# Lecture Notes in Mathematics

Edited by A. Dold and B. Eckmann



686

# Combinatorial Mathematics

Proceedings of the International Conference on Combinatorial Theory Canberra, August 16–27, 1977

Edited by
D. A. Holton and Jennifer Seberry

Springer-Verlag
Berlin Heidelberg New York

Australian Academy of Science
Canberra

**Editors**
D. A. Holton
Department of Mathematics
University of Melbourne
Parkville, Victoria 3052/Australia

Jennifer Seberry
Applied Mathematics Department
University of Sydney
Sydney. N. S. W. 2006/Australia

Distribution rights for Australia: Australian Academy of Science, Canberra

ISBN 0-85847-049-7 Australian Academy of Science Canberra

AMS Subject Classifications (1970): 05-04, 05A15, 05A17, 05A19, 05A99, 05B05, 05B15, 05B20, 05B25, 05B30, 05B40, 05B45, 05B99, 05C10, 05C15, 05C20, 05C25, 05C30, 05C35, 05C99, 15A24, 20B25, 20H15, 20M05, 50B30, 52A45, 62-XX, 62K10, 68A20, 94A10, 82A05

ISBN 3-540-08953-5 Springer-Verlag Berlin Heidelberg New York
ISBN 0-387-08953-5 Springer-Verlag New York Heidelberg Berlin


Printed in Germany

Printing and binding: Beltz Offsetdruck, Hemsbach/Bergstr.
2141/3140-543210

## PREFACE

The International Conference on Combinatorial Theory was held at the Australian National University from August 16-27 1977. The names of the eighty-nine participants are listed at the end of this volume.

This Conference was sponsored jointly by the International Mathematical Union and the Australian Academy of Science and was organised under the auspices of the Academy. Grants from the IMU and the Australian Government enabled us to invite a number of overseas specialists to the conference. With the exception of Professor Tutte, whose paper will appear elsewhere, the texts of the talks of the invited speakers appear in these Proceedings. We wish to thank our sponsors and the Australian Government for their support.

In addition to the invited addresses, three instructional series of talks were given. Professors Tutte and Bondy gave four lectures on the Reconstruction Conjecture, Professor Hughes gave four lectures on Designs and Professors Mullin and Vanstone gave four lectures on $(r, \lambda)$ Systems. The first two of these series will appear elsewhere. Professor Bondy's lectures will appear in the *Journal of Graph Theory* under the title "Graph reconstruction - a survey". This paper is coauthored by Professor R.L. Hemminger. The work by Professor Tutte is to appear in *Graph Theory and Related Topics*, the Proceedings of the Conference held in Waterloo in July 1977. Professor Hughes' material will appear in a book that he is currently writing. Only the material of Professor Mullin therefore, appears in these Proceedings.

At the conference there was a large number of contributed talks. Of these, twenty-eight appear in this volume. Papers which are given by title only in the Table of Contents will appear elsewhere.

It takes a great many people to make a conference the size of the present one run smoothly. We thank all those people who so willingly chaired sessions and refereed papers. Thanks too must go to the Australian National University, the Australian Academy of Science and the Canberra College of Advanced Education.

The ANU provided us with a number of lecture theatres, as well as library and other facilities. Neville Smythe of the ANU was a great help to us in the preconference organisation and in arranging typing and photocopying during the conference.

Considerable help was provided by the staff of the Academy. We particularly wish to express our thanks to Pat Tart and Beth Steward for their assistance which started many months before the conference. They were invaluable registering delegates, producing the daily newsletter, organising entertainment, and taking n + 1 jobs off our shoulders and executing them efficiently. Jack Deeble was a great help in the publication of these Proceedings.

At the Canberra CAE we were greatly helped by Peter O'Hallaron and Alan Brace. We thank Peter especially, for his liaison work between the College and the conference and his general assistance, particularly with regard to social events. We are grateful to the College for providing both lecture facilities for an afternoon session and transport for delegates during the conference.

We cannot let this opportunity go by of thanking Bernhard Neumann and Cheryl Praeger for the part they played in the running of the conference. The original idea of holding the conference was Bernhard's and he consistently gave his support throughout. Cheryl also was invaluable, especially in the early days of planning when the conference was on a very flimsy financial footing.

Finally we would like to thank Marjorie Funston, Helen Wort and Janet Midgley for their fine secretarial work in the periods of pressure before and after the conference.

D.A.H.

J.S.

TABLE OF CONTENTS

INVITED ADDRESSES



INSTRUCTIONAL LECTURE

## CONTRIBUTED PAPERS

PROBLEMS

## PAPERS BY TITLE ONLY

B. Alspach and R.J. Sutcliffe:
Vertex-transitive graphs of order 2p.

M. Deza and I.G. Rosenberg:
Generalized intersection pattern.

P. Eades:
A note on the Hadamard conjecture.

D. Glynn:
Constructing inverse planes.

I.P. Goulden and D.M. Jackson:
A unified treatment of sequence enumeration.

Richard Levingston:
Distance regular graphs and incidence structures.

N.J. Pullman:
Path number problems.

P. Robinson:
Orthogonal designs.

A. Rahilly:
Tangentially transitive projective planes.

P.J. Schellenberg, G.H.J. van Rees and S.A. Vanstone:
The existence of balanced tournament designs.

H.N.V. Temperley:
Potts problem on the plane triangular lattice.

W. Tutte:
The linear algebra of chromatic polynomials.

05Cxx

# REFLECTIONS ON THE LEGITIMATE DECK PROBLEM

J.A. Bondy

University of Waterloo
Waterloo, Ontario
Canada

ABSTRACT

We study the following problem: given a collection $\mathcal{H} = (H_i | 1 \le i \le n)$ of $n$ graphs, each on $n-1$ vertices, when does there exist a graph $G$ whose vertex-deleted subgraphs are the members of $\mathcal{H}$?

## 1. LEGITIMATE DECKS

A *deck* of $n$ cards is a collection $(H_i | 1 \le i \le n)$ of $n$ graphs, each having $n-1$ vertices. If there exists a graph $G$ with vertex set $\{1,2,\ldots,n\}$ such that

$$G_i \cong H_i \qquad (1 \le i \le n)$$

(where $G_i$ denotes the subgraph of $G$ obtained on deleting vertex $i$) the deck $(H_i | 1 \le i \le n)$ is said to be *legitimate*, and we call $G$ a *generator* of the deck. Decks which are not legitimate are, of course, *illegitimate*. The deck shown in figure 1(a) is legitimate: a generator is displayed in figure 1(b); but the deck of figure 2 is illegitimate, because we see from $H_1$ that every generator is acyclic, and from $H_2$ that no generator can possibly be so.

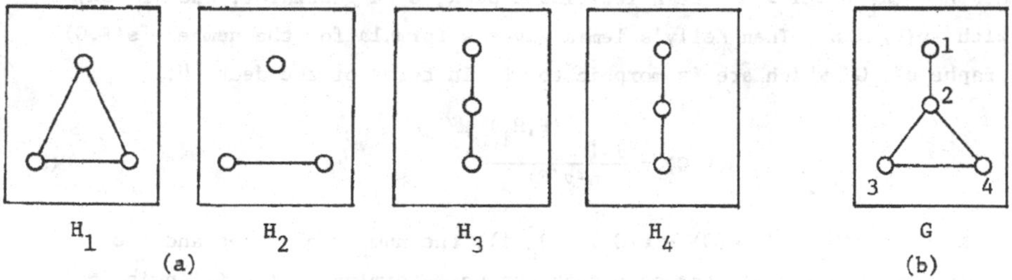


Figure 1

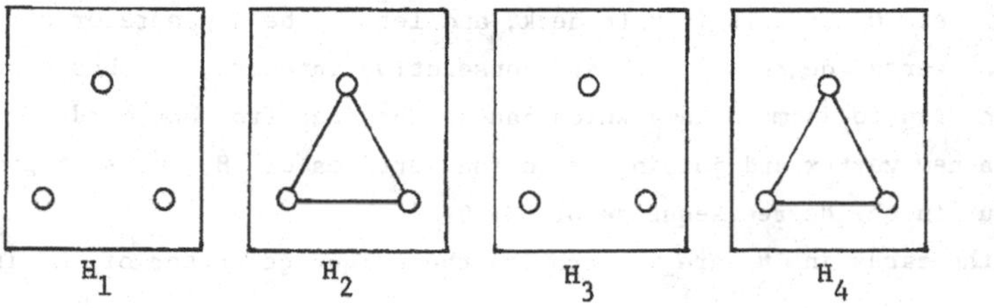


Figure 2

A less obvious example of an illegitimate deck is given in figure 3.

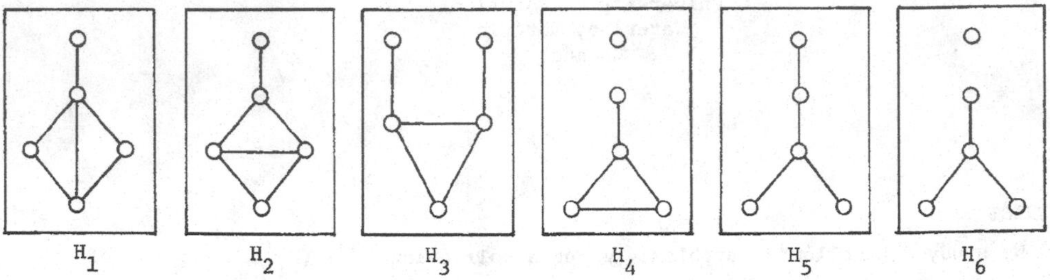

Figure 3

In the Reconstruction Conjecture [ 1 ], the problem is to show that no deck has more than one generator, up to isomorphism. The *Legitimate Deck Problem*, by contrast, seeks a characterization of those decks having at least one generator (in other words, legitimate decks). It was first mentioned by Harary [ 7 ] in 1968, more as an aside to the Reconstruction Conjecture than as a problem of independent interest. However, it does appear to be quite a basic question, having links with much existing graph theory.

This paper surveys the first few tentative steps which have been made towards an understanding of legitimate decks. Our notation and terminology is that of Bondy and Murty [ 2 ]. However, all graphs are assumed to be simple. Before proceeding, we make a couple of simple observations, based on a fundamental result in the theory of reconstruction known as Kelly's lemma [ 9 ].

Let $\mathcal{H} = (H_i | 1 \le i \le n)$ be a legitimate deck, G a generator, and F any graph with $\nu(F) < n$. Then Kelly's lemma gives a formula for the number $s(F,G)$ of subgraphs of G which are isomorphic to F in terms of the deck $\mathcal{H}$:

$$s(F,G) = \frac{\sum_{i=1}^{n} s(F,H_i)}{n-\nu(F)} \tag{1}$$

Since $s(K_2,G) = \varepsilon(G)$ and $\varepsilon(G) - \varepsilon(G_i) = d_G(i)$, the number of edges and the degree sequence of any generator of a deck can be determined. The following proposition is now easily established.

PROPOSITION. Let $\mathcal{H}$ be a legitimate deck, and let G be a generator of $\mathcal{H}$. Then

(i) if no two vertex degrees in G are consecutive integers, $\mathcal{H}$ has a unique generator (up to isomorphism) which can be obtained from any card $H_i$ by adding a new vertex and joining it to the vertices of $H_i$ whose degrees do not occur in the degree sequence of G;

(ii) if all the cards in $\mathcal{H}$ are isomorphic, the unique generator of $\mathcal{H}$ is vertex-transitive.

## 2. THE KELLY CONDITIONS

A variant of Kelly's lemma, this time involving induced subgraphs, can be invoked to yield a strong necessary condition for legitimacy. As before, let $\mathcal{H} = (H_i | 1 \le i \le n)$ be a legitimate deck, G a generator, and F any graph with $\nu(F) < n$. Then the number $s'(F,G)$ of induced subgraphs of G which are isomorphic to F is given by

$$s'(F,G) = \frac{\sum_{i=1}^{n} s'(F,H_i)}{n-\nu(F)} \tag{2}$$

Since the numbers $s'(F,G)$ must clearly be integers, we obtain the following condition on $\mathcal{H}$:

$$\text{(K1)} \qquad n-\nu(F) \mid \sum_{i=1}^{n} s'(F,H_i)$$

for each graph F with $\nu(F) < n$.

This condition appears to detect the vast majority of illegitimate decks; for instance, our deck of figure 3 fails (K1) when $F = K_{1,3}$. However, it is not as discriminating as might initially be supposed.

Let G be a vertex-transitive graph on a prime number p of vertices. Then, by (2)

$$s'(F,G) = \frac{\sum_{v \in V} s'(F,G_v)}{p-\nu(F)} = \frac{p \cdot s'(F,G_v)}{p-\nu(F)}$$

Since $s'(F,G)$ is an integer and $(p,p-\nu(F)) = 1$, we see that

$$p-\nu(F) \mid s'(F,G_v) \tag{3}$$

for each graph F with $\nu(F) < p$.

Consider, now, a deck of p cards, each of which is a vertex-deleted subgraph of a vertex-transitive graph. By (3), this deck satisfies (K1). However, it follows from our proposition that the deck is legitimate only when all of its cards are isomorphic. An example of an illegitimate deck formed in this way is given in figure 4.

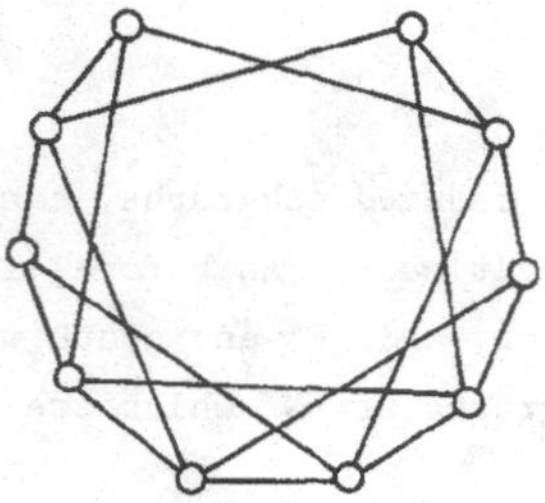

k copies

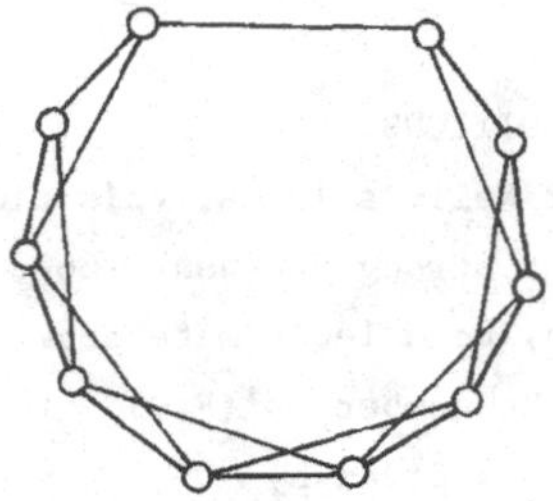

11-k copies

Figure 4

Another family of illegitimate decks which satisfy (K1) can be constructed from the star $K_{1,p}$, where, again, $p$ is a prime. Since

$$s'(F,K_{1,p}) = \frac{s'(F,K_p^c) + p\cdot s'(F,K_{1,p-1})}{p+1 - \nu(F)}$$

we have

$$p+1 - \nu(F) \mid s'(F,K_{1,p-1})$$

for each nonempty graph $F$ with $\nu(F) < p+1$. It now easily follows that the deck consisting of $k$ copies of $K_{1,p-1}$ and $p+1-k$ copies of $K_p^c$ satisfies (K1) for every $k$. This deck is legitimate, however, only when $k = 0$ or $k = p$ (provided that $p > 3$). Moreover, because

$$s'(F,G) = s'(F^c,G^c)$$

one obtains further illegitimate decks satisfying (K1) by taking $k$ copies of $K_1 + K_{p-1}$ and $p+1 - k$ copies of $K_p$ (where again $k \neq 0, p$ and $p > 3$). The above constructions can also be generalised to decks of $q$ and $q+1$ cards, respectively, where $q$ is any prime power.

These examples, due to Hafstrøm [5] and Jackson [8], amply demonstrate that further necessary conditions are required to supplement (K1). A natural one is:

(K2) $$s'(F,H_i) \leq \frac{\sum_{i=1}^{n} s'(F,H_i)}{n-\nu(F)}$$

for each graph $F$ with $\nu(F) < n$, and every $i$ $(1 \le i \le n)$. (In other words, no card in the deck can contain more copies of a graph than are to be found in a generator). Although condition (K2) eliminates many of our previous illegitimate decks, several families still remain; for instance, those derived from stars, with $k = p-1$ or $k = p+1$.

We remark, in concluding this section, that the obvious analogues of (K1) and (K2) for subgraphs (rather than induced subgraphs) are, in fact, subsumed in (K1) and (K2), and hence do not yield any new information.

## 3. THE SYMMETRIC ARRAY CONDITION

One unsatisfactory feature of the illegitimate decks derived from vertex-transitive graphs is that such decks include pairs of cards having very little in common, whereas, in a legitimate deck $(H_i | 1 \le i \le n)$ with generator $G$, any two cards $H_i$ and $H_j$ necessarily share the vertex-deleted subgraph $G_{ij} = G - \{i,j\}$. Furthermore, if $G$ has adjacency matrix $A = [a_{ij}]$, then

$$a_{ij} = \varepsilon(G) + \varepsilon(G_{ij}) - \varepsilon(G_i) - \varepsilon(G_j).$$

These observations prompted Randić [14] and Simpson [15] to formulate a quite different necessary condition for legitimacy, the so-called *symmetric array condition:* the vertex-deleted subgraphs of the cards $H_i$ can be arranged in a symmetric $n \times n$ array $[H_{ij}]$ such that

(SAC1) for $1 \le i \le n$, the vertex-deleted subgraphs of $H_i$ appear as the nondiagonal entries of row $i$;

(SAC1) if $A^* = [a^*_{ij}]$, where $a^*_{ii} = 0$ and $a^*_{ij} = \dfrac{\sum_{i=1}^{n} \varepsilon(H_i)}{n-2} + \varepsilon(H_{ij}) - \varepsilon(H_i) - \varepsilon(H_j)$

then $A^*$ is a symmetric (0,1)-matrix.

If a deck $(H_i | 1 \le i \le n)$ satisfies the symmetric array condition, then the graph $G^*$ whose adjacency matrix is $A^*$ is clearly a potential generator of the deck. The following result of Ramachandran [13] shows that $G^*$ does, moreover, possess some of the properties that one would demand of any generator.

THEOREM. Let $(H_i | 1 \le i \le n)$ be a deck which satisfies the symmetric array condition, and let $G^*$ denote the graph whose vertex set is $\{1,2,\ldots,n\}$ and whose adjacency matrix is the matrix $A^*$ of (SAC2). Then

(i) the degree sequence of $G^*$ is given by

$$d_{G^*}(i) = \frac{\sum_{i=1}^{n} \varepsilon(H_i)}{n-2} - \varepsilon(H_i)$$

(ii) the degree sequence of $G_i^*$ is the same as that of $H_i$ $(1 \le i \le n)$.

Despite this theorem, it is not difficult to find examples of illegitimate decks which satisfy both (SAC1) and (SAC2). We describe one class, due to Jackson [8].

Let $G$ be a vertex-deleted subgraph of a regular, but not vertex-transitive, graph on an even number $n$ of vertices. Then it follows from our proposition (section 1) that the deck consisting of $n$ copies of $G$ is illegitimate. However, symmetric arrays can readily be constructed from symmetric latin squares of order $n$ (which exist for all even $n$ [3]) and any such array automatically satisfies (SAC2). This construction is illustrated in figure 5.

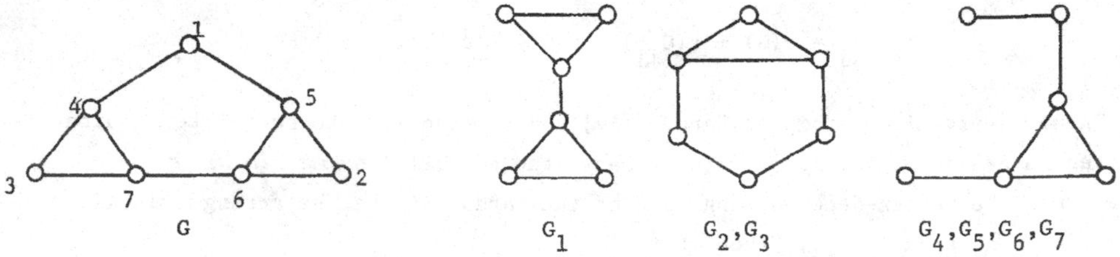


Latin Square

$$\begin{bmatrix} 1&5&2&6&3&7&4\\ 5&2&6&3&7&4&1\\ 2&6&3&7&4&1&5\\ 6&3&7&4&1&5&2\\ 3&7&4&1&5&2&6\\ 7&4&1&5&2&6&3\\ 4&1&5&2&6&3&7 \end{bmatrix}$$

Symmetric Array

$$\begin{bmatrix} * & G_5 & G_2 & G_6 & G_3 & G_7 & G_4 & G_1\\ G_5 & * & G_6 & G_3 & G_7 & G_4 & G_1 & G_2\\ G_2 & G_6 & * & G_7 & G_4 & G_1 & G_5 & G_3\\ G_6 & G_3 & G_7 & * & G_1 & G_5 & G_2 & G_4\\ G_3 & G_7 & G_4 & G_1 & * & G_2 & G_6 & G_5\\ G_7 & G_4 & G_1 & G_5 & G_2 & * & G_3 & G_6\\ G_4 & G_1 & G_5 & G_2 & G_6 & G_3 & * & G_7\\ G_1 & G_2 & G_3 & G_4 & G_5 & G_6 & G_7 & * \end{bmatrix}$$

$$\begin{bmatrix} 0&0&1&0&1&0&0&1\\ 0&0&0&1&0&0&1&1\\ 1&0&0&0&0&1&0&1\\ 0&1&0&0&1&0&1&0\\ 1&0&0&1&0&1&0&0\\ 0&0&1&0&1&0&1&0\\ 0&1&0&1&0&1&0&0\\ 1&1&1&0&0&0&0&0 \end{bmatrix}$$

A*

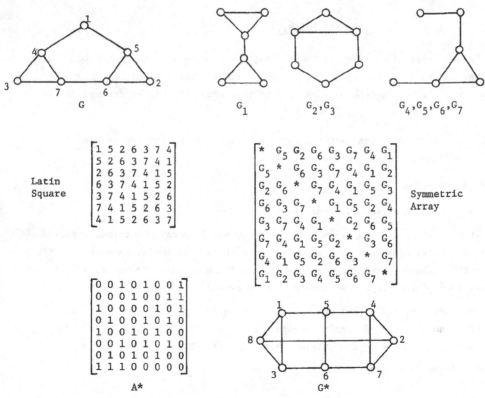

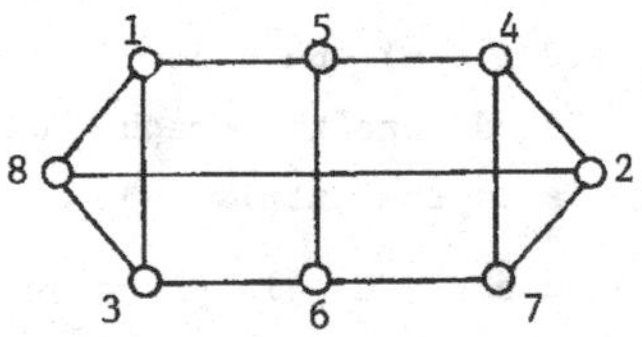


G*

Figure 5

As we have seen, neither the Kelly condition nor the symmetric array condition alone suffices to characterize legitimate decks. Although no illegitimate deck has yet been found to satisfy both conditions simultaneously, such decks must surely exist.

## 4. EXTREMAL CONDITIONS

In this section, we briefly indicate some links between the Legitimate Deck Problem and extremal graph theory.

As a simple example, consider a legitimate deck $\mathcal{H} = (H_i | 1 \le i \le n)$ in which all the cards $H_i$ are trees. Since each $H_i$ has $n-2$ edges, an application of Kelly's lemma shows that any generator $G$ of $\mathcal{H}$ must have $n$ edges. Now every graph on $n$ vertices and $n$ edges contains a cycle. Because each $H_i$ is a tree, we conclude that $G$ must itself be a cycle and, hence, that each $H_i$ must be a path. It follows that any deck in which all the cards are trees, but not all are paths, is necessarily illegitimate.

More generally, any extremal theorem of the form

$$s(F,G) = 0 \Rightarrow \varepsilon(G) \le t(F;\nu(G)) \tag{4}$$

can be employed to construct illegitimate decks, provided that

$$\frac{t(F;n-1)}{t(F;n)} > \frac{n-2}{n}$$

for some $n$. Under this latter condition, there exist decks $(H_i | 1 \le i \le n)$ such that $s(F,H_i) = 0$ for all $i$ and

$$\frac{\sum_{i=1}^{n} \varepsilon(H_i)}{n-2} > t(F;n)$$

By (4), such decks are illegitimate.

In a similar way, Ramsey theory can be tapped to yield examples of illegitimate decks. We present one small illustration.

Suppose that the deck of figure 6 were legitimate, with generator $G$. By Kelly's lemma

$$s'(K_3,G) = 1 \quad \text{and} \quad s'(K_3^c,G) = 0$$

However, an elementary result of Ramsey theory [4] states that, for any graph $G$ on 6 vertices

$$s'(K_3,G) + s'(K_3^c,G) \ge 2$$

The deck of figure 6 is therefore illegitimate.

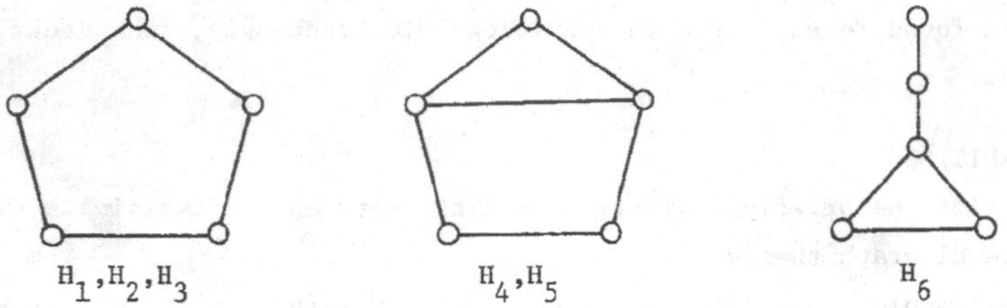


Figure 6

We have just scratched the surface here. The Legitimate Deck Problem must surely be linked in a similar manner to much of established graph theory.

## 5. THE SET RECONSTRUCTION PROBLEM

One interesting aspect of the Legitimate Deck Problem is to be found in its relationship to the Reconstruction Problem.

We define a *set reconstruction* of $G$ to be a graph $H$ such that (i) for any $v \in V(G)$ there is a $w \in V(H)$ with $H_w \cong G_v$, and (ii) for any $v \in V(H)$, there is a $w \in V(G)$ with $G_w \cong H_v$. In other words, a set reconstruction of a graph $G$ is a graph $H$ with the same set of vertex-deleted subgraphs as $G$. We then say that $G$ is *set-reconstructible* if every set reconstruction of $G$ is isomorphic to $G$. A set-reconstructible graph is also, of course, reconstructible, but not conversely: the graphs $G$ and $H$ of figure 7, although reconstructible, are not set-reconstructible.

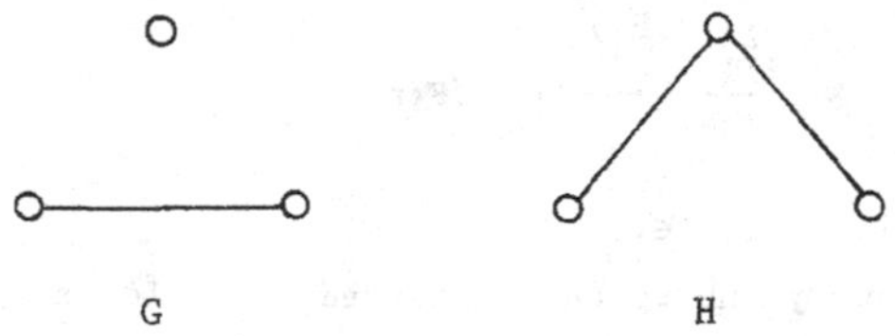


Figure 7

However, these two graphs are the only known exceptions, and the following conjecture has been proposed [6].

THE SET RECONSTRUCTION CONJECTURE. All finite simple graphs on at least four vertices are set-reconstructible.

This conjecture has been verified for graphs with up to nine vertices [12], and has been proved for various classes of graphs [11]. Also, it is known that certain parameters, such as the number of edges and the minimum degree, are set-reconstructible (although the only published proof [10; Proposition 2] is not completely watertight).

Now, where do legitimate decks come in? Let $\mathcal{G} = \{G_1, G_2, \ldots, G_m\}$ be an ordered set of $m$ cards, each on $n-1$ vertices (where $m \le n$) and call a vector $(x_1, x_2, \ldots, x_m)$ of positive integers whose sum is $n$ a *legitimate multiplier* for $\mathcal{G}$ if the deck consisting of $x_i$ copies of $G_i$ $(1 \le i \le m)$ is legitimate. Then the Set Reconstruction Conjecture divides naturally into two subconjectures:

CONJECTURE 1. *The Legitimate Multiplicity Conjecture.* Any ordered set $\{G_1, G_2, \ldots, G_m\}$ of $m$ cards, each on $n-1$ vertices, has at most one legitimate multiplier $(x_1, x_2, \ldots, x_m)$.

CONJECTURE 2. *The Reconstruction Conjecture.*

A proof of conjecture 1 would thus reduce the Set Reconstruction Conjecture to the standard Reconstruction Conjecture. We shall consider conjecture 1 for small values of $m$; the case $m = 1$ is, of course, trivial.

Let us assume that $\mathcal{G} = \{G_1, G_2, \ldots, G_m\}$ has a legitimate multiplier $(x_1, x_2, \ldots, x_m)$, and let $G$ be a graph with $x_i$ vertex-deleted subgraphs of type $G_i$ $(1 \le i \le m)$. We shall denote the numbers of vertices and edges and the minimum degree of $G$ by $\nu, \varepsilon$ and $\delta$, respectively; as remarked above, these parameters are set-reconstructible (that is, determined uniquely by the set $\mathcal{G}$). By the definition of a legitimate multiplier, we have

$$x_i > 0 \qquad (1 \le i \le m)$$

and

$$\sum_{i=1}^{m} x_i = \nu \tag{5}$$

Suppose that $G_i$ has $\varepsilon_i$ edges $(1 \le i \le m)$, where

$$\varepsilon_1 \ge \varepsilon_2 \ge \ldots \ge \varepsilon_m \tag{6}$$

Then Kelly's lemma, with $F = K_2$, gives

$$\sum_{i=1}^{m} \varepsilon_i x_i = \varepsilon(\nu-2) \tag{7}$$

Furthermore, by the proposition of section 1, we can assume that, for some $i$ $(1 \le i \le m)$

$$\varepsilon_i - \varepsilon_{i+1} = 1 \tag{8}$$

When $m = 2$, equations (5), (7), and (8) simplify to

$$x_1 + x_2 = \nu$$
$$\varepsilon_1 x_1 + \varepsilon_2 x_2 = \varepsilon(\nu-2)$$
$$\varepsilon_1 - \varepsilon_2 = 1$$

and these equations have the unique solution

$$x_1 = \varepsilon(\nu-2) - \nu\varepsilon_2$$
$$x_2 = \nu\varepsilon_1 - \varepsilon(\nu-2)$$

When $m = 3$, equations (5) and (7) become

$$x_1 + x_2 + x_3 = \nu \tag{9}$$

$$\varepsilon_1 x_1 + \varepsilon_2 x_2 + \varepsilon_3 x_3 = \varepsilon(\nu-2) \tag{10}$$

Let $\varepsilon_1 - \varepsilon_3 = k$. In view of (6) and (8), we can assume that $k \geq 1$, and that either $\varepsilon_1 - \varepsilon_2 = 1$ or $\varepsilon_2 - \varepsilon_3 = 1$. In fact, there is no loss of generality in assuming that $\varepsilon_2 - \varepsilon_3 = 1$, as can be seen by considering the complementary set $\{G_1^c, G_2^c, G_3^c\}$. Therefore, with the aid of (9), (10) simplifies to

$$kx_1 + x_2 = \varepsilon(\nu-2) - \nu\varepsilon_3 \tag{11}$$

We examine three cases, depending on the value of $k$.

If $k > 2$, then the number of vertices of degree $\delta-1$ or $\delta$ in $G_3$ is the same as the number of vertices of degree $\delta$ in $G$. But this number is precisely $x_1$. Therefore the system has a unique solution.

If $k = 2$, denote by $a$ the number of vertices of degree $\delta+2$ in $G_1$, and by $b$ the number of vertices of degree $\delta-1$ in $G_3$. Counting the number of edges of $G$ with one end of degree $\delta$ and the other of degree $\delta+2$ in two ways, we obtain

$$x_1(x_3-a) = x_3 b \tag{12}$$

Subtracting (9) from (11) and setting $c = \varepsilon(\nu-2) - \nu\varepsilon_2$ yields

$$x_1 - x_3 = c \tag{13}$$

We can solve (12) and (13) for $x_1$ and $x_3$ and obtain two quadratic equations:

$$x_1^2 - (a+b+c)x_1 + bc = 0$$

$$x_3^2 - (a+b-c)x_3 - ac = 0$$

Since $a$ and $b$ are both nonnegative, at least one of these equations has at most one positive root. Thus, again, the system has a unique solution.

Unfortunately, the case $k = 1$ appears to be far less tractable than the previous ones. When $k = 1$, we have $\varepsilon_1 = \varepsilon_2$ and, even though the graphs $G_1$ and $G_2$ are, by definition, nonisomorphic, they might be quite similar. Indeed, if the Reconstruction Conjecture were false, they could even have the same collection of vertex-deleted subgraphs. The problem of determining the multiplicities $x_1$ and $x_2$ would clearly be a difficult one under such circumstances.

REFERENCES


(1) J.A. Bondy and R.L. Hemminger, "Graph reconstruction - a survey", *J. Graph Theory* 1(1977), to appear.

(2) J.A. Bondy and U.S.R. Murty, *Graph Theory with Applications*, MacMillan, London and American Elsevier, New York, 1976.

(3) J. Dénes and A.D. Keedwell, *Latin Squares and their Applications*, Academic Press, New York, 1974.

(4) A.W. Goodman, "On sets of acquaintances and strangers at any party", *Amer. Math. Monthly* 66(1959), 778-783.

(5) U. Hafstrøm, personal communication, 1976.

(6) F. Harary, "On the reconstruction of a graph from a collection of subgraphs", in, *Theory of Graphs and its Applications*, (Proceedings of the Symposium held in Prague, 1964), edited by M. Fiedler, Czechoslovak Academy of Sciences, Prague, 1964, 47-52.

(7) F. Harary, "The four color conjecture and other graphical diseases", in *Proof Techniques in Graph Theory*, (Proceedings of the Second Ann Arbor Graph Theory Conference, Ann Arbor, Mich., 1968), edited by F. Harary, Academic Press, New York, 1969, 1-9.

(8) W. Jackson, "Legitimate decks", preprint, 1977.

(9) P.J. Kelly, "A congruence theorem for trees", *Pacific J. Math.* 7(1957), 961-968.

(10) B. Manvel, "On reconstruction of graphs", in *The Many Facets of Graph Theory*, (Proceedings of the Conference held at Western Michigan University, Kalamazoo, Mich., 1968), edited by G. Chartrand and S.F. Kapoor, Lecture Notes in Math., Vol. 110, Springer-Verlag, New York, 207-214.

REFERENCES con't...

(11) B. Manvel, "On reconstructing graphs from their sets of subgraphs", *J. Combinatorial Theory* (B), 21(1976), 156-165.

(12) B.D. McKay, "Computer reconstruction of small graphs", *J. Graph Theory* 1(1977), to appear.

(13) S. Ramachandran, "A test for legitimate decks", preprint, 1977.

(14) M. Randić, "On the reconstruction problem for graphs", preprint, 1977.

(15) J.E. Simpson, "Legitimate decks of graphs", *Notices Amer. Math. Soc.* 21(1974), A-39.

05C99

# SOME EXTREMAL PROBLEMS ON FAMILIES OF GRAPHS AND RELATED PROBLEMS

P. Erdös

Hungarian Academy of Science,
Budapest, Hungary

Let $G(n)$ be a graph of $n$ vertices, $G(n; \ell)$ a graph of $n$ vertices and $\ell$ edges. $f(n; G(m))$ is the smallest integer so that every $G(n; f(n; G(m))$ contains a subgraph isomorphic to $G(m)$. More generally let $G_1, \ldots$ be a finite or infinite family of finite graphs. $f(n; G_1, \ldots)$ is the smallest integer so that every $G(n; f(n; G_1, \ldots))$ contains one of the $G_k$ as a subgraph. Many papers have been published in the last few years on the determination or estimation of these functions. In one of my recent papers I give a far from complete list of papers dealing with extremal problems in graph theory. Bollobás is about to publish a comprehensive book on this subject which will also contain a very extensive list of references.

In this paper I first of all state a few of my favorite unsolved extremal problems. Then I prove the following theorems:

THEOREM 1. Assume that $G(n)$ does not contain a $C_{2k+1}$ for $3 \le k \le r$. Then the independence number of $G(n)$ is greater than $c_1 n^{1-1/r}$.

$C_k$ is a circuit of $k$ edges and the independence number of $G(n)$ is the cardinal number of the largest set of vertices no two of which are joined by an edge.

$K(m)$ is the complete graph of $m$ vertices. Denote by $K_{top}(m)$ an arbitrary subdivision of $K(m)$ (i.e. a topological complete $K(m)$. $K_{top}(3)$ is simply a circuit).

THEOREM 2. There is a function $f(c) > 0$ so that every $G(n; [c\, n^2])$ contains a $K_{top}(\ell)$ with $\ell \ge f(c)n^{\frac{1}{2}}$.

Before proving the theorems we will state several related conjectures.

P. Erdös, Some recent progress on extremal problems in graph theory, Proc. sixth Southeastern conference on combinatorics graph theory and computing 1975, *Utilitas* ***Math.*** *press*, 3-14, *Congress Num XIV*. We will refer to this paper as I.

For further problems see my paper: Problems and results in graph theory and combinatorial analysis, Proc. fifth British comb conference 1975, 169-192, *Utilitas Math. Cong. Num XV*. I refer to this paper as II. For some historical remarks see P. Erdös, Problems in number theory and combinatorics, Proc. sixth Manitoba conference on numerical math, *Congress Num. XVIII*, 35-58. For some further extremal and other problems see my paper, Some recent problems and results in graph theory combinatorics and number theory. Proc. seventh Southeastern conference. *Ut. Math. press*, 3-14,

(Congress Num. XVIII).

A weaker version of Theorem 2 was proved in: P. Erdős and A. Hajnal, On complete topological subgraphs of certain graphs, *Ann. Univ. Sci. Budapest*, 7(1969), 193-199. Theorem 2 is stated as a conjecture in this paper.

1. Simonovits and I conjectured that if G is bipartite (unless stated otherwise G is always bipartite) then there is a rational number $\alpha$, $1 \le \alpha < 2$ so that

(1.1) $$\lim_{n=\infty} f(n;\ G)/n^{\alpha} = c_{\alpha},\ 0 < c_{\alpha} < \infty.$$

We are very far from being able to prove (1). As a first step one should prove that to every bipartite G there is an $\alpha$ so that for every $\varepsilon > 0$ and $n > n_0(\varepsilon)$

(1.2) $$n^{\alpha-\varepsilon} < f(n;\ G) < n^{\alpha+\varepsilon}.$$

(1.2) perhaps will not be very hard to prove.

We further conjecture that to every rational $\alpha$, $1 \le \alpha < 2$, there is a G for which (1.1) is satisfied.

P. Erdős and M. Simonovits, Some extremal problems in graph theory, *Coll. Math. Soc. Bòljai* 4, Combinatorial theory and its applications (1969), 377 - 390, North Holland, see also II.

Nothing like (1) holds for hypergraphs. This follows from a result of Szemerédi and Rursa see II p. 179.

For non-bipartite graphs the results of Simonovits, Stone and myself cleared up the situation to some extent, though many problems remain.

[1] P. Erdős and A. Stone, "On the structure of linear graphs", *Bull. Amer. Math. Soc.* 52(1946), 1087-1091.

[2] P. Erdős and M. Simonovits, "A limit theorem is graph theory", *Studia Sci. Math. Hungar.* 1 (1966), 51-57.

2. Define V(G) as the minimum valency (or degree) of all the vertices of G. Put $V_1(G) = \max V(G')$ where the maximum is taken over all the subgraphs of G. Simonovits and I asked: Is it true that

(2.1) $$f(n:\ G) < c\, n^{3/2} \text{ if } V_1(G) = 2\ ?$$

We now expect that (2.1) is false, but can prove nothing.

Assume $V_1(G) = r$. A result of Rényi and myself implies $f(n:\ G) > c\, n^{2(1-1/r)}$. Define $\alpha_1(r)$ and $\alpha_2(r)$ as follows: For $V_1(G) = r$ and every $\varepsilon > 0$ if $n > n_0(\varepsilon)$,

$$n^{\alpha_1(r)-\epsilon} < f(n;\ G) < n^{\alpha_2(r)+\epsilon}.$$

Our result with Rényi implies $\alpha_1(r) \ge 2(1 - 1/r)$. Is this best possible? Is it true that for every r, $\alpha_2(r) < 2$ ? Unfortunately we do not know this even for $r = 2$.

[1] P. Erdős and A. Rényi, "On the evolution of random graphs", *Publ. Math. Inst. Hung. Acad. Sci.* 5 (1960), 17-67.

3. Denote by $D_n$ the graph of the n dimensional cube, it has $2^n$ vertices and $n\ 2^{n-1}$ edges, $D_2 = C_4$).

Simonovits and I proved $f(n;\ D_3) < c\ n^{8/5}$. Probably the exponent 8/5 is best possible, but we have not even been able to prove $f(n;\ D_3)/n^{3/2} \to \infty$.

Brown, V.T. Sós, Rényi and I proved

(3.1) $$f(n;\ C_4) = (\tfrac{1}{2} + o(1))n^{3/2}.$$

Let $\theta$ be a power of a prime. We also proved

(3.2) $$f(\theta^2 + \theta + 1;\ C_4) \ge \tfrac{1}{2}(p^3 + p) + p^2 + 1,$$

perhaps there is equality in (3.2). I proved in I that

(3.3) $$f(n;\ C_4) \le \tfrac{1}{2}\, n^{3/2} + \frac{n}{4} - (\frac{3}{16} + o(1))n^{\frac{1}{2}}\ .$$

I conjectured

(3.4) $$f(n;\ C_4) = \tfrac{1}{2}\, n^{3/2} + \frac{n}{4} + o(n).$$

It is not impossible that in (3.4) the error term is $O(n^{\frac{1}{2}})$.

K(u, v) is the complete bipartite graph of u white and v black vertices. Kövári, V.T. Sós, P. Turán and I proved

(3.5) $$f(n;\ K(r, r)) < c\ n^{2-1/r}.$$

Very likely the exponent in (3.5) is best possible. For $r = 2$ this is implied by (3.3) and Brown proved it for $r = 3$, but for $r > 3$ nothing is known.

Denote by G - e the subgraph of G from which the edge e has been omitted. Simonovits and I proved

(3.6) $$f(n;\ D_3 - e) < c n^{3/2}$$

and I proved

(3.7) $$f(n;\ K(r, r) - e) < c\ n^{1-1/(r-1)}.$$

Simonovits and I tried to characterize the graphs G with the property that for every proper subgraph G' .

(3.8) $$f(n;\ G')/f(n;\ G) \to 0.$$

We were of course unsuccessful, but in view of (3.6) and (3.7) it seemed to us that highly symmetric graphs are likely to satisfy (3.8).

Our paper with Simonovits is quoted in 1.

[1] W.G. Brown, "On graphs that do not contain a Thomsen graph", *Canad. Math. Bull.* 9 (1966), 281-285.

[2] P. Erdös, A. Rényi and V.T. Sós, "On a problem of graph theory", *Studia. Sci. Math. Hung.* 1 (1966), 215-235.

[3] T. Kövári, V.T. Sós, and P. Turán, "On a problem of K. Zarankievcz", *Coll. Math.* 3 (1954), 50-57.

[4] P. Erdös, "On an extremal problem in graph theory", *Coll. Math.* 13 (1964), 251-254.

4. We have

(4.1) $$\frac{1}{2\sqrt{2}} \le \lim f(n;\ C_3,\ C_4)/n^{3/2} \le \tfrac{1}{2} .$$

The lower bound is a result of Reiman and E. Klein (Mrs. Szekeres). The upper bound is (3.1). Determine the value of the limit in (4.1). I never managed to get anywhere with this question and cannot decide whether it is really difficult or whether I overlook a simple argument. I was never able to improve (4.1).

More generally, let $G_1, \ldots, G_k$ be a family of graphs some of which are bipartite. I hope and expect that

(4.2) $$\lim_{n=\infty} f(n;\ G_1,\ \ldots,\ G_k)/n^{\alpha} = c$$

Assume that the conjecture (1.1) holds and let $\alpha_i$ be the rational number for which

$$\lim_{n=\infty}\ f(n;\ G_i)/n^{\alpha_i} = c_i,\ 0 < c_i < \infty.$$

Perhaps $\alpha = \min_{1\le i\le k} \alpha_i$. I have of course no real evidence for this. I am sure that the situation changes completely for infinite families of graphs $\{G_k\}$, $1 \le k < \infty$. At the moment I do not know an example of an infinite family of graphs $\{G_k\}$, $1 \le k < \infty$ so that there is an $\alpha$ with $f(n;\ G_k)/n^{\alpha} \to \infty$ for every $k$, but for some $\beta < \alpha$

(4.3) $$f(n;\ G_1,\ \ldots\ )/n^{\beta} \to 0.$$

Probably the family of graphs $G$ with $V_1(G) \ge 3$ satisfies (4.3) for every $\beta > 1$. This is an old conjecture of Sauer and myself. If true then, since for these graphs $f(n;\ G) > c\, n^{4/3}$ by our result with Rényi stated in 2, this family would have the above property.

Our problem with Gauer is discussed in I. p. 10 and II p. 178.

[1] I. Reimann, "Über ein Problem von K. Zaranhievicz, *Acta Math. Acad. Sci. Hungar*, 9(1958), 269-278.

5. As far as I know G. Dirac was the first to investigate $f(n;\ K_{top}(m))$. Trivially $f(n;\ K_{top}(3)) = n$ and G. Dirac proved $f(n;\ K_{top}(4)) = 2n - 2$. He conjectured $f(n;\ K_{top}(5)) = 3n - 5$. It is surprising that this attractive conjecture is still open. Mader proved that $f(n;\ K_{top}(m)) \leq 2^{m-2}n$, he conjectured

$$f(n;\ K_{top}(m)) < c\, m^2 n. \tag{5.1}$$

(5.1) is probably rather deep. It is easy to see (as was of course known to Mader) that the conjecture if true is best possible - apart from the value of c.

Theorem 2 can be considered as proving (5.1) for large values of m, but it is very doubtful if it will help in proving (5.1). Before we prove our theorems we give a preliminary discussion and state some conjectures, some of which are in my opinion more interesting than the theorems. First of all it would be of interest to determine the largest f(c) for which Theorem 2 holds. I am sure that it will be a continuous strictly increasing function of c. It is not hard to prove that $f(c) \to 0$ as $c \to 0$ and $f(c) \to \infty$ as $c \to \frac{1}{2}$. It would of course be interesting to determine f(c) explicitly.

I am sure that the following strengthening of Theorem 2 holds.

Conjecture 1: Every $G(n;\ [c_1 n^{\frac{1}{2}}])$ contains $[c_2 n^{\frac{1}{2}}]$ vertices $x_1, \ldots, x_r$, $r = [c_2 n^{\frac{1}{2}}]$ so that $x_i$ and $x_j$, $1 \leq i < j \leq r$ are joined by vertex disjoint paths of length 2.

This conjecture is clearly connected with the following problem of perhaps greater independent interest.

Let $|S| = n$, $A_k \subset S$, $|A_k| > c\, n$, $1 \leq k \leq m$. Determine the largest $f(n, m, \varepsilon, c)$ so that there always are sets $A_{k_i}$, $1 \leq i \leq f(n, m, c, \varepsilon)$ for which for every $1 \leq i_1 < i_2 \leq f(n, m, c, \varepsilon)$

$$|A_{k_{i_1}} \cap A_{k_{i_2}}| > \varepsilon n .$$

$\varepsilon > 0$ can be chosen as small as we wish but must be independent of n and m. Observe that if $c > \frac{1}{2}$ then for sufficiently small $\varepsilon = \varepsilon(c)$, $f(n, m, c, \varepsilon) = m$. Thus the problem is of interest only for $c \leq \frac{1}{2}$.

The connection between this problem and the conjecture is easy to establish. First of all it is well known and easy to see that every $G(n;\ c\, n^2)$ contains a subgraph $G(N)$, $N > c_1 n$ each vertex of which has valency greater than $(2c + o(1)N$. (To prove the lemma omit successively the vertices of smallest valency).

Let the vertices of $G(N)$ be $x_1, \dots x_N$. The sets $A_k$ are the vertices joined to $x_k$. It is immediate that Conjecture 1 is a consequence of

Conjecture 2. For $n = m$ and $\varepsilon = \varepsilon(c)$ sufficiently small

$$f(n, m, c, \varepsilon) > \eta\, n^{\frac{1}{2}} \tag{5.2}$$

for some $\eta = \eta(c, \varepsilon) > 0$.

I can not even disprove

$$f(n, m, c, \varepsilon) \geq \eta\, m \tag{5.3}$$

for $m < n$ and $\eta = \eta(c, \varepsilon)$. On the other hand I can not prove (5.3) for $c = 1$ even for $c = \frac{1}{2}$.

Perhaps for every $m \leq 2^n$ and $\varepsilon = \varepsilon(\eta)$

$$f(n, m, c, \varepsilon) > m^{1-\eta}. \tag{5.4}$$

(5.4), if true, is best possible. To see this, let the A's be all subsets of S having at least cn elements, and let $m = 2^n - \sum_{0 \leq i < cn} \binom{n}{i}$. It is easy to see that in this case (5.4) can not be improved.

These conjectures have many connections with other interesting questions in graph theory. First of all an old conjecture of Kneser states as follows: Let $|S| = 2n + k$. The vertices of G are the $\binom{2n+k}{n}$ subsets of size n of S. Join two vertices if the corresponding n-sets are disjoint. Prove that the chromatic number $\chi(G)$ of G is $k + 2$. This conjecture has recently been proved by Lovász and Báráry in a surprisingly simple way. Their proofs will appear soon.

Define now a graph $G_{(n, m, \varepsilon)}$ as follows. Its vertices are the m sets $A_k \subset S$. Two A's are joined if $|A_{k_1} \cap A_{k_2}| < \varepsilon n$. Determine or estimate $\chi(G_{(n, m, \varepsilon)})$. (5.2) would follow from $\chi(G(n, m, \varepsilon)) < c\, n^{\frac{1}{2}}$. Perhaps very much more is true e.g. $\chi(G_{(n, m, \varepsilon)}) < c_1(\log m)^{c_2}$.

Ramsey's theorem can be used to obtain weaker inequalities than (5.2). Let $1/r > c > 1/r+1$. A simple argument shows that for sufficiently small $\varepsilon$ the largest independent set of $\overline{G}_{(n, m, \varepsilon)}$ is at most r. ($\overline{G}$ is the complementary graph of G, i.e. two vertices are joined in $\overline{G}$ if and only if they are not joined in G. A set of vertices is independent if no two of them are joined by an edge). Thus by a well known theorem of Szekeres and myself it contains a complete graph of size $> c\, n^{1/r}$. (for $r = 1$ the whole graph of course is complete). In other words (5.2) holds with $1/r$ instead of $\frac{1}{2}$.

Now we show that it is possible to obtain considerably stronger results. First of all assume $1/3 < c < \frac{1}{2}$. Clearly in our graph the largest independent set has size 2, but we can easily get some further information. Let $A_1, \ldots, A_5, |A_i| > cn$, $c > 1/3$ be any five of our sets. Then a simple argument shows that there is a set $S_1 \subset S$, $|S_1| > \varepsilon n$ which is contained in three of the A's. In other words every five points of our graph spans a triangle. Thus the complementary graph of our graph contains no triangle and no pentagon. But then by Theorem 1 it contains an independent set of size $> c\, n^{2/3}$ or our graph contains a complete graph of size greater than $cn^{2/3}$.

More generally assume $|A_k| > n/r{+}1\ (1 + \eta)$ for some $\eta > 0$ ($A_k \subset S$, $|S| = n$, $1 \le k \le m$). Join two sets $A_{k_1}$ and $A_{k_2}$ if $|A_{k_1} \cap A_{k_2}| > \varepsilon n$, $\varepsilon = \varepsilon(\eta)$ is sufficiently small. Then these graphs belonging to the set system the graphs depend on $\varepsilon$ has the following property: For every fixed $t = t_\varepsilon$ and $\ell \le t$ every set of $\ell(r + 1)$ vertices contains a $k(\ell + 1)$. I hope that for sufficiently large $t = t(r, \delta)$ this condition implies that our graph contains a complete graph of size $> m^{1-\delta}$. (For Conjecture 1 it suffices to prove this for $\delta = \frac{1}{2}$).


[1] G. Dirac, "In abstrakten Graphen vorhandene vollständige 4-Graphen und ihre Unterteilunge", *Math. Naohriohton* 22 (1960), 61-85;

for a very simple proof see:

[2] P. Erdös and L. Pósa, "On the maximal number of disjoint circuits of a graph", *Publicationes Math.* 9 (1962), 3-12, see p. 8.

[3] W. Mader, "Homomorphieeigenxhatten und mittlere Kantendichte von Graphen", *Math. Annalen* 174 (1967), 265-268.

[4] P. Erdös and G. Szekeres, "On a combinatorial problem in geometry, *Compositio Math.* 2 (1935), 463-470.


6. Now we prove Theorem 1. Let the vertices of our graph G(n) be $x_1, \ldots, x_m$. Denote by $S_i$ the set of those $x_j$'s which can be joined to $x_1$ by a path of length i but not by a shorter path ($S_0$ is defined to be $x_1$). Observe that the set $S_i$ is independent of the set $\bigcup_{j \ge i+2} S_j$ (i.e. no vertex of $S_i$ is joined (by an edge) to a vertex of $\bigcup_{j\ge i+2} S_j$). Observe further that for $1 \le i \le r$, $s_i$ is an independent set. For if two vertices of $S_i$ are joined then our G(n) contains an odd circuit of size $\le 2i + 1$, which contradicts our assumptions. Observe next that for some i, $0 \le i \le r - 1$,

$$(6.1) \qquad \frac{|S_i|}{|S_{i-1}|} < n^{1/r}.$$

(6.1) follows immediately from the fact that $S_{i_1} \cap S_{i_2} = \phi$ and that $|\bigcup_{i=0}^{r} S_i| \le n$. (In fact we can assume $|\bigcup_{i=1}^{r} S_i| < n$ for if not then $\max_{1\le i\le r} |S_i| \ge (n-1)/r$ which implies

Theorem 1). Let now $i \geq 0$ be the smallest index satisfying (6.1). We construct our large independent subset of $G(n)$ as follows: The vertices of $S_i$ will be in our large independent set. $G_1$ is the subgraph of $G$ spanned by those vertices of $G$ which are not in $\bigcup_{j=0}^{i+1} S_j$. Clearly by (1) and the minimum property of $i$

$$(6.2) \qquad \left|\bigcup_{j=0}^{i+1} S_j\right| < (n^{1/r} + 1)\,|S_i|$$

or $G_1$ has at least $n - (n^{1/r} + 1)\,|S_i|$ vertices and no vertex of $S_i$ is joined to any vertex of $G_1$. Repeat the same construction for $G_1$ and continue until all vertices are exhausted. The union of the $S_i$ belonging to the $G_i$ will be our large independent set of size $> (1 - \eta)n^{1-1/r}$ for every $\eta > 0$ if $n > n_0(\eta)$. This last statement easily follows from (6.1) and (6.2).

Probably the exponent $1 - 1/r$ cannot be improved this is known only for $r = 1$. I expect that $cn^{1-1/r}$ can be improved by a logarithmic factor but this also is known only for $r = 1$.

Assume now that $G(n)$ has girth greater than $2r + r$. (i.e. $G(n)$ has no circuit of length $\leq 2r + 2$). I cannot prove more than Theorem 1, i.e. I can only show that $G(n)$ has an independent set of size greater than $cn^{1-1/r}$. I wonder if the exponent $1 - 1/r$ is best possible. The case $r = 1$ is perhaps most interesting, i.e. $G(n)$ has no triangle and rectangle. Is there an independent set of size $> n^{\frac{1}{2}+\varepsilon}$? I do not know.

[1] P. Erdös, "Graph Theory and probability II", *Canad J. Math.* 13 (1961), 346-352.

For a penetrating and deep study of extremel problems on cycles in graphs see;

[2] J.A. Bondy and M. Simonovits, "Cycles of even length in graphs", *J. Combinatorial Theory* 16B (1974), 97-105.

[3] J.E. Graver and J. Yackel, "Some graph theoretic results associated with Ramsey's theorem", *J. Combinatorial Theory* 4 (1968), 125-175.

7. To finish our paper we now prove Theorem 2. First of all observe that Theorem 2. clearly holds for $c > \frac{1}{4}$. To see this observe that, by the lemma stated in 5, our $G(n; (\frac{1}{4} + \delta)n^2)$ contains a subgraph $G'$ of $N > c_1 n$ vertices each vertex of which has valency greater than $N(1+\delta)/2$. But then to every two vertices of $G_1$ there exist $\delta N > \delta c_1 n$ vertices which are joined to both of them. But then it is immediate that every set $y_1, \ldots, y_t$, $t = [\delta c_1 n]$ of vertices is a $K_{top}(t)$ i.e. any two are joined by vertex disjoint paths of length two. Thus Theorem 2 is proved for $c > \frac{1}{4}$.

Assume now that Theorem 2 is false. Let $C$ be the upper bound of the numbers for which Theorem 2 fails. In other words, for every $\varepsilon > 0$ there is an infinite sequence $n_1 < n_2 < \ldots$ and graphs $G(n_i; (C - \varepsilon)n_i^2)$ which do not contain a $K_{top}(\ell)$ for

$\ell > \eta n^{\frac{1}{2}}$, for any fixed $\eta$ if $n_i > n(\eta, \varepsilon)$, but no such sequence of graphs $G(n; (C + \varepsilon)n^2)$ exist. We now easily show that this assumption leads to a contradiction.

First of all our assumption means that there is an infinite sequence of integers $n_1 < \ldots$ so that there is a graph $G(n_i; (C - o(1)n_i^2)$ the largest $K_{top}(\ell)$ of which satisfies $\ell/n_i^{\frac{1}{2}} \to 0$ and that $C$ is the largest number with this property. Further by the trivial lemma stated in 5, we can assume that every vertex of our $G$ has valency not less than $(2C - o(1))n_i^2$. Our assumption implies that there is a sequence $\eta_i \to 0$ and $\ell_i \to \infty$ so that our $G(n_i; (C - o(1)n_i^2)$ has the property that we can omit $[\eta_i n_i]$ of its vertices, so that in the remaining graph $G'(n_i - [\eta\, n_i]) = G_i'$ there are two vertices which can not be joined by a path of length less than $k_i$. To see this, observe that if our statement would be false then for sufficiently small $\eta$ every set of $[\eta\, n_i^{\frac{1}{2}}] = \ell$ sets of vertices of our $G(n_i)$ would be a $K_{top}(\ell)$.

To arrive at the contradiction let $y_1$ and $y_2$ be two vertices of our $G_i'$ which can not be joined by a path of length less than $k_i$. Observe that every vertex of our $G_i'$ has valency not less than $(2C - o(1) - \eta_i)n_i = (2C - o(1))n_i$. Denote by $S_1^{(j)}$, respectively $S_2^{(j)}$, the set of vertices which can be joined to $y_1$, respectively $y_2$, with $i$ but not with fewer edges. Clearly for every $t \le [\frac{k_i-1}{2}]$ the two sets $\bigcup_{j=0}^{t} S_1^{(j)}$ and $\bigcup_{j=0}^{t} S_2^{(j)}$ are disjoint. $(S_1^{(0)} = y_1,\ S_2^{(0)} = y_2)$ (Otherwise there would be a path of length less than $k_i$ joining $y_1$ and $y_2$). Without loss of generality we can thus assume

$$(7.1) \qquad |S_1^{(t)}| < \frac{n_i}{2}\,,\quad |S_1^{(1)}| > (2C - o(1)n_i.$$

From (7.1) we obtain that there is an $2 \le r < t$ for which

$$(7.2) \qquad |S_1^{(r)}| < \frac{n_i}{2(t-1)}\ .$$

Let now $G_i^{(r)}$ be the subgraph of $G_i'$ spanned by the vertices of $\bigcup_{j=0}^{r-1} S_1^{(j)}$. The valency of every one of its vertices is at least $(2C - o(1) - \frac{1}{2(t-1)})n_i = (2C - o(1))n_i$ (since the vertices not in $G_i^{(r)}$ which are joined to a vertex of $G_i^{(r)}$ are all in $S_1^{(r)}$ which implies our statement by (7.2)).

The sequences of graphs $G_i^{(r)}$ establish our contradiction. The i-th graph has by (7.1) and (7.2) more than $(2C - o(1))n_i$ and fewer than $\frac{n_i}{2}$ vertices each of which has valency not less than $(2C - o(1))n_i$ and the largest $K_{top}(\ell)$ of it is $o(n_i^{\frac{1}{2}})$. This contradicts the maximality property of $C$ and hence Theorem 2 is proved.

05B05, 15A24

# INTEGRAL PROPERTIES OF COMBINATORIAL MATRICES*

Marshall Hall, Jr.

California Institute of Technology

## 1. INTRODUCTION

The incidence matrix $A$ of a symmetric block design $D$, with parameters, $v,k,\lambda$ where $v > k > \lambda > 0$ and $k(k-1) = \lambda(v-1)$ satisfies

(1.1) $$AA^T = (k-\lambda)I + \lambda J$$

$A^T$ being the transpose of $A$, and $J$ the matrix of all ones. It also satisfies

(1.2) $$A^TA = (k-\lambda)I + \lambda J, \quad AJ = kJ, \quad JA = kJ.$$

An integral matrix $A$ satisfying all relations in (1.1) and (1.2) is the incidence matrix of a design, but there exist integral matrices satisfying (1.1) but not (1.2). These have been investigated [3, 4, 5, 6] and are treated in section 2. Integral matrices $A$ satisfying

(1.3) $$AA^T = mI$$

include Hadamard matrices and are also related to (1.1).

Given an $r$ rowed matrix $X$ the problem of finding a matrix $A$ with these as its first rows and satisfying (1.1), (1.2) or (1.3) is called the *completion problem.* $X$ must satisfy certain obvious necessary conditions. In Hall-Ryser [5] it has been shown that over the rational field that if there are any rational solutions of (1.1) or (1.3) then a rational completion always exists. For an integral start $X$ it is always possible to complete up to 7 remaining rows [2, 4, 7] but not 8. These results are discussed and summarized in section 3.

Conditions for the existence of rational matrices satisfying these combinatorial relations are well known. The transition to integral matrices is a difficult major step and should be the subject of much further study.

## 2. THE INCIDENCE EQUATION.

Let $A$ be a $v$ by $v$ real matrix satisfying

---

*This research was supported in part by NSF Grant MPS-72-0535A02.

$$AA^T = (k-\lambda)I + \lambda J. \tag{2.1}$$

Here $A^T$ is the transpose of $A$, and $J$ is the $v$ by $v$ matrix all of whose entries are 1's. Furthermore we suppose that $v, k, \lambda$ are integers satisfying

$$v > k > \lambda > 0, \; k(k-1) = \lambda(v-1). \tag{2.2}$$

If $v, k, \lambda$ satisfy these conditions we call (2.1) the *incidence equation* for $A$.

For the elementary theory of block designs see [1] Chapter 10. If $D$ is a symmetric $v, k, \lambda$ block design, then we may assume that (2.2) holds. Let $P_1,\ldots,P_v$ be the points of $D$, and $B_1,\ldots,B_v$ the blocks of $D$. Then the incidence matrix $A$ of $D$ is defined as

$$A = [a_{ij}], \; i, j = 1,\ldots,v$$

$$a_{ij} = 1 \text{ if } P_i \in B_j \quad . \tag{2.3}$$
$$a_{ij} = 0 \text{ if } P_i \notin B_j$$

Here $A$ satisfies (2.1) and also the further relations

$$A^TA = (k-\lambda)I + \lambda J, \; AJ = kJ, \; JA = kJ. \tag{2.4}$$

If $A$ is a non-singular real matrix satisfying (2.1) and either the second or third relation of (2.4) then it is known [1, p.104] that (2.2) and the other relations in (2.4) hold. Such a matrix $A$ we call a normal solution.

In (2.1) if we multiply any column of $A$ by $-1$ the relation still holds. If this is done so that the column sums of $A$ are non-negative we say that $A$ is in *normalized form*.

In order that (2.1) have a rational solution the Bruck-Ryser-Chowla conditions must hold [1, p.107].

I *If* $v$ *is even* $k-\lambda$ *is a square*

II *If* $v$ *is odd then*

$$z^2 = (k-\lambda)x^2 + (-1)^{\frac{v-1}{2}}\lambda y^2$$

*has a solution in integers* $x, y, z$ not all zero.

We shall always assume that these conditions hold.

For solutions A of (2.1) which do not in general satisfy (2.4) the following notation is appropriate:

(2.5) $$s_i = \sum_r a_{ri} .$$

Here $s_i$ is the $i^{th}$ column sum of A.

(2.6) $$A^T A = [t_{ij}] .$$

Thus $t_{ij}$ is the inner product of the $i^{th}$ and $j^{th}$ columns of A.

Multiply (2.1) on the left by $A^{-1}$ to obtain

(2.7) $$A^T = (k-\lambda)A^{-1} + \lambda A^{-1}J .$$

Multiply this on the right by J and use (2.2) to obtain

(2.8) $$A^T J = (k-\lambda+\lambda v)A^{-1}J = k^2 A^{-1}J.$$

Multiplying (2.7) by $k^2A$ on the right and replace $k^2A^{-1}J$ by $A^TJ$, we find

(2.9) $$k^2 A^T A = k^2(k-\lambda)I + \lambda A^T JA .$$

By direct calculation

(2.10) $$A^T JA = [s_i s_j] .$$

Thus

(2.11) $$k^2[t_{ij}] = k^2(k-\lambda)I + \lambda[s_i s_j] .$$

This is equivalent to

(2.12) $$k^2 t_{ij} = k^2(k-\lambda)\delta_{ij} + \lambda s_i s_j, \quad i,j = 1,\ldots,v .$$

A further easy calculation is

$$JAA^TJ = (s^2 +\ldots+ s_v^2)J \tag{2.13}$$

$$J[(k-\lambda)I + \lambda J]J = k^2vJ$$

and we conclude

$$s_1^2 +\ldots+ s_v^2 = k^2v \; . \tag{2.14}$$

The relations established so far are solely algebraic consequences of (2.1) and (2.2). Let us now make the additional assumption that A is an integral matrix. Two lemmas are easily established.

LEMMA A. *Suppose that* A *is an integral matrix satisfying* (2.1) *and* (2.2). *Then* $s_i \equiv t_{ii}$ (mod 2) *and* $s_i \leq t_{ii}$. *If* $s_i = t_{ii}$ *then* $a_{ri} = 0$ *or* 1 *for* $r = 1,\ldots,v$ .

PROOF: Since $t_{ii} = \Sigma a_{ri}^2$ and $a_{ri} \equiv a_{ri}^2$ (mod 2) we have $s_i \equiv t_{ii}$ (mod 2). Also $a_{ri} \leq a_{ri}^2$ with equality only if $a_{ri} = 0$ or 1 so that $s_i \leq t_{ii}$ with equality only if every $a_{ri} = 0$ or 1.

LEMMA B. *Suppose that* A *is an integral matrix satisfying* (2.1) *and* (2.2). *Then we cannot have an* $s_i$ *lying strictly between* k *and* $k(k-\lambda)/\lambda$.

PROOF: The inequality $s_i \leq t_{ii}$ of Lemma A when applied to (2.12) gives

$$k^2s_i \leq k^2t_{ii} = k^2(k-\lambda) + \lambda s_i^2 \; ,$$

$$0 \leq \lambda s_i^2 - k^2s_i + k^2(k-\lambda) \; . \tag{2.15}$$

As the roots of $\lambda x^2 - k^2x + k^2(k-\lambda)$ are $x = k$ and $x = k(k-\lambda)/\lambda$ and $\lambda > 0$ any value of $s_i$ strictly between k and $k(k-\lambda)/\lambda$ will violate (2.15).

Two results due to Ryser [6] follow from the above.

THEOREM 2.1 (Ryser). *Let* A *be a* v *by* v *integral matrix satisfying* $AA^T = A^TA = (k-\lambda)I + \lambda J$ *where* $v > k > \lambda > 0$ *and* $k(k-1) = \lambda(v-1)$. *Then* A *or* -A *is the incidence matrix of a symmetric design.*

THEOREM 2.2 (Ryser). *Let* $A$ *be a* $v$ *by* $v$ *integral matrix satisfying* $AA^T = (k-\lambda)I + \lambda J$ *where* $v > k > \lambda > 0$ *and* $k(k-1) = \lambda(v-1)$ *and suppose that* $k-\lambda$ *is odd and that* $(k,\lambda)$ *is squarefree. Then for* $A$ *in normalized form,* $A$ *is the incidence matrix of a symmetric block design.*

For Theorem (2.1) as

$$t_{ij} = (k-\lambda)\delta_{ij} + \lambda$$

(2.16)
$$k^2 t_{ij} = k^2(k-\lambda)\delta_{ij} + \lambda s_i s_j$$

Here $s_i^2 = k^2$ and $s_i s_j = k^2$ so that either $s_i = k$ in every case or $s_i = -k$ in every case. In this latter situation replace $A$ by $-A$ so that $s_i = k$, but as $t_{ii} = k$ from Lemma A every entry of $A$ is $0$ or $1$ and $A$ is the incidence matrix of a design.

For Theorem 2.2 since $k^2 t_{ii} = k^2(k-\lambda) + \lambda s_i^2$ as $(k,\lambda)$ is squarefree it can be shown that $k$ divides $s_i$, so that

(2.17) $$s_i = ku_i, \quad i = 1,\ldots,v .$$

From (2.14) it now follows that

(2.18) $$u_1^2 + \ldots + u_v^2 = v .$$

If some $u_i = 0$, then $s_i = 0$ and from (2.12) $t_{ii} = k-\lambda$, but from Lemma A $s_i \equiv t_{ii} \pmod 2$ or $0 \equiv k-\lambda \pmod 2$ which is not possible if $k-\lambda$ is odd. With $A$ in normalized form $s_i \geq 0$ and as $s_i > 0$, $u_i > 0$ and from (2.18) $u_1 = \ldots = u_v = 1$, whence $s_1 = \ldots = s_v = k$ and again $A$ is a $0$, $1$ matrix and so the incidence matrix of a design.

We say that an integral matrix $A$ satisfying (2.1) is of type I if it is the incidence matrix of a design and of type II if it is not. For Type II we must have either $k-\lambda$ even or $(k,\lambda)$ divisible by a squared factor.

Let us consider the case of projective planes with parameters $v = n^2 + n + 1$, $k = n + 1$, $\lambda = 1$. Since $(k,\lambda) = 1$ we need only consider cases where $k-\lambda = n$ is even. From the Bruck-Ryser conditions it follows that either $n \equiv 0 \pmod 4$ or $n \equiv 2 \pmod 4$ and $n = x^2 + y^2$. From Lemma B either $s_i \leq k = n + 1$ or $s_i \geq (n+1)n$. As $s_i = ku_i = (n+1)u_i$, either $u_i = 0$, $1$ or $u_i \geq n+1$. Since

(2.19) $$u_1^2 + \ldots + u_v^2 = v = n^2 + n + 1$$

we either have $u_1 = \ldots = u_v = 1$ giving a Type I solution or a Type II solution, numbering the $u$'s appropriately with

(2.20) $$u_1 = n, \quad u_2 = \ldots = u_{n+2} = 1, \quad u_{n+3} = \ldots u_v = 0.$$

Thus

(2.21) $$s_1 = n(n+1), \quad s_2 = \ldots = s_{n+2} = n + 1, \quad s_{n+3} = \ldots = s_v = 0\ .$$

It now follows that

(2.22) $$t_{11} = n(n+1), \quad t_{ii} = n+1, \quad i = 2,\ldots n+2, \quad t_{ii} = n, \quad i = n+2,\ldots,v$$

and indeed that

(2.23) $$A^T A \;=\; \begin{array}{|c|ccc|cc|} & & n+1 & & n^2-1 & \\ \hline n^2+n, & n, & n, \ldots & n & 0 & 0 \\ \hline n, & n+1, & 1, \ldots & 1 & & \\ n & 1 & n+1, & & 0 & \\ \vdots & \vdots & & & & \\ n & & & n+1 & & \\ \hline 0 & & & & n & 0 \\ \vdots & & & & & \ddots \\ 0 & & & 0 & & n \end{array}$$

From this and Lemma A the first $n+2$ columns of $A$ consist of 0's and 1's and $A$ has the shape

(2.24) $$A \;=\; \begin{array}{|c|ccccc|ccc|} & & & n+1 & & & & n^2-1 & \\ 0 & 1 & 1 & 1 & \ldots & 1 & 0 & \ldots & 0 \\ \hline 1 & 1 & 0 & . & . & 0 & & & \\ 1 & 1 & & & & 0 & & \ldots & \\ \vdots & \vdots & & . & . & . & & & \\ 1 & 1 & 0 & . & . & . & & & \\ 1 & 0 & 1 & & & 0 & & \ldots & \\ 1 & & 1 & & & \vdots & & & \\ 1 & & 1 & & & 0 & & & \\ 1 & 0 & & & & \vdots & & & \\ 1 & & & & & 1 & & & \end{array}$$

Here the first column of $A$ has a $0$ in the first row and $1$ in the remaining $n^2+n$ rows. Each of the next $n+1$ columns has a $1$ in the first row and $n$ further 1's, different columns using different sets of $n$ rows. This much of $A$ is forced by (2.23) and the observations of Lemma A. Let $\bar{A}$ be the matrix obtained from $A$ by deleting the first row and column. Then immediately

$$\bar{A}\bar{A}^T = nI_{n^2+n} \quad . \tag{2.25}$$

Conversely an integral $\bar{A}$ satisfying (2.25) with its first $n+1$ columns as in (2.24) can be bordered as in (2.24) to give an integral $A$ satisfying $AA^T = nI + J$. One way of constructing such an $\bar{A}$ is to take

$$\bar{A} = \left|\begin{array}{ccccccc} 1 & 0 & 0 & \cdot & & 0 & \\ 1 & & & \overset{n-1}{\boxed{X_1}} & & & \\ \vdots & & & & & & \\ 1 & 0 & & & \overset{n-1}{\boxed{X_2}} & & \\ 0 & 1 & & & & & \\ & 1 & & & & \ddots & \\ 0 & 1 & & & & & \overset{n-1}{\boxed{X_{n+1}}} \\ 0 & & \vdots & 0 & & & \\ 0 & & 1 & & & & \end{array}\right| \tag{2.26}$$

Here for $i = 1,\ldots,n+1$

$$\underline{X_i} = \left|\begin{array}{cc} 1 & \\ \vdots & X_i \\ 1 & \end{array}\right| \; , \quad \underline{X_i X_i^T} = nI_n \; .$$

For example with $n = 10$, we may take $\underline{X_i} = X$

$$X = \left|\begin{array}{rrrrrrrrrr} 1 & 1 & 1 & 1 & 1 & 1 & 1 & 1 & 1 & 1 \\ 1 & 1 & -1 & 0 & 0 & 0 & 2 & -1 & -1 & -1 \\ 1 & 1 & 0 & -1 & 0 & 0 & -1 & -2 & 1 & 1 \\ 1 & 1 & 0 & 0 & -1 & 0 & -1 & 1 & -2 & 1 \\ 1 & 1 & 0 & 0 & 0 & -1 & -1 & 1 & 1 & -2 \\ 1 & -1 & 1 & 1 & 1 & 1 & -1 & -1 & -1 & -1 \\ 1 & -1 & 2 & -1 & -1 & -1 & 1 & 0 & 0 & 0 \\ 1 & -1 & -1 & -2 & 1 & 1 & 0 & 1 & 0 & 0 \\ 1 & -1 & -1 & 1 & -2 & 1 & 0 & 0 & 1 & 0 \\ 1 & -1 & -1 & 1 & 1 & -2 & 0 & 0 & 0 & 1 \end{array}\right| \tag{2.27}$$

But this structure is not necessary. For example with $n = 10$ we may take two columns of 10 ones in $\bar{A}$ and complete them to a submatrix $A_{20}$ of dimension 20 with $A_{20}A_{20}^T = 10I_{20}$ as follows

(2.28)

$A_{20}$ =

| | 1 | 2 | 3 | 4 | 5 | 6 | 7 | 8 | 9 | 10 | 11 | 12 | 13 | 14 | 15 | 16 | 17 | 18 | 19 | 20 |
|---|---|---|---|---|---|---|---|---|---|---|---|---|---|---|---|---|---|---|---|---|
| 1 | 1 | | 1 | | 2 | 2 | | | | | | | | | | | | | | |
| 2 | 1 | | 1 | | -2 | 1 | 1 | 1 | 1 | | | | | | | | | | | |
| 3 | 1 | | 1 | | | -1 | -2 | | | 1 | 1 | | | | | | | | | |
| 4 | 1 | | 1 | | | -1 | 1 | -2 | | 1 | | 1 | | | | | | | | |
| 5 | 1 | | 1 | | | -1 | | | -1 | -2 | -1 | -1 | | | | | | | | |
| 6 | 1 | | -1 | | | | | | | | | | 2 | -1 | 1 | 1 | 1 | | | |
| 7 | 1 | | -1 | | | | | | | | | | 1 | 1 | -2 | -1 | | 1 | | |
| 8 | 1 | | -1 | | | | | | | | | | -1 | 2 | 1 | | 1 | -1 | | |
| 9 | 1 | | -1 | | | | | | | | | | -1 | -1 | | | -1 | | 2 | 1 |
| 10 | 1 | | -1 | | | | | | | | | | -1 | -1 | | | -1 | | -2 | -1 |
| 11 | | 1 | | 1 | | | 1 | 1 | -2 | | 1 | 1 | | | | | | | | |
| 12 | | 1 | | 1 | | | -1 | | 1 | -1 | -1 | 2 | | | | | | | | |
| 13 | | 1 | | 1 | | | | -1 | 1 | -1 | 2 | -1 | | | | | | | | |
| 14 | | 1 | | 1 | | | | | | 1 | -1 | -1 | | | | | | | -1 | 2 |
| 15 | | 1 | | 1 | | | | | | 1 | -1 | -1 | | | | | | | 1 | -2 |
| 16 | | | 1 | -1 | 1 | -1 | 1 | 1 | 1 | | | | | | 1 | -1 | | 1 | | |
| 17 | | 1 | | -1 | -1 | 1 | -1 | -1 | -1 | | | | | | 1 | -1 | | 1 | | |
| 18 | | 1 | | -1 | | | | | | | | | 1 | 1 | | 1 | -2 | -1 | | |
| 19 | | 1 | | -1 | | | | | | | | | -1 | | -1 | 2 | 1 | 1 | | |
| 20 | | 1 | | -1 | | | | | | | | | | -1 | -1 | -1 | 1 | -2 | | |

In (2.24) the columns of $A$ are arranged so that $s_1 = n(n+1)$, $s_2 = \ldots s_{n+2} = n+1$, $s_{n+3} = \ldots = s_v = 0$. Let us rearrange so that $s_1 = \ldots = s_{n+1} = n+1$, $s_{n+2} = n(n+1)$, $s_{n+3} = \ldots = s_v = 0$. Now let $U = [u_{ij}]$ be an orthogonal matrix and consider the matrix $A^* = AU$. Here $a^*_{ij} = \sum_t a_{it}u_{tj}$ and so for column sums $s^*_j = \sum_i a^*_{ij} = \sum_{i,t} a_{it}u_{tj} = \sum_t s_t u_{tj}$. Hence if $U$ is of the shape

(2.29)

$$U = \left|\begin{array}{c|c} I_{n+1} & 0 \\ \hline 0 & \frac{1}{n} H_{n^2} \end{array}\right|$$

where $H_{n^2}$ is an Hadamard matrix of order $n^2$ with its first row consisting entirely of 1's then for $A^*$, $s_1^* = \dots s_{n+1}^* = n+1$ and for $j > n+2$ $s_j^* = \sum_{t=n+2}^{v} s_t u_{tj} = n(n+1)\cdot\frac{1}{n} + 0 = n+1$ since $s_{n+2} = n(n+1)$ and $s_j = 0$ for $j > n+2$ while $u_{n+2,j} = \frac{1}{n}$.

In this case the matrix $A^* = AU$ satisfies the incidence equation since $A^*A^{*T} = AUU^TA^T = AA^T = nI + J$ and as every column sum of $A^*$ is $n+1$ we have $JA^* = (n+1)J$ and so $A^*$ is normal. As $n$ is even $n^2 \equiv 0 \pmod 4$ and it is to be expected that an Hadamard matrix $H_{n^2}$ exists. For $n = 10$, a number of Hadamards $H_{100}$ are known. However thedenominator of $A^* = AU$ is not $n$ but is at most $\frac{1}{2}n$, since in a row of $A$ beyond the first $n+1$ columns the square sum and so the sum of elements is even. Hence the inner product with a column of $\pm$ 1's in an Hadamard matrix will be even.

In this way a number of normal matrices $A$ of denominator 5 have been found satisfying the incidence equations for the plane of order 10.

3. COMPLETION PROPERTIES.

Let $A$ be a rational $n$ by $n$ matrix (or perhaps $v$ by $v$ matrix) and let $X$ be the $r$ by $n$ matrix consisting of the first $r$ rows of $A$, while $Y$ is the $s$ by $n$ matrix $(r+s = n)$ consisting of the remaining rows of $A$. Thus

(3.1) $$A = \left|\begin{matrix} X \\ Y \end{matrix}\right| .$$

The equations of interest here are

(3.2) $$AA^T = mI_n,$$

or

(3.3) $$AA^T = (k-\lambda)I + \lambda J = A^TA$$

$$AJ = JA = kJ,\ v > k > \lambda > 0,\ k(k-1) = \lambda(v-1).$$

If the matrix $X$ is given then $A$ of (3.1) is called a completion and we are interested in completions satisfying (3.2) or (3.3).

For the existence of rational solutions of (3.3) the Bruck-Ryser-Chowla conditions given in the previous section are both necessary and sufficient. Sufficiency involves the deep Hasse-Minkowski theory of quadratic forms. For (3.2) necessary conditions are

(1) If $n$ is odd, $m$ is a square

(2) If $n \equiv 2 \pmod 4$, $m = a^2 + b^2$ for integers $a, b$

(3) If $n \equiv 0 \pmod 4$, $m$ is positive.

These conditions, necessary for the existence of rational solutions of (3.2) are in fact sufficient for the existence of integral solutions.

Two general theorems are relevant here. The first comes directly from Hall-Ryser [5].

THEOREM 3.1. *Suppose that* $A$ *is a non singular square matrix of order* $n$ *such that*

$$AA^T = \left|\begin{array}{c|c} D_1 & 0 \\ \hline 0 & D_2 \end{array}\right| = D_1 \oplus D_2$$

*where* $D_1$ *is of order* $r$, $D_2$ *of order* $s$ *and* $r+s = n$. *Let* $X$ *be an arbitrary matrix of size* $r$ *by* $n$ *such that* $XX^T = D_1$. *Then there is an* $n$ *by* $n$ *matrix* $Z$ *having* $X$ *as its first* $r$ *rows such that* $ZZ^T = D_1 \oplus D_2$. *This result holds for all fields* $F$ *of characteristic not* 2.

The next is a slight generalization of a theorem in [5].

THEOREM 3.2. *Suppose that* $AA^T = D_1 \oplus D_2$ *where* $A$ *is of order* $n$ *and non-singular and* $D_1$ *and* $D_2$ *are of order* $r$ *and* $s = n-r$. *Suppose further that* $X$ *and* $Y$ *are* $r$ *by* $n$ *matrices such that* $XX^T = YY^T = D_1$. *Then there exists an orthogonal matrix* $U$ *of order* $n$ *such that* $XU = Y$. *This result holds for all fields* $F$ *of characteristic not* 2.

Let us consider first the completion problem for equation (3.2). A necessary condition on the $r$ by $n$ matrix $X$ is

$$(3.4) \qquad XX^T = mI_r .$$

Assuming the appropriate necessary condition (1), (2) or (3) for the existence of some rational A satisfying (3.2) it follows [5] that over the rational field a completion always exists. If X is integral and $r+s = n$, and $s \leq 7$ then an integral completion exists from Hall [2] and Verheiden [7]. If X has at least 3 rows of $\pm 1$'s, and $s \leq 7$ there is a completion to an Hadamard matrix. This is false for $s = 8$ as the example $X = [1,1,1,1,1,1,1,1]$ shows. Here $AA^T = 9I$ has solutions, but a row of square sum 9 has its sum odd and so cannot be orthogonal to the vector X. Verheiden's proof rests on the deep fact that an integral quadratic form of determinant 1 on at most 7 variables is integrally equivalent to a sum of squares. This conclusion is false for 8 variables. For the case of $AA^T = 9I$ above there is a completion of denominator 2 and Verheiden shows that there is always a completion with denominator a power of 2.

Now let us consider (3.3) and suppose that X is an r by v matrix. It is necessary that S satisfy

$$XX^T = (k-\lambda)I_r + \lambda J_{rr}, \quad XJ = kJ_{rv}. \tag{3.5}$$

Here $J_{rr}$ and $J_{rv}$ are r by r and r by v matrices of all 1's. Assuming the necessary Bruck-Ryser-Chowla conditions, it is shown in Hall-Ryser [5] that over the rational field a completion always exists. An integral matrix X satisfying (3.5) is always a zero-one matrix since for any row of X the square sum of the entries is k and the sum of the entries is also k. It has been shown by Hall [4] that an integral completion always exists if there are at most four rows remaining and Verheiden [7] has extended this to seven rows. With $v = 11$, $k = 5$, $\lambda = 2$ the following example shows that the result is not true for 8 remaining rows

$$X = \begin{vmatrix} 1 & 1 & 1 & 1 & 1 & 0 & 0 & 0 & 0 & 0 & 0 \\ 1 & 1 & 0 & 0 & 0 & 1 & 1 & 1 & 0 & 0 & 0 \\ 1 & 1 & 0 & 0 & 0 & 0 & 0 & 0 & 1 & 1 & 1 \end{vmatrix} . \tag{3.6}$$

In this case there is a completion of denominator 2. For (3.3) it is not known what general conditions on denominators are.

## References

[1] Marshall Hall, Jr., *Combinatorial Theory*, John Wiley, New York and London, 1967.

[2] Marshall Hall, Jr., *Integral matrices* A *for which* $AA^T = mI$. Academic Press, to appear.

[3] Marshall Hall, Jr., Matrices satisfying the incidence equation, *Proceedings of the Fifth Hungarian Congress on Combinatorics*, 1976.

[4] Marshall Hall, Jr., Combinatorial Completions, *Proceedings of the Cambridge Combinatorial Conference*, 1977.

[5] M. Hall and H.J. Ryser, Normal completions of incidence matrices, *Amer. J. of Math.* 76(1954), 581-589.

[6] H.J. Ryser, Matrices with integer elements in combinatorial investigations, *Amer. J. Math.* 74(1952), 769-773.

[7] E. Verheiden, Integral and rational completions of combinatorial matrices, I, II *Combinatorial Theory* (A), to appear.

# A CLASS OF THREE-DESIGNS

Haim Hanani
Department of Mathematics,
Technion, Haifa, Israel

ABSTRACT

It is proved that balanced 3-designs $B_3[k, \lambda, v]$ exist for $k = 5$, $\lambda = 30$ and every $v \geq 5$.

## 1. INTRODUCTION

Let $v \geq k \geq t \geq 2$ and $\lambda$ be positive integers. A *balanced* t-*design* $B_t[k, \lambda; v]$ is a pair $(X, \mathcal{B})$, where X is a set of *points* and $\mathcal{B}$ a family of not necessarily distinct subsets $B_i$ - called *blocks* - of X, satisfying the following conditions:

(i) $|X| = v$;
(ii) $|B_i| = k$ for every $B_i \in \mathcal{B}$;
(iii) every t-subset of X is contained in exactly $\lambda$ blocks of $\mathcal{B}$.

A well-known theorem (see e.g. [4]) states:

THEOREM 1.1. A necessary condition for the existence of a balanced t-design $B_t[k, \lambda; v]$ is that

$$\lambda\binom{v-h}{t-h} \equiv 0 \left(\mathrm{mod}\binom{k-h}{t-h}\right), \qquad h = 0, 1, \ldots, t-1 .$$

Proof. $\lambda\binom{v-h}{t-h}/\binom{k-h}{t-h}$ is the number of blocks which contain h fixed points of X.

Balanced 2-designs $B_2[k, \lambda; v]$ are known as *balanced incomplete block designs* (BIBD) $B[k, \lambda; v]$. The BIBD's are discussed extensively in [5].

With regard to balanced 3-designs it is known [3] that for $k = 4$ (and every $\lambda$) the necessary condition of Theorem 1.1 is also sufficient. More explicitly:

THEOREM 1.2. A necessary and sufficient condition for the existence of a balanced 3-design $B_3[4, \lambda; v]$ is that $\lambda v \equiv 0 \pmod 2$, $\lambda(v-1)(v-2) \equiv 0 \pmod 3$ and $\lambda v(v-1)(v-2) \equiv 0 \pmod 8$.

A class of 3-designs has also been constructed by Alltop [1].

It will be proved in this paper that the condition of Theorem 1.1 is sufficient also for $k = 5$ and $\lambda = 30$, in other words it will be proved that for every integer $v \geq 5$ there exists a design $B_3[5, 30; v]$.

## 2. NOTATION

The notation is basically the same as in [9], namely:

lower case letters (a,k,m,...) will denote points or integers;
capital letters (B,K,C,...) will denote sets of points or sets of integers;
script capital letters ($\mathcal{B}$,$\mathcal{G}$,...) will denote families of sets;
q denotes exclusively a prime-power (an integer which is a power of a prime);
$|S|$ denotes the cardinality of the set S;

I(n) denotes the set of non-negative integers smaller than n, e.g. I(5) = {0,1,2,3,4};

Z(n) denotes the cycle of residues mod n;

GF(q) denotes Galois field of order q;

Z(p,x), when p is a prime, denotes Z(p) with the additional information that x is the primitive root used;

GF(q,f(x)=0) denotes GF(q) with the additional information that x is the primitive element used;

whenever the blocks are written within brackets < > and X = Z(p,x) or X = GF(q,f(x)=0), then the points are denoted by *exponents* of x and so the symbol $\alpha$ denotes the point $x^\alpha$; for the element 0 the symbol $\emptyset$ is used;

whenever the blocks are written within braces { }, then the points are denoted by the elements of Z(p,x) or GF(q,f(x)=0), respectively;

when X = Y × Z, then the points are denoted by a symbol (a,b) where a is an exponent of an element, or an element in Y, and b an exponent of an element, or an element in Z, depending on the brackets of the block;

in case of transversal designs, X = Y × Z where Y denotes the set of points in a group and Z the family of groups; in such case a semicolon is used in the symbol (a;b);

the words "mod q" after a block denote that all the elements of the block should be taken cyclically by adding to them all the residues of Z(q) or all the elements of GF(q) respectively;

if S is a set of integers, then $S + 1 = \{s+1 : s \in S\}$.

## 3. TRANSVERSAL t-DESIGNS

Let $s \geq t$, $r$ and $\lambda$ be positive integers. A *transversal t-design* $T_t[s, \lambda; r]$ is a triple $(X, G, P)$, where $X$ is a set of *points*, $G$ - a family of subsets of $X$ called *groups*, which form a partition of $X$, and $P$ a family of subsets of $X$ called *blocks*, satisfying the following conditions:

(i) $|G_i| = r$ for every $G_i \in G$;

(ii) $|G| = s$;

(iii) $|G_i \cap B_j| = 1$ for every $G_i \in G$ and every $B_j \in P$;

(iv) every t-subset of $X$, such that each of its points is contained in a distinct group, is contained in exactly $\lambda$ blocks of $P$.

It follows immediately that in a transversal t-design $T_t[s, \lambda; r]$, $|X| = sr$, $|B_j| = s$ for every $B_j \in P$, and $|P| = \lambda r^t$.

The set of integers $r$ for which transversal t-designs $T_t[s, \lambda; r]$ exist will be denoted by $T_t(s, \lambda)$.

Transversal 2-designs $T_2[s, \lambda; r]$ are known simply as transversal designs $T[s, \lambda; r]$, (see e.g. [5]).

Clearly we have:

LEMMA 3.1. If $\lambda'$ divides $\lambda$, then $T_t(s, \lambda') \subset T_t(s, \lambda)$.

LEMMA 3.2. If $s \leq s'$, then $T_t(s', \lambda) \subset T_t(s, \lambda)$.

LEMMA 3.3. If $\{r, r'\} \subset T_t(s, 1)$, then also $rr' \in T_t(s, 1)$.

Proof. For every block $B$ of $T_t[s, 1; r]$ consider the elements of $B$ as the groups of $T_t[s, 1; r']$ and form on them the blocks of $T_t[s, 1; r']$.

LEMMA 3.4. For every $r > 0$, $r \in T_t(t+1, 1)$ holds.

Proof. $X = I(r) \times I(t+1)$. $P = \{(a_i; i) : i \in I(t+1), \sum a_i \equiv 0 (\mathrm{mod}\ r)\}$.

For $t = 3$ we prove further

LEMMA 3.5. If $q$ is a prime-power, then $q \in T_3(q+1, 1)$.

Proof. $X = GF(q, f(x) = 0) \times I(q) \cup \{(\infty_\alpha) : \alpha \in GF(q)\}$.

$P = \{(\infty_\alpha), (0; 0), (\alpha x^{2i} + \beta x^i; i+1) : i = 0, 1, \ldots q-2\} \mathrm{mod}(q;-), \alpha \in GF(q), \beta \in GF(q)$.

From Lemmas 3.2-3.5 follows the equivalent of MacNeish's theorem for transversal 3-designs:

THEOREM 3.1. If $r = \Pi q_j$, where $q_j$ are powers of distinct primes and $s = 1 + \max(3, \min q_j)$, then $r \in T_3(s, 1)$.

We shall not develop here further the theory of transversal 3-designs and we shall prove only two lemmas which will be applied subsequently.

LEMMA 3.6. $4 \in T_3(6, 1)$.

Proof. $X = (Z(3, 2) \cup \{\infty\}) \times (Z(5, 2) \cup \{\infty\})$.

$$P = \langle(\infty; \infty), (\infty; \emptyset), (\infty; 0), (\infty; 1), (\infty; 2), (\infty; 3)\rangle,$$

$$\langle(\emptyset, \infty), (\emptyset; \emptyset), (\emptyset; 0), (\emptyset; 1), (\emptyset, 2), (\emptyset, 3)\rangle \mathrm{ mod}(3; -),$$

$$\langle(\infty; \infty), (\infty; \emptyset), (0; 0), (0; 2), (1; 1), (1; 3)\rangle \mathrm{ mod}(3; 5),$$

$$\langle(\emptyset; \infty), (\emptyset; \emptyset), (0; 1), (0; 3), (1; 0), (1; 2)\rangle \mathrm{ mod}(3; 5),$$

$$\langle(\infty; \alpha), (\infty; \alpha+2), (\alpha; \alpha+1), (\alpha; \alpha+3), (\alpha+1; \infty), (\alpha+1; \emptyset)\rangle \mathrm{ mod}(3; 5), \quad \alpha = 0, 1.$$

LEMMA 3.7. $3 \in T_3(5, 2)$.

Proof. $X = (Z(2) \cup \{\infty\}) \times Z(5, 2)$.

$$P = \langle(\emptyset; \emptyset), (\emptyset; 0), (\emptyset; 1), (\emptyset; 2), (\emptyset; 3)\rangle \mathrm{ mod}(2; -), \text{ twice},$$

$$\langle(\infty; \emptyset), (\emptyset; 0), (\emptyset; 2), (0; 1), (0; 3)\rangle \mathrm{ mod}(2; 5),$$

$$\langle(\infty; \emptyset), (\infty; \alpha), (\infty; \alpha+2), (\emptyset, \alpha+1), (\emptyset, \alpha+3)\rangle \mathrm{ mod}(2; 5), \alpha = 0, 1,$$

$$\langle(\infty; \emptyset), (\emptyset; 0), (\emptyset; \varepsilon), (0; 2), (0; \varepsilon+2)\rangle \mathrm{ mod}(2; 5), \varepsilon = \pm 1.$$

## 4. PAIRWISE BALANCED t-DESIGNS

Let t, v and $\lambda$ be positive integers and K a set of positive integers. A *pairwise balanced* t-*design* $B_t[K, \lambda; v]$ is a pair $(X, \mathcal{B})$, where X is a set of points and $\mathcal{B}$ a family of blocks (subsets of X), satisfying the following conditions:

(i) $|X| = v$;

(ii) $|B_i| \in K$ for every $B_i \in \mathcal{B}$;

(iii) every t-subset of X is contained in exactly $\lambda$ blocks of $\mathcal{B}$.

A pairwise balanced t-design $B_t[K, \lambda; v]$, where $K = \{k\}$ consists of exactly one integer is a balanced t-design $B_t[k, \lambda; v]$.

The set of integers v for which pairwise balanced t-designs $B_t[K, \lambda; v]$ exist, will be denoted by $B_t(K, \lambda)$. Similarly the set of integers v for which balanced t-designs $B_t[k, \lambda; v]$ exist will be denoted by $B_t(k, \lambda)$.

The following lemmas are evident.

LEMMA 4.1. $K \subset B_t(K, 1)$.

LEMMA 4.2. If $K' \subset K$, then $B_t(K', \lambda) \subset B_t(K, \lambda)$.

LEMMA 4.3. If $\lambda'$ divides $\lambda$, then $B_t(K, \lambda') \subset B_t(K, \lambda)$.

And more generally:

LEMMA 4.4. $B_t(K, \lambda) \cap B_t(K, \lambda') \subset B_t(K, n\lambda+n'\lambda')$, where $\lambda$ and $\lambda'$ are any positive integers, and n and n' any non-negative integers.

Further we have:

LEMMA 4.5. If $v \in B_t(K', \lambda')$ and $K' \subset B_t(K, \lambda)$, then $v \in B_t(K, \lambda\lambda')$ holds.

The following special case of Lemma 4.5 will be most useful.

LEMMA 4.6. If $v \in B_t(K, 1)$ and $K \subset B_t(k, \lambda)$, then $v \in B_t(k, \lambda)$ holds.

Taking as blocks all the distinct (k-1)-subsets of a k-set we obtain:

LEMMA 4.7. $k \in B_t(k-1, k-t)$.

Further, applying Lemma 4.5, it follows:

LEMMA 4.8. $B_t(k, \lambda) \subset B_t(k-1, (k-t)\lambda)$.

Deleting one point from a design $B_t[k, \lambda; v]$ we obtain a design $B_t[\{k, k-1\}, \lambda; v-1]$. Applying Lemmas 4.7 and 4.5 it follows:

LEMMA 4.9. $B_t(k, \lambda) - 1 \subset B_t(k-1, (k-t)\lambda)$.

In a similar way, by deleting one point from a design $B_t[k, \lambda; v]$ and considering only those blocks which contained the deleted point, we obtain:

LEMMA 4.10. $B_t(k, \lambda) - 1 \subset B_{t-1}(k-1, \lambda)$.

## 5. FINITE PLANES [4]

Let q be a prime-power and d a positive integer. Consider the field $GF(q^d)$ and extend it to $F = GF(q^d) \cup \{\infty\}$. We introduce the linear transformation

$$\eta = T(\xi) = (\alpha\xi+\beta)/(\gamma\xi+\delta),\ \{\alpha,\beta,\gamma,\delta\} \subset GF(q^d),\ \{\xi,\eta\} \subset F,\ \alpha\delta-\beta\gamma \neq 0 .$$

The linear transformations are known to be one-one and to form a group.

The cross ratio $(\xi, \xi_2, \xi_3, \xi_4) = \dfrac{\xi - \xi_2}{\xi - \xi_4}\Big/\dfrac{\xi_3 - \xi_2}{\xi_3 - \xi_4}$ is the image of $\xi$ under the linear transformation, which carries $\xi_2$, $\xi_3$ and $\xi_4$, respectively, into the elements 0, 1 and $\infty$ of $GF(q) \cup \{\infty\}$.

A subset C of F is a *circle* if $(\xi_1,\xi_2,\xi_3,\xi_4) \in GF(q)$ whenever $\{\xi_1,\xi_2,\xi_3,\xi_4\} \subset C$, and if no set properly containing C has this property.

A linear transformation transforms circles into circles; also for any two circles there exists a linear transformation transforming one of them into the other, and for any three distinct elements of F there exists exactly one circle containing them. Further, observing that the set $GF(q) \cup \{\infty\}$ forms a circle, we deduce that every circle has exactly q+1 elements.

The extended field F with the system of circles on it forms a *finite inversive geometry* IG(q, d). IG(q, d) is clearly a balanced 3-design $B_3[q+1, 1; q^d+1]$, the circles serving as blocks.

Considering in IG(q, d) the circles which contain the element $\infty$ and deleting this element, a *finite affine geometry* AG(q, d) is obtained in which the truncated circles serve as *lines*. Each line has clearly q elements and - by Lemma 4.10 - a finite affine geometry AG(q, d) is a BIBD $B[q, 1; q^d]$, (see e.g. [2, p.167-179]).

In sequel we shall limit ourselves to *finite inversive planes* and *finite affine planes* (i.e. the respective geometries with d = 2). It is known [2, l.c.] that the blocks of $B[q, 1; q^2]$ (the lines of AG(q, 2)) can be partitioned into q+1 subfamilies, each consisting of q disjoint blocks. We thus obtain the following:

THEOREM 5.1. Let q be a prime-power, then $q^2 + 1 \in B_3(q+1, 1)$; furthermore, the design $B_3[q+1, 1; q^2+1]$ can be constructed in such a way, that for a given point x of the design, x is contained in a class of q blocks, which - when the point x is omitted

- are disjoint.

Consider a finite inversive plane IG(q, 2) and the related AG(q, 2). If L is any line of AG(q, 2) and C any circle of IG(q, 2), then either $L \subset C$ or $|L \cap C| \le 2$. Accordingly, if we delete t parallel lines from IG(q, 2), then the size of the remaining - partly truncated - blocks varies between q - 2t and q + 1. Considering the density of the prime-powers among the integers and Lemma 4.5, we obtain [4]:

THEOREM 5.2. For every integers $v \ge k \ge 3$, $v \in B(\{n : k \le n \le q[(q+k)/2] - 1\}, 1)$ holds, where, for $k \ge 11$, q is the smallest prime-power satisfying $q \ge 2k - 1$, and for $3 \le k \le 10$, q = 23.

For specific values of k this result may be considerably improved. We are especially interested in the case k = 5 and we prove

THEOREM 5.3. For every $v \ge 5$, $v \in B_3(K_5, 1)$ holds, where $K_5 = \{5,\dots,40,83,\dots,86\}$.

Proof. By Theorem 5.2 and Lemma 4.5 it suffices to prove our theorem for $5 \le v \le 321$, and by Lemma 4.1 for $v \notin K_5$. For $228 \le v \le 321$ delete appropriate number of lines and points from IG(19,2),

for $160 \le v \le 227$ - from IG(16,2),
for $117 \le v \le 159$ - from IG(13,2),
for $87 \le v \le 116$ - from IG(11,2),
for $63 \le v \le 82$ - from IG(9, 2),
for $48 \le v \le 62$ - from IG(8, 2),
for $v \in \{41,42,43,44,46,47\}$ - from IG(7, 2)

and for v = 45 delete from IG(7, 2) 5 points, no 4 of which are on a circle.

## 6. BALANCED 3-DESIGNS

We prove:

THEOREM 6.1. Let q be a prime-power. If $v+1 \in B_3(q+1, \lambda)$, then $qv+1 \in B_3(q+1, \lambda)$.

Proof. Let $X = I(q) \times I(v) \cup \{x\}$ and choose an additional point y. Form a balanced 3-design $B_3[q+1, \lambda; v+1]$ on the set $I(v) \cup \{y\}$ and denote by $A'$ the subfamily of blocks containing the point y with the point y deleted, and by $A$ the subfamily of all the other blocks of the 3-design $B_3[q+1, \lambda; v+1]$.

For every block $A' = \{a_i' : i \in I(q)\} \in A'$ construct - by Theorem 5.1 - on the set $I(q) \times A' \cup \{x\}$ a design $B_3[q+1, 1; q^2+1]$ such that the sets $I(q) \times \{a_i'\} \cup \{x\}$ be among its blocks; these blocks have to be taken exactly $\lambda$ times altogether. For every block $A \in A$ construct - by Lemma 3.5 - a transversal 3-design $T_3[q+1, 1; q]$ on the set $I(q) \times A$.

In the case q = 5 an additional result can be obtained, namely:

LEMMA 6.1. $22 \in B_3(6, 1)$.

Proof. $X = Z(2) \times Z(11, 2)$.

$$\mathcal{B} = < (\emptyset,\emptyset),(\emptyset,2\alpha),(\emptyset,2\alpha+7),(0,\emptyset),(0,2\alpha+2),(0,2\alpha+5) > \mathrm{mod}(-,11),\ \alpha = 0,1,2,3,4,$$
$$< (\emptyset,0),(\emptyset,2),(\emptyset,4),(\emptyset,6),(\emptyset,8),(0,\emptyset) > \mathrm{mod}(-,11),$$
$$< (\emptyset,\emptyset),(0,1),(0,3),(0,5),(0,7),(0,9) > \mathrm{mod}(-,11).$$

THEOREM 6.2. If $v+1 \in B_3(6, \lambda)$, then $4v+2 \in B_3(6, \lambda)$.

Proof. Let $X = I(4) \times I(v) \cup \{x_1, x_2\}$. Form a balanced 3-design $B_3[6, \lambda; v+1]$ on the set $I(v) \cup \{y\}$ - where y is some additional point - and denote by $\mathcal{A}'$ the subfamily of blocks containing the point y with the point y deleted and by $\mathcal{A}$ the subfamily of all the other blocks of the 3-design $B_3[6, \lambda; v+1]$. For every block $A' = \{a_i' : i \in I(5)\} \in \mathcal{A}'$ form - by Lemma 6.1 - on the set $I(4) \times A' \cup \{x_1, x_2\}$ a design $B_3[6, 1; 22]$, such that the sets $I(4) \times \{a_i'\} \cup \{x_1, x_2\}$ be among its blocks; these blocks have to be taken exactly $\lambda$ times altogether. For every block $A \in \mathcal{A}$ form - by Lemma 3.6 - a transversal 3-design $T_3[6, 1; 4]$ on the set $I(4) \times A$.

We are now able to prove the main theorem.

THEOREM 6.3. For every $v \geq 5$, $v \in B_3(5, 30)$ holds.

Proof. By Theorem 5.3 and Lemma 4.6 it suffices to prove that $K_5 \subset B_3(5, 30)$. By Lemmas 4.8 and 4.9 if $v \in B_3(6, 10)$ then $\{v, v-1\} \subset B_3(5, 30)$. The existence of designs $B_3[6, 10; v]$ for some values of v - considering Lemma 4.3 - is proved in Table 1. The construction of $B_3[5, 30; v]$ for other values of $v \notin \{39, 40\}$ is given in Table 2. It remains to prove $\{39, 40\} \subset B_3(5, 30)$. For this purpose we prove the existence of two auxiliary designs as follows:

1) Let a set $X = Z(3, 2) \times Z(5, 2)$ be given. By $P[5, 6; 15]$ we define a design which is basically the same as $B_3[5, 6; 15]$ with the difference that the 3-subsets of X, $Z(3) \times \{i\}$, $i \in Z(5)$ do not appear in any block at all. We prove the existence of $P[5, 6; 15]$ by construction of its blocks as follows:

$$\mathcal{B} = < (\emptyset,\emptyset),(\emptyset,\alpha),(\emptyset,\alpha+2),(\beta,\alpha),(\beta+1,\alpha+2) > \mathrm{mod}(3,5),\ \alpha = 0,1,\ \beta = 0,1,$$
$$< (\beta,\emptyset),(\beta,\alpha),(\beta,\alpha+2),(\beta+1,\alpha),(\beta+1,\alpha+2) > \mathrm{mod}(3,5),\ \alpha = 0,1,\ \beta = 0,1,$$
$$< (\emptyset,\emptyset),(\beta,0),(\beta,2\alpha+1),(\beta+1,2),(\beta+1,2\alpha-1) > \mathrm{mod}(3,5),\ \alpha = 0,1,\ \beta = 0,1,$$
$$< (\emptyset,\emptyset),(\emptyset,\alpha),(\emptyset,\alpha+2),(0,\alpha+\varepsilon),(1,\alpha+\varepsilon) > \mathrm{mod}(3,5),\ \alpha = 0,1,\ \varepsilon = \pm 1,$$
$$< (\emptyset,\emptyset),(0,\alpha),(0,\alpha+2),(1,\alpha),(1,\alpha+2) > \mathrm{mod}(3,5),\ \alpha = 0,1.$$

2) In a similar way we define $P[5, 6; 16]$. Here the set is $X = Z(3,2) \times Z(5,2) \cup \{(\infty)\}$ and the design $P[5, 6; 16]$ is basically the same as $B_3[5, 6; 16]$ with the difference that the 3-subsets of $Z(3) \times \{i\} \cup \{(\infty)\}$ do not appear in any block at all. The blocks of $P[5, 6; 16]$ are as follows:

$\mathcal{B} = \langle\ (\infty),\ (\beta,0),(\beta,2\alpha+1),(\beta+1,2),(\beta+1,2\alpha+3)\ \rangle \bmod(3,5),\ \alpha=0,1,\ \beta=0,1,$

$\langle\ (\infty),\ (\beta,0),(\beta,2),(\beta+1,1),(\beta+1,3)\ \rangle \bmod(3,5),\ \beta=0,1,$

$\langle\ (\emptyset,\emptyset),(\emptyset,\alpha),(\emptyset,\alpha+2),(\beta,\alpha),(\beta+1,\alpha+2)\ \rangle \bmod(3,5),\ \alpha=0,1,\ \beta=0,1,$

$\langle\ (\emptyset,\emptyset),(0,\gamma),(0,\gamma+1),(1,\gamma),(1,\gamma+1)\ \rangle \bmod(3,5),\ \gamma=0,1,2,3,$

$\langle\ (\emptyset,\emptyset),(0,\alpha)(0,\alpha+2),(1,\alpha),(1,\alpha+2)\ \rangle \bmod(3,5),\ \alpha=0,1,$ twice

$\langle\ (\alpha,\emptyset),(\alpha,\alpha),(\alpha,\alpha+2),(\alpha+1,\alpha+1),(\alpha+1,\alpha+3)\ \rangle \bmod(3,5),\ \alpha=0,1,$

$\langle\ (\emptyset,\emptyset),(0,1),(0,3),(1,0),(1,2)\ \rangle \bmod(3,5),$

$\langle\ (\emptyset,\emptyset),(\emptyset,0),(\emptyset,1),(\emptyset,2),(\emptyset,3)\ \rangle \bmod(3,-),$ 3 times.

We prove now that $v \in B_3(5,30)$ for $v \in \{39,40\}$. The set $X$ of points of the designs is defined as follows:

for $v = 39$, $X = I(3) \times I(13)$,

for $v = 40$, $X = I(3) \times I(13) \cup \{\infty\}$.

Let $y$ be an additional point. On $I(13) \cup \{y\}$ form a design $B_3[6,5;14]$ as in Table 1 (insert there $y$ instead of $\infty$). Consider the family of blocks which contain the point $y$, with the point $y$ deleted, and denote the subfamily with $\alpha=0$ by $\mathcal{A}'$ and the subfamily with $\alpha \in \{1,2\}$ by $\mathcal{A}''$. Further denote by $\mathcal{A}$ the subfamily of all other blocks of this design. Clearly, the blocks of $\mathcal{A}' \cup \mathcal{A}''$ are of size 5 and the blocks of $\mathcal{A}$ of size 6. For every block $A' \in \mathcal{A}'$ form:

for $v = 39$, a design $B_3[5,6;15]$ on $I(3) \times A'$,

for $v = 40$, a design $B_3[5,6;16]$ on $I(3) \times A' \cup \{\infty\}$,

as in Table 2. For every block $A'' \in \mathcal{A}''$ form:

for $v = 39$, a design $P[5,6;15]$ on $I(3) \times A''$,

for $v = 40$, a design $P[5,6;16]$ on $I(3) \times A'' \cup \{\infty\}$,

as above. Further, for every block $A \in \mathcal{A}$ form on $A$ a design $B_3[5,3;6]$ by Lemma 4.7 and for every block $B$ of this design form $T_3[5,2;3]$ on $I(3) \times B$ by Lemma 3.7.

Table 1

| v | λ | $B_3[6,\lambda;v]$ |
|---|---|---|
| 8 | 10 | X=I(8). $\mathcal{B}$=All 6-point subsets of I(8) |
| 10 | 5 | X=Z(2) × Z(5,2). |
| | | $\mathcal{B}$ = < (∅,∅), (∅,0), (∅,2), (0,∅), (0,1), (0,3) >mod(2,5), |
| | | < (∅,0), (∅,1), (∅,2), (∅,3), (0,α), (0,α+2) >mod(2,5), α=0,1. |
| 12 | 2 | X=Z(11,2) ∪ {∞}. |
| | | $\mathcal{B}$ = < ∞,0,2,4,6,8 >mod 11, < ∅,1,3,5,7,9 >mod 11. |
| 14 | 5 | X=Z(13,2) ∪ {∞}. |
| | | $\mathcal{B}$ = < ∞,∅,α,α+3,α+6,α+9 >mod 13, α=0,1,2, |
| | | < β,β+1,β+4,β+5,β+8,β+9 >mod 13, β=0,1,2,3. |
| 18 | 5 | X=Z(17,3) ∪ {∞}. |
| | | $\mathcal{B}$ = < ∞,∅,α,α+4,α+8,α+12 >mod 17, α=0,1,2,3, |
| | | < β,β+1,β+2,β+8,β+9,β+10 >mod 17, β=0,1,...,7. |
| 20 | 10 | X=Z(19,2) ∪ {∞}. |
| | | $\mathcal{B}$ = < ∞,∅,α,α+1,α+9,α+10 >mod 19, α=0,1,...,8, |
| | | < α,α+1,α+5,α+9,α+10,α+14 >mod 19, α=0,1,...,8, |
| | | < β,β+2,β+6,β+8,β+12,β+14 >mod 19, β=0,1,...,5, |
| | | < γ,γ+3,γ+6,γ+9,γ+12,γ+15 >mod 19, γ=0,1,2, twice. |
| 22 | 1 | Lemma 6.1. |
| 24 | 10 | X=Z(23,5) ∪ {∞}. |
| | | $\mathcal{B}$ = < ∞,∅,α,α+2,α+11,α+13 >mod 23, α=0,1,...,10, |
| | | < α,α+1,α+2,α+11,α+12,α+13 >mod 23, α=0,1,...,10, |
| | | < α,α+β+1,α+β+4,α+11,α+β+12,α+β+15 >mod 23,α=0,1,...,10,β=0,1 |
| 26 | 1 | Theorem 5.1. |
| 30 | 10 | Theorem 6.2 and 8 ε $B_3(6,10)$ as above. |
| 36 | 10 | Theorem 6.1 and 8 ε $B_3(6,10)$ as above. |
| 38 | 5 | Theorem 6.2 and 10 ε $B_3(6,5)$ as above. |

Table 1 (continued)

| $v$ | $\lambda$ | $B_3[6,\lambda;v]$ |
|---|---|---|
| 84 | 10 | $X=Z(83,2)\cup\{\infty\}$.<br>$\mathcal{B} = <\infty,\emptyset,\alpha,\alpha+3,\alpha+41,\alpha+44>\bmod 83,\ \alpha=0,1,\ldots,40,$<br>$<\alpha,\alpha+9,\alpha+19,\alpha+41,\alpha+50,\alpha+60>\bmod 83,\ \alpha=0,1,\ldots,40,$<br>$<\alpha,\alpha+4,\alpha+10,\alpha+41,\alpha+45,\alpha+51>\bmod 83,\ \alpha=0,1,\ldots,40,$<br>$<\alpha,\alpha+8,\alpha+16,\alpha+41,\alpha+49,\alpha+57>\bmod 83,\ \alpha=0,1,\ldots,40,$<br>$<\alpha,\alpha+1,\alpha+20,\alpha+41,\alpha+42,\alpha+61>\bmod 83,\ \alpha=0,1,\ldots,40,$<br>$<\alpha,\alpha+2,\alpha+11,\alpha+41,\alpha+43,\alpha+52>\bmod 83,\ \alpha=0,1,\ldots,40,$<br>$<\alpha,\alpha+8,\alpha+18,\alpha+41,\alpha+49,\alpha+59>\bmod 83,\ \alpha=0,1,\ldots,40,$ twice,<br>$<\alpha,\alpha+1,\alpha+4,\alpha+41,\alpha+42,\alpha+45>\bmod 83,\ \alpha=0,1,\ldots,40,$ twice,<br>$<\alpha,\alpha+2,\alpha+17,\alpha+41,\alpha+43,\alpha+58>\bmod 83,\ \alpha=0,1,\ldots,40,$ twice,<br>$<\alpha,\alpha+3,\alpha+14,\alpha+41,\alpha+44,\alpha+55>\bmod 83,\ \alpha=0,1,\ldots,40,$ twice. |
| 86 | 1 | Theorem 6.2 and $22\in B_3(6,1)$ by Lemma 6.1. |

Table 2

| $v$ | $\lambda$ | $B_3[5,\lambda;v]$ |
|---|---|---|
| 6 | 3 | Lemma 4.7. |
| 15 | 6 | $X=Z(3,2)\times Z(5,2)$.<br>$\mathcal{B} = <(\emptyset,\emptyset),(\emptyset,0),(\emptyset,1),(\emptyset,2),(\emptyset,3)>\bmod(3,-),$<br>$<(\emptyset,\emptyset),(\emptyset,\alpha),(\emptyset,\alpha+2),(\beta,\alpha+1),(\beta,\alpha+3)>\bmod(3,5),\alpha=0,1,\ \beta=0,1,$<br>$<(\emptyset,\emptyset),(\emptyset,\alpha),(\emptyset,\alpha+2),(0,\emptyset),(1,\emptyset)>\bmod(3,5),\ \alpha=0,1,$<br>$<(\emptyset,\emptyset),(\emptyset,\alpha),(\emptyset,\alpha+2),(\beta,\alpha),(\beta+1,\alpha+2)>\bmod(3,5),\alpha=0,1,\ \beta=0,1,$<br>$<(\emptyset,\emptyset),(0,\alpha),(0,\alpha+2),(1,\alpha),(1,\alpha+2)>\bmod(3,5),\ \alpha=0,1,$<br>$<(\emptyset,\alpha+2\beta+1),(0,\alpha),(0,\alpha+2),(1,\alpha),(1,\alpha+2)>\bmod(3,5),\alpha=0,1,\beta=0,1,$<br>$<(\emptyset,\emptyset),(\beta,0),(\beta,2),(\beta+1,1),(\beta+1,3)>\bmod(3,5),\ \beta=0,1.$ |
| 16 | 6 | $X=GF(16,x^4=x+1)$.<br>$\mathcal{B} = <1,4,7,10,13>\bmod 16,$<br>$<\emptyset,3\alpha+2,3\alpha+3,3\alpha+4,3\alpha+5>\bmod 16,\ \alpha=0,1,2,3,4,$<br>$<\emptyset,3\alpha,3\alpha+2,3\alpha+4,3\alpha+6>\bmod 16,\ \alpha=0,1,2,3,4,$<br>$<\emptyset,3\alpha+\beta,3\alpha+\beta+1,3\alpha+\beta+3,3\alpha+\beta+5>\bmod 16,\ \alpha=0,1,2,3,4,\ \beta=0,1.$ |

Table 2 (continued)

| v | λ | $B_3[5,\lambda;v]$ |
|---|---|---|
| 27 | 6 | $X=GF(27,x^3=x+2)$. |
| | | $\mathcal{B} = < \emptyset,\alpha,\alpha+1,\alpha+2,\alpha+20 >$ mod 27, $\alpha=0,1,\ldots,25$, |
| | | $< \emptyset,\alpha,\alpha+2,\alpha+3,\alpha+7 >$ mod 27, $\alpha=0,1,\ldots,25$, |
| | | $< \emptyset,\beta,\beta+1,\beta+13,\beta+14 >$ mod 27, $\beta=0,1,\ldots,12$. |
| 28 | 30 | $X=GF(27,x^3=x+2) \cup \{\infty\}$. |
| | | $\mathcal{B}$ = Blocks of $B_3[5,6;27]$ on GF(27), 4 times, |
| | | $< \emptyset,\alpha,\alpha+1,\alpha+3,\alpha+15 >$ mod 27, $\alpha=0,1,\ldots,25$, |
| | | $< \infty,\beta,\beta+\gamma+1,\beta+13,\beta+\gamma+14 >$ mod 27, $\beta=0,1,\ldots,12$, $\gamma=0,1,2,3$, |
| | | $< \infty,\beta,\beta+2,\beta+13,\beta+15 >$ mod 27, $\beta=0,1,\ldots,12$, |
| | | $< \emptyset,\beta,\beta+1,\beta+13,\beta+14 >$ mod 27, $\beta=0,1,\ldots,12$. |
| 31 | 6 | $X=Z(31,3)$. |
| | | $\mathcal{B} = < \alpha,\alpha+1,\alpha+2,\alpha+3,\alpha+5 >$ mod 31, $\alpha=0,1,\ldots,29$, |
| | | $< \alpha,\alpha+3,\alpha+4,\alpha+14,\alpha+15 >$ mod 31, $\alpha=0,1,\ldots,29$, |
| | | $< \emptyset,\beta,\beta+3,\beta+15,\beta+18 >$ mod 31, $\beta=0,1,\ldots,14$, |
| | | $< \gamma,\gamma+6,\gamma+12,\gamma+18,\gamma+24 >$ mod 31, $\gamma=0,1,\ldots,5$, twice. |
| 32 | 2 | $X=GF(32,x^5=x^2+1)$. |
| | | $\mathcal{B} = < \alpha,\alpha+1,\alpha+5,\alpha+21,\alpha+23 >$ mod 32, $\alpha=0,1,\ldots,30$. |
| 33 | 15 | Theorem 6.1 and $9 \in B_3(5,15)$ by Lemma 4.9 and $10 \in B_3(6,5)$ as in Table 1. |
| 34 | 15 | $X=Z(2) \times Z(17,3)$. |
| | | $\mathcal{B} = < (\emptyset,\emptyset),(0,\alpha),(0,\alpha+1),(0,\alpha+2),(0,\alpha+4) >$ mod(2,17), $\alpha=0,1,\ldots,15$, 4 times, |
| | | $< (\emptyset,\emptyset),(0,\emptyset),(0,\alpha),(0,\alpha+1),(0,\alpha+7) >$ mod(2,17), $\alpha=0,1,\ldots,15$, 3 times, |
| | | $< (\emptyset,\emptyset),(\emptyset,\alpha),(\emptyset,\alpha+1),(0,\emptyset),(0,\alpha+4) >$ mod(2,17), $\alpha=0,1,\ldots,15$, |
| | | $< (\emptyset,\emptyset),(\emptyset,\alpha),(\emptyset,\alpha+5),(0,\emptyset),(0,\alpha+4) >$ mod(2,17), $\alpha=0,1,\ldots,15$, |
| | | $< (\emptyset,\emptyset),(\emptyset,\beta),(\emptyset,\beta+8),(0,\beta+1),(0,\beta+9) >$ mod(2,17), $\beta=0,1,\ldots,7$, 4 times, |
| | | $< (\emptyset,\emptyset),(\emptyset,\beta),(\emptyset,\beta+8),(0,\beta+4),(0,\beta+12) >$ mod(2,17), $\beta=0,1,\ldots,7$, 5 times, |
| | | $< (\emptyset,\emptyset),(\emptyset,\beta),(\emptyset,\beta+8),(0,\beta+5),(0,\beta+13) >$ mod(2,17), $\beta=0,1,\ldots,7$, 3 times, |
| | | $< (\emptyset,\emptyset),(\emptyset,\beta),(\emptyset,\beta+8),(0,\beta+6),(0,\beta+14) >$ mod(2,17), $\beta=0,1,\ldots,7$, twice, |
| | | $< (\emptyset,\emptyset),(\emptyset,\beta),(\emptyset,\beta+8),(0,\beta+7),(0,\beta+15) >$ mod(2,17), $\beta=0,1,\ldots,7$. |

REFERENCES

[1] W.O. Alltop, "Some 3-designs and a 4-design", *J. Comb. Th. Ser. A* 11 (1971), 190-195.

[2] M. Hall, Jr., *Combinatorial Theory*, Blaisdell, Waltham, Mass., 1967.

[3] H. Hanani, "On some tactical configurations", *Canadian J. Math.* 15 (1963), 702-722.

[4] H. Hanani, "Truncated finite planes", Combinatorics, *Proc. Symp. in Pure Maths*, A.M.S. XIX (1971), 115-120.

[5] H. Hanani, "Balanced incomplete block designs and related designs", *Discrete Math.* 11 (1975), 255-369.

05C99

# ISOMORPHIC FACTORISATIONS III: COMPLETE MULTIPARTITE GRAPHS

Frank Harary
Department of Mathematics,
University of Michigan,
Ann Arbor, Michigan 48109,
UNITED STATES OF AMERICA

Robert W. Robinson and Nicholas C. Wormald
Department of Mathematics,
University of Newcastle,
New South Wales, 2308,
AUSTRALIA

ABSTRACT

In the first paper of this series we showed that a factorisation of the complete graph $K_p$ into t isomorphic subgraphs exists whenever the Divisibility Condition holds, that is, the number of lines is divisible by t. Our present objective is to investigate for complete multipartite graphs the extent to which the Divisibility Condition implies the existence of an isomorphic factorisation. We find that this is indeed the situation for all complete bipartite graphs but not for all k-partite graphs when $k \geq 3$.

## 1. INTRODUCTION

An *isomorphic factorisation* of a graph $G = (V,E)$ is a partition $\{E_1,\dots,E_t\}$ of the line set E such that the spanning subgraphs $(V,E_1),(V,E_2),\dots,(V,E_t)$ are all isomorphic to each other. In this case, if $H \cong (V,E_1)$ we say H *divides* G and write either $H|G$ or $H \in G/t$. Also if G has an isomorphic factorisation into exactly t isomorphic subgraphs we say that G *is divisible* by t and write $t|G$. To avoid a triviality we always take $t > 1$.

For given t and given G having exactly q lines, an obvious necessary condition for the divisibility of G by t is that t divide q. This is called the *Divisibility Condition* for G and t.

In [2] it is shown that whenever G is a complete graph the Divisibility Condition is sufficient for $t|G$, the divisibility of G by t. Our present object is to investigate the sufficiency of the Divisibility Condition for complete multipartite graphs. The *complete r-partite graph* of type $(n_1,n_2,\dots,n_r)$ is the complement of the disjoint union $K(n_1) \cup \dots \cup K(n_r)$ of complete graphs and is denoted by $K(n_1,n_2,\dots,n_r)$. As usual, the complete 2-partite graphs are called *bipartite* while the 3-partite graphs are called *tripartite*. We begin the next section by verifying that the Divisibility Condition is always sufficient for complete bipartite graphs. Then we show that for complete tripartite graphs the Divisibility Condition is not sufficient for any odd t, while for t = 2 and 4 the Divisibility Condition is sufficient. A further class of examples of insufficiency of the Divisibility Condition is provided in Section 3 for t = 2 and complete r-partite graphs for all r which are divisible by 4. We conclude by discussing some of the problems which are left open concerning the sufficiency of the Divisibility Condition for complete multipartite graphs. Attention is drawn to a related notion of divisibility for numerical partitions which is implicit in all of our systematic counterexamples. In the positive direction, we conjecture that for even t the Divisibility Condition is sufficient for all complete tripartite graphs.

## 2. COMPLETE BIPARTITE AND TRIPARTITE GRAPHS

In contrast to the difficult problem of verifying the Divisibility Conjecture for complete graphs [2], the analogue for complete bipartite graphs is trivial.

Theorem 1. *If $t|mn$ then $K(m,n)$ is divisible by $t$.*

Proof. Since $t|mn$, we have $t = rs$ for some $r,s$ such that $r|m$ and $s|n$. Then $K(m/r,n/s)$, together with the required number $m(1-1/r) + n(1-1/s)$ of isolated points, divides $K(m,n)$. □

Consider now the factorisation of the complete tripartite graph $K(m,n,s)$ into t factors. In this case, the Divisibility Condition is that $t|(mn+ns+sm)$. We will show that this condition is sufficient when t is 2 or 4, but not sufficient when t is odd.

Let A, B and C be pairwise disjoint point sets. It is convenient for construction purposes to write $K(A,B,C)$ for the complete tripartite graph with parts A, B and C.

Theorem 2. *If m, n and s are positive integers such that $mn+ns+sm$ is even, then the complete tripartite graph $K(m,n,s)$ is divisible by 2.*

Proof. Suppose $mn+ns+sm$ is even. Then at least two of m, n and s are even, so we assume that m and n are both even. Let $A_1,A_2,B_1,B_2$ and C be pairwise disjoint point sets such that $A_1$ and $A_2$ each have cardinality $m/2$, $B_1$ and $B_2$ each have cardinality $n/2$, and C has cardinality s, and let $A = A_1 \cup A_2$ and $B = B_1 \cup B_2$.

We now construct an element of $K(A,B,C)/2$. Define $G_1$ to be the spanning subgraph of $K(A,B,C)$ containing just those lines which join $A_1$ with C, $B_2$ with C or $B_1$ with A. Then let $G_2$ be the spanning subgraph of $K(A,B,C)$ containing those lines which join $A_2$ with C, $B_1$ with C or $B_2$ with A. It is clear that $G_1$ and $G_2$ form a factorisation of $K(A,B,C)$. Furthermore, there is an isomorphism between $G_1$ and $G_2$ induced by interchanging $A_1$ with $A_2$ and $B_1$ with $B_2$. Hence $G_1$ divides $K(A,B,C)$, so $K(m,n,s)$ is divisible by 2. □

The graphs $G_1$ and $G_2$ are illustrated in Figure 1. Here each letter represents a point set and each line between two sets represents the inclusion of all lines joining the two sets.

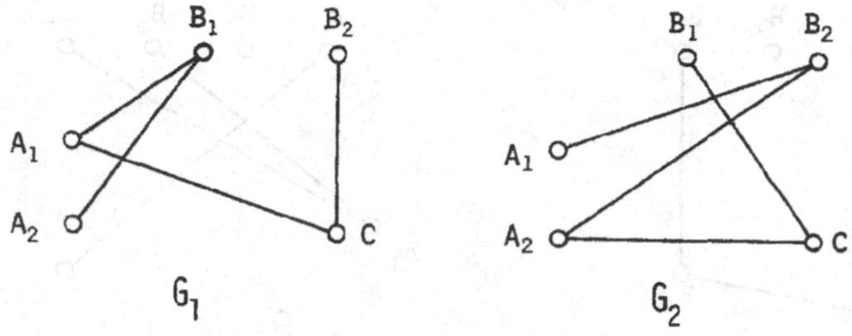


Figure 1. Two graphs in a factorisation of a complete tripartite graph.

We now extend the idea used for t = 2 to prove the sufficiency of the Divisibility Condition for complete tripartite graphs when t = 4.

Theorem 3. *If m, n and s are positive integers such that mn+ns+sm is divisible by 4, then the complete tripartite graph K(m,n,s) is divisible by 4.*

Proof. Suppose mn+ns+sm is divisible by 4. Then either m, n and s are all even, or one is odd. If one is odd, then the remaining two are either both divisible by 4 or both oddly even (divisible by 2 but not by 4). Alternatively, if m, n and s are all even, we can choose two which are both divisible by 4 or both oddly even. We can therefore assume without loss of generality that $m/2 = a$ and $n/2 = b$, where a and b are integers, a+b is even and $a \leq b$.

We will construct a graph in K(m,n,s)/4 by making use of seven disjoint point sets: $A_1$ and $A_2$ each containing a points, $B_1$ and $B_3$ each containing $(b+a)/2$ points, $B_2$ and $B_4$ each containing $(b-a)/2$ points, and C containing s points. Note that if b = a, then $B_2$ and $B_4$ are both empty. Let $A = A_1 \cup A_2$ and $B = B_1 \cup B_2 \cup B_3 \cup B_4$. Then A has cardinality m, B has cardinality n and C has cardinality s. Let $G_1$ be the spanning subgraph of K(A,B,C) containing just those lines which join $A_1$ with $B_1$ or $B_2$, or C with $A_2$ or $B_4$. Similarly, define the spanning subgraphs $G_2$, $G_3$ and $G_4$ of K(A,B,C) as follows: $G_2$ contains all lines joining $A_1$ with $B_3$ or $B_4$, or C with $B_1$; $G_3$ contains all lines joining $A_2$ with $B_1$ or $B_2$, or C with $B_3$; and $G_4$ contains all lines joining $A_2$ with $B_3$ or $B_4$, or C with $A_1$ or $B_2$. The line sets of $G_1, G_2, G_3$ and $G_4$ partition the lines of K(A,B,C). Furthermore, as $A_1 \cup B_2$, $B_1$, $A_2 \cup B_4$ and $B_3$ all have cardinality $(a+b)/2$, clearly $G_1, G_2, G_3$ and $G_4$ are all isomorphic to $K(a,b) \cup K((a+b)/2,s)$. Hence the latter graph is in K(m,n,s)/4. □

The graphs $G_1, G_2, G_3$ and $G_4$ are illustrated in Figure 2 following the same conventions as in Figure 1. Note that by combining $G_1$ and $G_3$ into the one graph, we obtain the element of K(m,n,s)/2 which was constructed in the proof of Theorem 2.

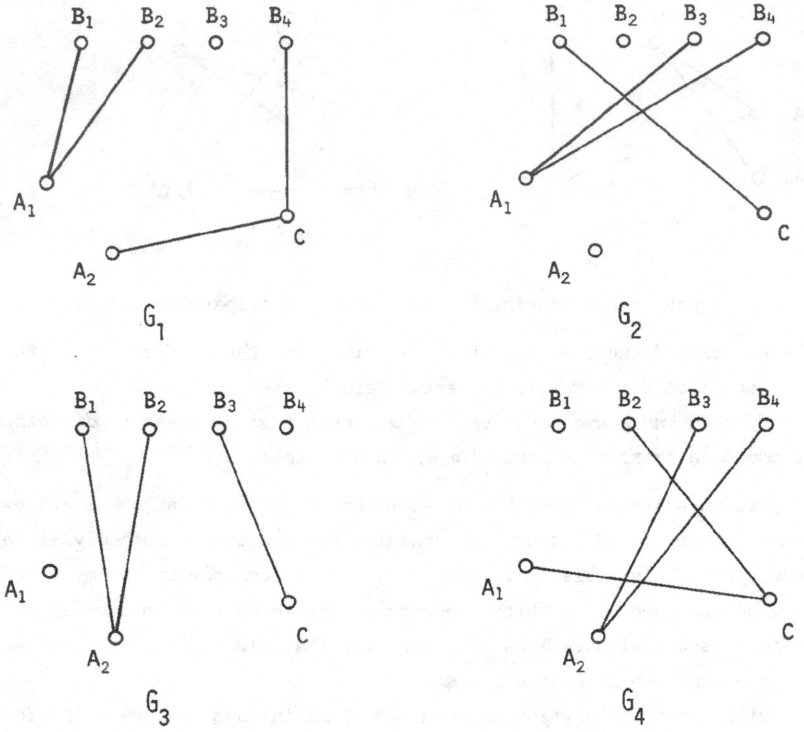


Figure 2. Four graphs in a factorisation of a complete tripartite graph.

The next theorem will provide our first example in which the Divisibility Condition fails to be sufficient.

Theorem 4. *If $t > 1$ is odd and $m \geq t(t+1)$ then $K(1,1,m)/t$ is empty.*

Proof. Let $t > 1$ be odd and $m \geq t(t+1)$. Suppose $K(1,1,m)/t$ is nonempty; let $H \in K(1,1,m)/t$. Then by the Divisibility Condition, $t \mid (2m+1)$.

It is convenient to consider a copy of $K(1,1,m)$ as $K(A,B,C)$ with point sets $A = \{u_1\}$, $B = \{u_2\}$ and $C = \{u_3,u_4,\ldots,u_{m+2}\}$. As $H \in K(1,1,m)/t$, there is some factorisation $H_1,H_2,\ldots,H_t$ of $K(A,B,C)$ such that $H_i \cong H$ for each $i$. We fix the latter isomorphisms, so that a given point of $H$ is associated with a unique point of $H_i$ for each $i$. For $1 \leq i \leq t$ let $d_{i1}$ and $d_{i2}$ denote the degrees in $H_i$ of the points $u_1$ and $u_2$ respectively. Since $u_1$ and $u_2$ each have degree $m+1$ in $K(A,B,C)$, the sum over the $t$ graphs of these degrees is

$$(1) \qquad \sum_{i=1}^{t} d_{i1} = \sum_{i=1}^{t} d_{i2} = m+1.$$

Let $r$ be the maximum degree among the points of $H$. Then $d_{i1} \le r$ for each $i$, so by (1) we have $tr \ge m+1$. As $t > 1$ and $m \ge t(t+1)$, $m/t > 2$ and hence $r > 2$. Let $v$ be a point in $H$ of degree $r$. Then for each $i$, the point of $H_i$ associated with $v$ must be $u_1$ or $u_2$, because all other points have degree 2 in $K(A,B,C)$. So if $v$ coincides with $u_1$ in $k$ of the graphs $H_i$, then it coincides $t-k$ times with $u_2$. We assume without loss of generality that $k \le t-k$, and hence $k < t-k$ as $t$ is odd. Thus, $t-k \ge (t+1)/2$, so $d_{i2} = r$ for at least $(t+1)/2$ values of $i$. Hence by (1), $r(t+1)/2 \le m+1$, so we have the upper bound

$$r \le 2(m+1)/(t+1). \tag{2}$$

We can also assume without loss of generality that $d_{i1} = r$ for $1 \le i \le k$. Suppose $d_{i1} \le 2$ for $k+1 \le i \le t$. Then

$$\begin{aligned}\sum_{i=1}^{t} d_{i1} &\le kr + 2(t-k) \\ &\le 2k(m+1)/(t+1)+2t && \text{by (2)} \\ &\le (t-1)(m+1)/(t+1)+2t && \text{as } k \le (t-1)/2 \\ &< m+1 && \text{as } m \ge t(t+1)\end{aligned}$$

which contradicts (1). Thus $d_{i1} > 2$ for some $i > k$, so some point $w \ne v$ of $H$ has degree $s \ge 3$. As in the case of $v$, $w$ must therefore be associated with either $u_1$ or $u_2$ in each of the $H_i$. This means that in each of the $H_i$, $u_1$ and $u_2$ represent the points $v$ and $w$ in some order. Hence, $d_{i2} = s$ for $1 \le i \le k$, and $d_{i1} = s$ otherwise. Equation (1) now implies that $kr+(t-k)s=ks+(t-k)r$, and hence $r = s$ since $k \ne t-k$. But (1) further implies that $t|m+1$, which is impossible because $t > 1$ and $t|2m+1$. It follows from this contradiction that $K(1,1,m)/t$ is empty. □

<u>Corollary 4a</u>. *The Divisibility Condition is sufficient for the existence of a graph in $K(m,n,s)/t$ when $t$ is 2 or 4, but not when $t$ is an odd integer greater than 1.*

## 3. COMPLETE r-PARTITE GRAPHS FOR r ≥ 4

We have seen that the Divisibility Condition is not in general sufficient for a complete tripartite graph to be divisible by $t$. We will extend this result to cover complete $r$-partite graphs when $r$ is divisible by 4.

To refer to some specific multipartite graphs, we need the following notation. The graph $K(r:a,b)$ is the complete $(r+1)$-partite graph on $ar+b$ points, where one of the parts has cardinality $b$ and each of the other parts has cardinality $a$.

<u>Theorem 5</u>. *If $4|(r+1)$ and $m > r+1$, then the complete $(r+1)$-partite graph $K(r:1,m)$ is not divisible by 2.*

Proof. For $K(r:1,m)$ to be divisible by 2, the Divisibility Condition implies that $m$ is odd. Now $K(r:1,m)$ has $r$ points of degree $m+r-1$ and $m$ points of degree $r$. Let $G$ be the graph $K(r:1,m)$ with point set $1,2,\ldots, m+r$, such that the points $1,2,\ldots,r$ all have degree $m+r-1$. Suppose $G$ is divisible by 2, and let $H_1$ and $H_2$ be the subgraphs of $G$ corresponding to a factorisation of $G$ into 2, so that $H_1 \cong H_2$. As $4|r+1$, $r$ must be odd, and hence $m+r-1$ is odd. Thus, whenever $i \le r$, the point $i$ has degree greater than $(m+r-1)/2$ in precisely one of $H_1$ and $H_2$. Moreover, since $m > r+1$ we have $(m+r-1)/2 > r$. Hence, none of the points $r+1, r+2,\ldots, m$ has degree greater than $(m+r-1)/2$ in either $H_1$ or $H_2$. It follows that if $k$ is the number of points of degree greater than $(m+r-1)/2$ in $H_1$, $2k = r$. This contradicts the hypothesis that $4|r+1$, so we conclude that $G/2$ is empty. □

Corollary 5a. *When $4|r$, the Divisibility Condition is insufficient for the existence of an isomorphic factorisation of a complete r-partite graph into two factors.*

The cases when $t > 2$ or $r$ is not divisible by 4 are more difficult to study. Corollary 5a certainly does not cover all instances of the failure of the Divisibility Condition. For example, $K(1,1,1,2,5)/2$ is empty; we omit the complex proof as its method does not appear to be susceptible to generalisation.

## 4. UNSOLVED PROBLEMS

We have been examining various values of $r$ and $t$ in an attempt to find precisely when the Divisibility Condition is sufficient for the existence of an isomorphic factorisation of a complete r-partite graph into $t$ factors. The sufficiency was shown when $r = 2$ for all $t$, and when $r = 3$ for $t = 2$ and 4, but we found that the Divisibility Condition is insufficient for $r = 3$ and $t$ odd. We make the following conjecture based on these results.

Tripartite Conjecture. Consider a complete tripartite graph $K = K(m,n,s)$ and an integer $t > 1$. If for all $m,n$ and $s$ the Divisibility Condition for $K$ and $t$ implies the existence of a graph in $K/t$, then $t$ is even, and conversely.

Reward. The first-named author offers U.S. $50 for settling this conjecture.

S. Quinn has just proved the Tripartite Conjecture for $t = 6$. His proof involves six cases with a separate construction for each and does not appear to generalize readily.

We have seen in Theorem 4 that for complete tripartite graphs $K_{m,n,s}$, the Divisibility Condition is not sufficient for any odd $t \ge 3$ when $n = s = 1$. However for certain values of $m,n,s$, the Divisibility Condition will be sufficient. This suggests another interesting open problem.

Tripartite Question. For precisely which values of m,n,s and t does the Divisibility Condition suffice for complete tripartite graphs?

We have only been concerned with the existence of graphs in K/t where K is a complete multipartite graph. A more difficult question is the determination of the entire set K/t.

Construction Question. When a complete multipartite graph K is divisible by t, what are all the graphs in K/t?

In particular, what is the set of graphs G in K(m,n)/2, K(m,n,s)/2 and K(m,n,s)/4? As the structure of self-complementary graphs $G \in K_p/2$ is entirely known, it may be quite tractable to determine fully the two sets above with t=2.

Our knowledge of the existence of isomorphic factorisations of complete r-partite graphs with $r \geq 4$ is scanty. It is not known whether there are t and r with $r \geq 4$ such that the Divisibility Condition is sufficient. In a slightly different direction, consider multipartite graphs in which the parts are of equal size, and call these *equipartite* graphs. We know of no case in which the Divisibility Condition fails to be sufficient for a complete equipartite graph. When each part has cardinality 1, a complete equipartite graph is simply an ordinary complete graph, and the sufficiency of the Divisibility Condition is the main theorem of [2]. We conjecture the following generalization of the Divisibility Theorem for complete graphs.

Equipartite Conjecture. For all complete equipartite graphs the Divisibility Condition implies the existence of an isomorphic factorisation.

Reward. U.S. $10 is offered for the first proof or disproof of this conjecture.

We next consider the divisibility of complete multipartite graphs from another angle. Let $\pi$ be an ordered partition of the positive integer n into non-negative parts so that zero entries are allowed. The partition $\pi$ is *divisible* by t if there is some partition $\pi'$ such that $\pi$ can be expressed as a sum of t copies of $\pi'$ with different orderings of $\pi'$ admitted. We then write $t|\pi$ and $\pi'|\pi$.

Let $\pi$ be the partition of 2q obtained as the degree sequence of a graph G with q lines. Suppose G is divisible by t and let $H \in G/t$. It is immediate that $\pi$ is divisible by t, and if $\pi'$ is the partition of H then $\pi'|\pi$. Thus, in order that a graph be divisible by t it is necessary that its partition be divisible by t. Our main sources of counterexamples to the sufficiency of the Divisibility Condition arise from this fact. In the proofs of Theorems 4 and 5 we have essentially shown that the partitions of the graphs involved are not divisible by t. Thus, Theorems 4 and 5 provide examples of a number t and a graph G such that the number q of lines is divisible by t but the partition $\pi$ of G is not. Since $\pi$ is a partition of 2q, the following number theoretic question is of importance in determining when a graph with q lines has an isomorphic factorisation into t factors.

Partition Question. For which partitions $\pi$ of 2q does $t|q$ imply $t|\pi$?

In this connection we note that a graph G is not necessarily divisible by t if its partition $\pi$ is divisible by t. To see this, let $G = K(1,1,7)$ and $t = 3$. Then if $\pi' = 4,2,2,1,1,0,0,0,0$ we have $\pi'|\pi$ and so $3|\pi$, but G is not divisible by 3, although $\pi'$ is a graphical partition. Thus, even a complete solution to this Partition Question will not supply a sufficient condition for a graph to be divisible by t.

Yamamoto, Ikeda, Shige-eda, Ushio and Hamada [4] have specified precisely which stars $K_{1,s}$ (unfortunately calling them "claws") divide into $K_{m,n}$, as well as those which divide $K_p$. These two results suggest at once the problem of characterizing those stars $K_{1,s}$ which divide a complete multipartite graph $K = K(n_1,n_2,\ldots,n_r)$. They also suggest the more general question of the existence of isomorphic factorisations $F|K$ where F is a given forest, which may of course consist of a tree together with enough isolated points to make F a spanning subgraph of K.

REFERENCES

1. F. Harary, *Graph Theory*. Addison-Wesley, Reading, Mass., 1969.

2. F. Harary, R.W. Robinson and N.C. Wormald, Isomorphic Factorisations I: Complete graphs, *Trans. Amer. Math. Soc.*, to appear.

3. F. Harary and W.D. Wallis, Isomorphic factorisations II: Combinatorial designs, *Proceedings of the Eighth Southeastern Conference on Combinatorics, Graph Theory and Computing*, Utilitas Math. Publ., Winnipeg, to appear.

4. S. Yamamoto, H. Ikeda, S. Shige-eda, K. Ushio and N. Hamada, On claw-decomposition of complete graphs and complete bigraphs, *Hiroshima Math. J.* 5 (1975) 33-42.

05B05, 05B25, 05B30

# BIPLANES AND SEMI-BIPLANES

Daniel Hughes

Westfield College (University of London)
London NW3 7ST

## 1. BIPLANES

It was noticed some time ago, perhaps by Marshall Hall, that while there exists an infinite number of symmetric 2-designs with $\lambda = 1$ (i.e., finite projective planes), there was no other value of $\lambda$ for which an infinite number of symmetric 2-designs was known. This inspired an interest in the problem for $\lambda = 2$. Such a design is called a *biplane* and can be defined synthetically as a finite collection of *points* and *blocks*, with an *incidence relation* (or, alternatively, blocks can be thought of as subsets of points) satisfying:

(1) two distinct points are in exactly two common blocks;

(2) two distinct blocks contain exactly two common points.

(Notice that these definitions make natural sense in the infinite case, and we can indeed speak of infinite biplanes.) We also demand:

(3) the total number of points, $v$, is greater than the number of points on any block, and every block contains more than two points.

It is then easy to prove:

Theorem 1. The number of points on any block is a constant $k$, and $v = 1 + k(k-1)/2$. Also, the number of blocks $= v$ and the number of blocks on any point is $k$.

Biplanes are know to exist for $k = 3,4,5,6,9,11$ and $13$, and for $k < 16$ there are no other possible values of $k$. (In fact, there is exactly one biplane for $k = 3,4$ and $5$, exactly three for $k = 6$, exactly four for $k = 9$, at least three for $k = 11$ and at least two for $k = 13$.) For no other value of $k$ is a biplane known at this time.

A *Singer group* of a symmetric 2-design is a group of automorphisms regular on the points (and hence on the blocks). Singer groups exist for the biplanes with $k = 3,4$ and $5$, for at least two (and possibly all) of the biplanes with $k = 6$, and for exactly one of the biplanes with $k = 9$. Baumert's search ([1]) shows that there are no other *cyclic* Singer groups for biplanes with $k \leqslant 100$.

L.J. Dickey and the author have made a computer search for Singer groups for biplanes with $n = k - 2 \leqslant 5000$, using the Honeywell 6060 at the University of Waterloo. We broke the work into several levels, each of which provides additional information about a more restrictive class of biplanes.

Level 1. The Bruck-Ryser-Chowla theorem says that if a biplane with $v$ points exists, then

(a) if $v$ is even, then $n = k - 2$ is a square;

(b) if $v$ is odd, then there is a non-trivial solution in integers for

$$x^2 = ny^2 + (-1)^{\frac{v-1}{2}} 2z^2$$

(here, as always, $n = k - 2$).

Since the solubility of diophantine equations of the type above depends completely upon certain Legendre symbols, it was very easy to carry this test out. About 1100 numbers $n \leqslant 5000$ passed the Bruck-Ryser-Chowla test.

Level 2. Certain tests due to Hughes ([4]) assert that if a biplane with $v$ points has an automorphism of odd prime order $p$ fixing no points, and if $v$ is odd, then there is a non-trivial solution in integers for

$$x^2 = ny^2 + (-1)^{\varepsilon} 2pz^2$$

where $\varepsilon = (v/p - 1)/2$.

The same programs as in the Level 1 search are applicable here. About 760 of the 1100 possible biplane parameters from Level 1 passed this test. The parameters which do *not* pass the test give us instances where, if a biplane exists then it can have no Singer group whatever.

Level 3. If an abelian Singer group exists then Hall's multiplier theorem ([3]) enables us to compute additional automorphisms of the biplane. Alternatively (and this was the approach we used) it enabled us to compute the number of elements of each order dividing $v$ which must be in a "difference set". This was a very efficient test, and rejected almost all the remaining 760 numbers, leaving only about 4 values of $n$. (In fact we kept improving this test, since it turned out to be capable of considerable refinement.)

Level 4. The remaining 4 cases were rejected by various *ad hoc* tests.

The conclusion is:

Theorem 2. If a biplane exists with an abelian Singer group, then its block size $k$ satisfies $k \leqslant 9$ or $k \geqslant 5003$.

This might be interpreted as strong evidence for the non-existence of infinitely many biplanes.

## 2. SEMI-BIPLANES

In Beukenhout's scheme ([2]) of generalized Coxeter diagrams, one of the simplest diagrams is

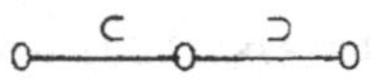

.

Using Beukenhout's recipes, this diagram describes exactly the class of structures which we shall call *semi-biplanes*, i.e.: $\mathcal{B}$ is a collection of *points* and certain subsets of points called *blocks*, satisfying:

(1) two distinct points of $\mathcal{B}$ are in exactly 0 or 2 common blocks of $\mathcal{B}$;

(2) two distinct blocks of $\mathcal{B}$ contain exactly 0 or 2 common points of $\mathcal{B}$;

(3) $\mathcal{B}$ is connected (in the graph-theoretic sense).

We can also insist on

(4) every block of $\mathcal{B}$ contains at least 3 points.

It is then easy to prove:

Theorem 3. If v is the number of points in a finite semi-biplane, then v is also the number of blocks. If k is the number of points on one block of $\mathcal{B}$, then every block of $\mathcal{B}$ contains k points and every point of $\mathcal{B}$ is on k blocks.

So a finite semi-biplane has *parameters* (v,k).

In addition, Beukenhout's schemes focusses particular interest on the automorphism groups of $\mathcal{B}$ which are *chain-transitive:* that is, transitive on the ordered triples (P,Q,y), where P,Q are points and y is a block on P and Q. So if $\mathcal{B}$ is a biplane, chain-transitivity is somewhat stronger than 2-transitivity, while if $\mathcal{B}$ is a semi-biplane which is not a biplane, then it is somewhat weaker, but still as strong as might be reasonably demanded.

Since there are only finitely many known finite biplanes, at first it seemed possible that the same might be true of semi-biplanes. That this is not the case was shown by L.J.Dickey and the author:

Theorem 4. If A is the incidence matrix of a semi-biplane, then so is

$$B = \begin{bmatrix} A & I \\ I & A^t \end{bmatrix} .$$

If the biplane associated with A has parameters (v,k), then that associated with B has parameters (2v,k + 1).

The proof is trivial. This gives an infinite class of semi-biplanes but in general these lack chain-transitive groups.

Subsequently Douglas Leonard, Richard Wilson and the author found another family (or perhaps three families) of semi-biplanes.

Let $\mathcal{P}$ be a finite projective plane of order q, and let $\alpha$ be an involutory automorphism of $\mathcal{P}$. Construct $\mathcal{B} = \mathcal{B}(\mathcal{P},\alpha)$ as follows: the points of $\mathcal{B}$ are the unordered pairs $(P,P^\alpha)$ for each point P of $\mathcal{P}$ such that $P \neq P^\alpha$; the blocks of $\mathcal{B}$ are the unordered pairs $(\ell,\ell^\alpha)$ where similarly $\ell$ is a line of $\mathcal{P}$ such that $\ell \neq \ell^\alpha$. The point $(P,P^\alpha)$ is on the block $(\ell,\ell^\alpha)$ is and only if P is on $\ell$ or $P$ is on $\ell^\alpha$ . Then it is easy to see:

Theorem 5. $\mathcal{B}(\mathcal{P},\alpha)$ is a semi-biplane, and has the following parameters:

(a) if $\alpha$ is an elation (so $q$ is even), then $v = q^2/2$, $k = q$.

(b) if $\alpha$ is a homology (so $q$ is odd), then $v = (q^2-1)/2$, $k = q$ .

(c) if $\alpha$ is a Baer involution (so $q$ is a square), then $v = (q^2-\sqrt{q})/2$, $k = q$

In addition, we have:

Theorem 6. If $G$ represents the automorphism group of $\mathcal{B}(\mathcal{P},\alpha)$ inherited from Aut $\mathcal{P}$ , then $G$ is exactly the centralizer in Aut $\mathcal{P}$ of $\alpha$ , mod $\langle\alpha\rangle$ .

Theorem 7. If $\mathcal{P}$ is Desarguesian, then

(a) $G$ is induced by the set of 3 by 3 lower triangular matrices

$$\begin{pmatrix} a & 0 & 0 \\ b & c & 0 \\ d & c & f \end{pmatrix}, \quad acf \neq 0$$

plus elements representing automorphism of $GF(q)$.

(b) $G$ is $PGL(2,q)$ extended by a cyclic group (of homologies) of order $q - 1$, plus elements representing automorphisms of $GF(q)$.

(c) $G$ is $P\Gamma L(3,\sqrt{q})$ extended by the automorphism of order 2 of $GF(q)$.

In cases (b) and (c) $G$ is non-soluble (except in case (b) when $q = 3$), and $G$ is chain-transitive in all three cases.

It is worth noting that the semi-biplanes $\mathcal{B}(\mathcal{P},\alpha)$ are all divisible (i.e., "group-divisible").

BIBLIOGRAPHY

(1) L. Baumert, Cyclic difference sets, Springer-Verlag, Lecture Notes in Mathematics, 182 (1971).

(2) F. Beukenhout, On generalized Coxeter diagrams, (unpublished but widely distributed).

(3) M. Hall, Jr., Cyclic projective planes, Duke Math. J., Vol. 14 (1947), 1079-1090.

(4) D.R. Hughes, Collineations and generalized incidence matrices, Trans. Amer. Math. Soc., Vol. 86 (1957), 284-296.

05B05, 05B30

# NEAR-SELF-COMPLEMENTARY DESIGNS AND A METHOD OF MIXED SUMS

R.C. Mullin and D. Stinson

University of Waterloo

ABSTRACT. An important class of BIBDs is that of (strongly) self-complementary-designs, designs which are invariant under complementation. Their parameter sets satisfy the relation $v = 2k$, yet for $k$ odd, there are an infinity of parameter sets which cannot be realized as self-complementary designs. For these parameters the idea of near-self-complementary designs is introduced. These designs have many aspects similar to self-complementary designs.

An extension of Bose's method of mixed differences is introduced and is applied to show the residuality of certain near-self-complementary designs.

1. INTRODUCTION.

A *balanced incomplete block design* BIBD $(v,b,r,k,\lambda)$ is a pair $(V,F)$ where $V$ is a v-set of objects called varieties, $F$ is a family of k-subsets of $V$, $b$ in number, which has the property that each variety occurs in precisely $r$ of these subsets and each pair of distinct varieties occurs in precisely $\lambda$ of these subsets. An important subclass of these designs is the *self-complementary or strongly self-complementary* designs, which are invariant under complementation. (Some authors refer to self-complementary designs as those isomorphic to their complements, for this reason, the option "strongly self-complementary design" is given as an alternative; for sake of simplicity, we use self-complementary or SCD). An SCD $(v,b,r,k,\lambda)$ is *simple* if $(b,r,\lambda) = 1$. Simple self-complementary designs enjoy the following properties.

(i) Simple self-complementary designs are affine resolvable, that is, any block meets all blocks except itself and its complement in precisely $k/2$ varieties [5].

(ii) Simple self-complementary designs are residual designs (cf. [3]).

(iii) Simple self-complementary designs are 3-designs, that is, every triple of varieties occurs in $\lambda_3$ blocks, where $\lambda_3$ is independent of the triple chosen. (This property is valid for any SCD design).

(iv) There exists an SCD $(4t+4, 8t+6, 4t+3, 2t+2, 2t+1)$ if and only if there exists an Hadamard matrix $H_{4t+4}$ [4].

Clearly a necessary condition for the existence of an SCD $(v,b,r,k,\lambda)$ is that $v = 2k$. This yields parameters of the form $(2x+2, t(4x+2), t(2x+1), x+1, tx)$.

However, if $k$ is odd and $t = (b,r,\lambda)$ is also odd, then it is known [5] that no SCD exists. Because of this deficiency, the following definition is given. A BIBD $D$ is *near-self-complementary* (NSC) if there exists an involutory mapping $\phi$ defined on the blocks of $D$ such that (i) $|B \cap \phi B| = 1$ and (ii) $|B \cup \phi B| = v-1$, for all blocks $B$ of $D$. $\phi B$ is the *near-complement* of $B$. An NSC $(v,b,r,k,\lambda)$ is simple if $(b,r,\lambda) = 1$.

It is evident from the definition that in any NSC $(v,b,r,k,\lambda)$ the relation $v = 2k$ holds, hence the set of parameters again has the form $(2x+2, t(4x+2), t(2x+1), x+1, tx)$. Since $(t(4x+2), t(2x+1), tx) = t$, for simple NSC designs the parameters have the form $(2x+2, 4x+2, 2x+1, x+1, x)$. Not surprisingly, the designs have different properties in the cases of $x$ even and $x$ odd. In either case the designs are quasi-residual [3], and some aspects of residuality are discussed in a later section.

## 2. PROPERTIES OF SIMPLE NSC DESIGNS WITH ODD BLOCK SIZE

The most interesting of the cases is that in which $x$ is even, or $k$ is odd, since no SCD can exist in this case. Letting $x = 2s$, the parameters become $(4s+2, 8s+2, 4s+1, 2s+1, 2s)$.

THEOREM 2.1. In an NSC $(4s+2, 8s+2, 4s+1, 2s+1, 2s)$ any block meets any block other than itself or its near-complement in either $s$ or $s+1$ elements.

PROOF. The result follows from a standard argument involving intersection numbers. (See, for example [5]). □

A variety of an NSC $D$ is said to be an *infinite element* if for all $B \in D$, $\infty$ is in $B \cup \phi B$. Since there are $r$ pairs $\{B,\phi B\}$, each infinite element occurs in precisely one of $\{B,\phi B\}$.

THEOREM 2.2. In an NSC $(4s+2, 8s+2, 4s+1, 2s+1, 2s)$ there are either one or two infinite elements.

PROOF. Since there are $4s+2$ varieties and only $4s+1$ pairs of blocks, there is at least one infinite element. Let us now assume that there are at least three infinite elements $\infty_1, \infty_2, \infty_3$. Let $F_i$ denote the set of blocks containing $\infty_i$ for $i = 1,2,3$. Let $|F_1 \cap F_2 \cap F_3| = \alpha$. Then $|F_1 \cap F_2 \cap \bar{F}_3| = |F_1 \cap \bar{F}_2 \cap F_3| = |\bar{F}_1 \cap F_2 \cap F_3|$ where $\bar{F}_i$ is the complement of $F_i$ in the block set of the design. Let $\beta$ denote the common value of these cardinalities. Further let $\gamma = |F_1 \cap (F_2 \cup F_3)|$. Then $|F_1| = \alpha + 2\beta + \gamma = 4s + 1$, and $\alpha + \gamma = |F_2 \cap F_3| = \lambda = 2s$ hence $2\beta = 2s + 1$, which is clearly impossible. □

Regrettably the NSC designs do not share the balance property with respect to triples that the SC designs possess. However the number of blocks containing a fixed triple cannot vary greatly within such a design, as is shown below.

THEOREM 2.3. In an NSC (4s+2, 8s+2, 4s+1, 2s+1, 2s) with one infinite element every triple of distinct varieties occurs in either s-2, s-1, s, or s+1 blocks.

PROOF. If A is a subset of varieties, then let $S_A$ denote the set of blocks of the design D which contain A. Let u, v, w denote three distinct varieties of D. Usually if a block contains u and not v or w, then its near-complement will contain v and w but not u.

This fails only if one of u, v or w is repeated or omitted from the pair $\{B,\phi B\}$. It is easily established that in any NSC with one infinite element, every non-infinite variety is contained in the intersection of precisely one near-complementary pair and is omitted from the union of precisely one near-complementary pair. Let $\delta_1(x,y)$ denote the number of near-complementary block pairs which contain u in one block, x in the other block, and omit y from both, $\delta_2(x,y)$ denote the number of complementary block pairs which contain u in both blocks and which are such that uxy occur together in one of the blocks, $\delta_3(x,y)$ denote the number of near-complementary block pairs such that ux is in one block and xy is in the other, and $\delta_4(x,y)$ be the number of block pairs in which u occurs in neither but x and y occur together in one of the blocks.

Let $\delta = \delta_1(v,w) + \delta_1(w,v) + \delta_2(v,w) - \delta_3(v,w) - \delta_3(w,v) - \delta_4(v,w)$. Then

$$|S_u - \{S_{uv} \cup S_{uw}\}| = |S_{vw} - S_u| + \delta(v,w).$$

Since each element is repeated at most once and omitted at most once $0 \le \delta_i \le 1$ for $i = 1,2,3,4$. Hence $-3 \le \delta \le 3$.

Now $|S_u - \{S_{uv} \cup S_{uw}\}| = |S_u| - |S_{uv}| - |S_{uw}| + |S_{uvw}| = r - 2\lambda + |S_{uvw}|$ and $|S_{vw} - S_u| = |S_{vw}| - |S_{uvw}| = \lambda - |S_{uvw}|$. This yields $|S_{uvw}| = t + (\delta-1)/2$. □

COROLLARY. If T is a triple of varieties which contains the infinite element, then T occurs in either s -1 or s blocks.

This follows from the fact that if $u = \infty$, then $\delta_2 = \delta_4 = 0$. □

In the case of two infinite elements it can again be shown that any triple of varieties again occurs in s-2, s-1, s or s+1 blocks, and a triple containing both infinite elements can only occur in s-1 or s blocks. The proof is similar to the above.

The authors know of only one such design with two infinite elements. It is listed below

| | |
|---|---|
| $B_1 : \infty_1 \infty_2\ 1$ | $B_6 :\ 1\ 2\ 3$ |
| $B_2 : \infty_1 \infty_2\ 2$ | $B_7 :\ 1\ 2\ 4$ |
| $B_3 : \infty_1\ 1\ 3$ | $B_8 : \infty_2\ 1\ 4$ |
| $B_4 : \infty_1\ 3\ 4$ | $B_9 : \infty_2\ 2\ 3$ |
| $B_5 : \infty_1\ 2\ 4$ | $B_{10} : \infty_2\ 3\ 4$ |

This design can also be re-partitioned to yield a design with just one infinite element as follows. $(B_1,B_9)$ $(B_2,B_7)$ $(B_3,B_8)$ $(B_4,B_6)$ $(B_5,B_{10})$.

For larger values of the parameters, no repartitioning of any such NSC design is possible in view of theorem 2.1.

## 3. PROPERTIES OF SIMPLE NSC DESIGNS WITH EVEN BLOCK SIZE

Nearly self-complementary designs with even block size are of less interest since their parameters coincide with those of self-complementary designs and the latter are known to exist for all possible parameter sets provided that Hadamard matrices $H_{4n}$ exist for all positive $n$. Moreover the properties of NSC designs are weaker for these parameter sets.

These properties are listed below (without proofs, since these are analogous to those of the previous section).

THEOREM 3.1. In any NSC $(4s+4,\ 8s+6,\ 4s+3,\ 2s+2,\ 2s+1)$ there are either one, two or three infinite elements.

THEOREM 3.2. In any NSC $(4s+4,\ 8s+6,\ 4s+3,\ 2s+2,\ 2s+1)$ with either one or two infinite elements, every triple of distinct varieties occurs in either $s-1$, $s$ or $s+1$ blocks. In any NSC $(4s+4,\ 8s+6,\ 4s+3,\ 2s+2,\ 2s+1)$ with three infinite elements every triple of distinct varieties occurs in either $s-2$, $s-1$, $s$, $s+1$ or $s+2$ blocks.

## 4. CYCLIC NSC DESIGNS

In this section a standard method [1] for obtaining certain NSC designs based on cyclic groups (cyclic designs) is discussed. As usual $Z_n$ denotes the cyclic group of order $n$, and $Z_n^*$ denotes the non-zero elements of $Z_n$.

THEOREM 4.1. Let $G = Z_n$ where $n = 2s+1$. If one can find a pair of blocks of the form $A^* = \{\infty,0\} \cup A$ and $B^* = \{0\} \cup B$ where $A$ and $B$ are $s-1$ and $s$-subsets of $G^*$ such that $A \cap B = \phi$ and the differences of $\{0\} \cup A$ and $\{0\} \cup B$ are symmetrically repeated (each occurring $s$ times), then the translates of $A^*$ and $B^*$ form an NSC $(2s+2,\ 4s+2,\ 2s+1,\ s+1,\ s)$.

PROOF. That the configuration is a BIBD with the required parameters is immediate from the method of differences. The appropriate block pairs are $\{A^*+\theta, B^*+\theta\}$, $\theta \in G$. □

EXAMPLES.

CASE 1 (s EVEN)

| | | |
|---|---|---|
| ∞ 0 2 | 0 3 4 | mod 5, |
| ∞ 0 2 4 5 | 0 3 6 7 8 | mod 9, |
| ∞ 0 2 3 4 8 11 | 0 5 6 7 9 10 12 | mod 13. |

CASE 2 (s ODD)

| | | |
|---|---|---|
| ∞ 0 | 0 2 | mod 3, |
| ∞ 0 2 6 | 0 3 4 5 | mod 7. |

Note that the general parameters (2s+2, 4s+2, 2s+1, s+1, s) are parameters of the designs derived from the symmetric designs (4s+3, 4s+3, 2s+1, 2s+1, s) which exist if and only if there exists an Hadamard matrix $H_{4s+4}$ (see, for example, [4]). It is well known (see, for example, [2, p.256] that a quasi-residual design ("design with the parameters of a residual design") is not necessarily a residual design. However we shall show that every cyclic NSC design is residual, hence the existence of such a design implies the existence of an Hadamard matrix. In particular, if $k$ is odd, say $k = 2s+1$, then the corresponding Hadamard matrix has order $8s+4$. For this reason, the existence of cyclic NSC designs with odd values for $k$ could prove useful in the theory of Hadamard matrices. To prove the cited result, we will use a new approach to the method of mixed differences [1].

## 5. A GENERALIZATION OF THE METHOD OF MIXED DIFFERENCES

It will be assumed here (as it was above) that the reader is familiar with the contents of [1]. The method will be extended here in terms of rings.

Let $R$ be a finite ring of order $n$, $R = \{0,r,s,\dots,t\}$. Consider $m$ "copies' of $R$,

$$R_1 = \{0_1,r_1,s_1,\dots,t_1\}$$
$$R_2 = \{0_2,r_2,s_2,\dots,t_2\}$$
$$\vdots$$
$$R_m = \{0_m,r_m,s_m,\dots,t_m\}.$$

Given two elements $x_i$ and $y_i$ from the *ith* copy, we define the *pure difference* $x_i - y_i$ as the element $(x-y)_i$. Thus pure differences are the natural differences operating within the *ith* copy. Now to each copy $R_k$ of $R$ assign an invertible element $w(k)$ of $R$, called the *weight* of $k$. By the *exterior combination* $x_i \ominus y_j$ of $x_i$ and $y_j$ (where $x_i \in R_i$ and $y_j \in R_j$, $i \neq j$) we mean the quantity $(w(i)x - w(j)y)_{ij}$. Let $S = \{s_i, t_j, \ldots, v_\ell\}$ be a subset of $V = \bigcup_{i=1}^{m} R_i$. If one forms the set of blocks $S \oplus \theta$ for $\theta \in R$ where

$$S \oplus \theta = \{(s + (w(i))^{-1}\theta)_i, (t + (w(j))^{-1}\theta)_j, \ldots, (v + (w(\ell))^{-1}\theta)_\ell\},$$

then $S$ is said to be developed through $R$.

THEOREM 5.1. Let $m$ copies $R_1, R_2, \ldots, R_m$ of a ring $R$ be given. Let $w(1), w(2), \ldots, w(m)$ be a set of corresponding invertible weights, also be given. If one can find a set of $t$ blocks $B_1, B_2, \ldots, B_t$ each of size $k$ with elements in $V = \bigcup_{i=1}^{m} R_i$ such that

(i) the non-zero pure differences are symmetrically repeated, each occurring $\lambda$ times, and

(ii) the exterior combinations $x_{ij}$ are symmetrically repeated, each occurring $\lambda$ times,

then the blocks $B_1, B_2, \ldots, B_t$ when developed through $R$, form BIBD $(mn, ms, r, k, \lambda)$ for an appropriate value of $r$.

PROOF. Since the pure differences are symmetrically repeated $\lambda$ times, each pair $\{x_i, y_i\}$, $x \neq y$ occurs in precisely $\lambda$ blocks. Suppose that $x_i$ and $y_j$ are given with $i \neq j$. Let $d_{ij}$ denote the exterior combination, so that $d = w(i)x - w(j)y$. Now $d_{ij}$ is represented as an exterior sum $\lambda$ times in the set of blocks $B_1, B_2, \ldots, B_t$. Let $u_i \ominus v_j$ be such a representation in $B_\ell$. Then

$$w(i)u - w(j)v = w(i)x - w(j)y,$$
$$y = v + (w(j))^{-1} w(i)(x-u).$$

However there exists a unique $\theta \in R$ such that $u + (w(i))^{-1}\theta = x$, namely $\theta = w(i)(x-u)$. Then $B_\ell + \theta$ contains (corresponding to $\{u_i, v_j\}$ the pair $\{x_i, (v + (w(j))^{-1} w(i)(x-u))_j\} = \{x_i, y_j\}$. Hence the pair $\{x_i, y_j\}$ occurs in precisely $\lambda$ blocks of the developed set. □

If all weights $w(i) = 1$ in the above, then the result is the standard method of mixed differences. The case of interest here is that of two copies of $Z_n$ (for appropriate $n$) and weights $1, -1$. Thus we look at mixed sums. Examples of this case are given below. (The blocks here can also be employed for mixed differences if minor modifications are made).

EXAMPLE 1. BIBD (10,30,9,3,2)

$$\text{Initial blocks } (0_1,1_2,4_2)\ (0_1,2_2,3_2)$$
$$(0_1,0_2,1_2)\ (0_1,2_2,4_2)\ (0_1,3_1,0_2)$$
$$(0_1,1_1,2_1) \text{ modulo } 5$$

EXAMPLE 2. BIBD (16,80,15,3,2)

$$\text{Initial blocks } (0_1,1_2,7_2)\ (0_1,2_2,6_2)\ (0_1,3_2,5_2)$$
$$(0_1,0_2,1_2)\ (0_1,2_2,7_2)\ (0_1,3_2,6_2)\ (0_1,4_2,5_2)$$
$$(0_1,4_1,4_2)\ (0_1,1_1,3_1)\ (0_1,1_1,3_1) \text{ modulo } 8$$

EXAMPLE 3. BIBD (22,154,21,3,2)

$$\text{Initial blocks } (0_1,1_2,10_2)\ (0_1,2_2,9_2)\ (0_1,3_2,8_2)$$
$$(0_1,4_2,7_2)\ (0_1,5_2,6_2)\ (0_1,0_2,1_2)\ (0_1,2_2,10_2)$$
$$(0_1,3_2,9_2)\ (0_1,4_2,8_2)\ (0_1,5_2,7_2)\ (0_1,6_1,0_2)$$
$$(0_1,1_1,2_1)\ (0_1,2_1,5_1)\ (0_1,3_1,7_1) \text{ modulo } 11.$$

It is also easy to construct such BIBD's modulo 14, 17, 20, 23, 26, 29 and probably for any modulus $\equiv 2 \pmod 3$.

## 6. THE RESIDUALITY OF CYCLIC NSC DESIGNS

As mentioned in section 4, cyclic NSC designs are generated by initial blocks of the form $A^* = \{\infty,0\} \cup A$ and $B^* = \{0\} \cup B$ where $A \cap B = \phi$.

THEOREM 6.1. Let D be a cyclic NSC (2x+2, 4x+2, 2x+1, x+1, x) design. Then D is the residual of a symmetric (4x+3, 4x+3, 2x+1, 2x+1, x) design.

PROOF. D is defined by the cyclic group $G = Z_{2x+1}$. View G as the additive group of the ring R of integers modulo 2x+1. Take two copies $R_1$ and $R_2$ of R, and weights $w(1) = 1$ and $w(2) = -1$. Using the notation employed above, let $A'$ denote $\{0\} \cup A$. Let $B^c$ denote the complement of $B^*$ in R. We adopt the convention that if S is a subset of R, then $S_i$ is the corresponding set in $R_i$, $i = 1,2$. It is readily verified that in $A'$ and $B^c$ the differences are symmetrically repeated, each occurring x-1 times. Consider the set of blocks

$$\{\alpha = A_1^* \cup B_2^c,\ \beta = B_1^* \cup A_2'\}.$$

Let $S_i \oplus S_j$ denote the multiset of exterior sums formed between $S_i$ and $S_j$. Since addition is commutative,

$$B_1^* \oplus A_2' = A_1' \oplus B_2^*.$$

Moreover it is clear that

$$(X \oplus Y) \;\&\; (X \oplus W) = X \oplus (Y \cup W)$$

for disjoint sets $Y$ and $W$, where & denotes multiset union.

Hence

$$\begin{aligned}(A_1' \oplus B_2^c) \;\&\; (B_1^* \oplus A_2') &= (A_1' \oplus B_2^c) \;\&\; (A_1' \oplus B_2^*)\\ &= A_1' \oplus (B_2^c \cup B_2^*)\\ &= A_1' \oplus R_2,\end{aligned}$$

and the mixed sums of $\alpha$ and $\beta$ are symmetrically repeated, each occurring $|A_1'| = x$ times. Hence $\alpha$ and $\beta$ developed through $R$, together with $B$, a block consisting of the elements of $R_2$ form the required design. □

As an example of the previous theorem, the design

$$\infty\ 0\ 2 \qquad 2\ 3\ 4 \qquad \text{mod } 5$$

is embedded in

$$\begin{array}{lllll|lllll} & & & & & 0_2 & 1_2 & 2_2 & 3_2 & 4_2 \\ \infty & 0_1 & 2_1 & 0_2 & 1_2 & 2_1 & 3_1 & 4_1 & 0_2 & 2_2 \\ \infty & 1_1 & 3_1 & 4_2 & 0_2 & 3_1 & 4_1 & 0_1 & 4_2 & 1_2 \\ \infty & 2_1 & 4_1 & 3_2 & 4_2 & 4_1 & 0_1 & 1_1 & 3_2 & 0_2 \\ \infty & 3_1 & 0_1 & 2_2 & 3_2 & 0_1 & 1_1 & 2_1 & 2_2 & 4_2 \\ \infty & 4_1 & 1_1 & 1_2 & 2_2 & 1_1 & 2_1 & 3_1 & 1_2 & 3_2 \end{array}$$

as a residual design.

In conclusion, the authors reiterate the fact that NSC designs are in a position to contribute to the theory of Hadamard matrices. We ask if all simple NSC designs are residual.

REFERENCES


[1] R.C. Bose, "On the construction of balanced incomplete block designs", *Ann. Eugenics*, 9 (1939), 353-399.

[2] M. Hall, Jr., *"Combinatorial Theory"*, Ginn [Blaisdell], Waltham, Mass. (1967).

[3] R.C. Mullin, "Resolvable designs and geometroids", *Utilitas Math.* 5 (1974), 137-149.

[4] P.J. Schellenberg, "On balanced Room squares and complete Howell rotations",

*Aequationes Math.* 9 (1973), 75-90.

5] R.G. Stanton and D.A. Sprott, "Block intersections in balanced incomplete block designs", *Canad. Math. Bull.* 7 (1964), 539-548.

05A17, 62-XX

# RECENT PROGRESS AND UNSOLVED PROBLEMS IN DOMINANCE THEORY

T.V. Narayana
Department of Mathematics
University of Alberta
Edmonton, Candada

DEDICATED TO R. PYKE

ABSTRACT

A survey of (unsolved) combinatorial, algebraic and statistical problems which have arisen since 1950 and are closely related to dominance or majorization.

## 1. INTRODUCTION AND BACKGROUND

Domination and dominance structure in combinatorics is often a special case of what Hardy, Littlewood and Polya (1952) call majorization. However with the rapid development of combinatorics and also the study of statistical configurations in recent years, which was stimulated by the availability of computers since (may I say?) 1950, it seems appropriate to retain the term "dominance" to distinguish that subdomain where integers and combinatorics are essentially involved - of the larger domain of problems where majorization is useful. The word "dominance", implying a pecking-order for the social sciences, was introduced by Landau (1953) who studied conditions for a score structure in the simplest round-robin (RR) tournament. If T represents the ordered scores of n players in a RR, where no ties are possible and the winner scores 1 point while the loser has 0, then Landau proved that T is a score vector if and only if T dominates $[0,\binom{2}{2},\binom{3}{2},\ldots,\binom{n}{2}]$ i.e.

$$t_1 \geq 0,$$

[1a] $$t_1 + t_2 \geq 1.$$

$$\vdots$$

$$t_1 + \ldots + t_{n-1} \geq \binom{n-1}{2},$$

$$t_1 + \ldots + t_n \geq \binom{n}{2}.$$

Of course the last inequality here becomes an equality, as obviously

[1b] $$w(T) = t_1 + \ldots + t_n = \binom{n}{2}.$$

One standard reference to graphs and tournaments is my colleague J.W. Moon's (1968) book, [cf. also his paper with Pullman (1970)]. Moon's book refers to many proofs and extensions of Landau's Theorem. A notable exception is the constructive and brief proof by Brauer, Gentry and Shaw (1968) which can be slightly shortened through dominance considerations. A more general definition of domination was given independently by Narayana (1955) in studying what is now generally called the Young lattice as popularized by Berge (1968). Here $(t_1,\ldots,t_n)$ dominates $(t_1',\ldots,t_n')$ if and only if

[1c] $$\sum_{j=1}^{r} t_j \geq \sum_{j=1}^{r} t_j' \quad (r = 1,2,\ldots,n).$$

In a sequence of papers, A. Young (1873-1940) developed group representations and essentially studied the combinatorics of dominance in connection with his famous Young tableaux or Young chains. As Young's collected works are soon to be published, I refer to it as well as the celebrated Hook Theorem due to Frame, Robinson and Thrall. More recently, Kreweras (1965) has given an elegant simultaneous treatment of both the problems of Young and Simon Newcomb. Thus, even combinatorially, dominance has had a long history; also, if $(t_1, t_2,\ldots,t_n,\ldots)$ are random variables [1c] represents "the sequence of partial sums of random variables". Such sequences play an important role in probability theory and for combinatorial results in this direction we cite the "ballot problem" and its generalization due to Takacs (1967). R. Pyke (1973) in his Jeffrey-Williams address touches on the role played by dominance in empirical processes in discussing Steck's (1971) results. We shall not treat these aspects of dominance in probability theory; nor shall we discuss the many other definitions of "dominance structure" possible even in tournaments, e.g. as developed by W.Maurer (1975) or by J. Zidek (1969), not to speak of other branches of mathematics!

Finally to indicate a class of combinatorial problems to which dominance as in [1c] applies, (many of which are unsolved), I refer to the delightful problem brought to my attention by Wynne (1976). The problem here is to assign integer weights to $n$ directors on a board so that (i) different subsets have different total weights (no ties possible in voting) and (ii) every subset of size $x$ always has more weight than every subset of size $x-1$ $(x=1,\ldots,n)$. The subsets [satisfying (ii)] may be called non-distorting in that every majority beats a minority. A solution to this problem is given in Table I; but the great difficulty is to either show such a solution is a minimal-sum solution or obtain a counter-example showing that such a solution is not minimal-sum. Indeed, in all such problems we can always replace minimal-sum by the weaker minimal-dominance; but even this simpler problem appears difficult. For other such problems, I refer to Erdos (1955).

TABLE I

Non-distorting, Tie-avoiding Integer Vote Weights $W_m$

| | | | | | | | | | | |
|---|---|---|---|---|---|---|---|---|---|---|
| members, m | 1 | 2 | 3 | 4 | 5 | 6 | 7 | 8 | 9 | 10 |
| totals, $S_m$ | 1 | 3 | 9 | 21 | 51 | 117 | 271 | 607 | 1363 | 3013 |
| column vectors of vote weights $[W_m]$ | 1 | 2 | 4 | 7 | 13 | 24 | 46 | 88 | 172 | 337 |
| | | 1 | 3 | 6 | 12 | 23 | 45 | 87 | 171 | 336 |
| | | | 2 | 5 | 11 | 22 | 44 | 86 | 170 | 335 |
| | | | | 3 | 9 | 20 | 42 | 84 | 168 | 333 |
| NOTE | | | | | 6 | 17 | 39 | 81 | 165 | 330 |
| The underlined values along | | | | | | 11 | 33 | 75 | 159 | 324 |
| the diagonal of vector elements | | | | | | | 22 | 64 | 148 | 313 |
| are the $I_m$ values - where | | | | | | | | 42 | 126 | 291 |
| $I_1 = I_2 = 1$ and | | | | | | | | | 84 | 249 |
| $I_m = 2I_{m-1} - [\mathrm{mod}_2(m-1)]I_{[m/2]-1}$ for $m \geq 3$ | | | | | | | | | | 165 |

## 2. RECENT PROGRESS

Pride of place with regard to the most important recent work surely goes to G. Kreweras' 1965, 1967 contributions, and in particular to his oft rediscovered Dominance Theorem. Almost every continent I visit has at least one proof of '1-dominance chains' and Australia is no exception (see Pitman 1972). These independent proofs by Epanechnikov, Steck, Mohanty, Sarkadi and perhaps others should be mentioned and the further statistical applications--notably by G.P. Steck--will be cited.

Let $A = (a_1,\ldots,a_k)$, $B = (b_1,\ldots,b_k)$ be integer, non-negative non-increasing sequences such that A dominates B, i.e.

$$[2a] \qquad a_i \geq b_i \quad (i=1,\ldots,k)$$

If $A \neq B$, in how many ways can we reach A from B through Young chains? In Kreweras' solution we note that some a's and b's might equal zero, and if $d = w(A) - w(B)$, a typical Young chain is

[2b] $$Y_d = A > Y_{d-1} > \dots > Y_1 > Y_0 = B.$$

Kreweras' response to this question is

[2c] $$Y(A,B) = [w(A)-w(B)]!\,\|\,[(a_i-i)-(b_j-j)]\#!\,\|$$

where the (i,j) element of the $k \times k$ determinant is given on the R.H.S. and $z\#! = (z!)^{-1}$. This formula is a generalisation of both Young's formula where $B = (0,\dots,0)$ and (in an infinite determinant form) of the Hook Theorem. He now introduces more general sequences than [2c] where we assume as usual A dominates B. For every positive integer r, we can interpose sets of non-increasing sequences $N_1,\dots,N_r$ ($N_i = n_{i1}, n_{i2},\dots,n_{ik}$; $i=1,\dots,r$) satisfying

[2d] $$A \geq N_1 \geq \dots \geq N_r \geq B,$$

where $A \geq N_1$ means A dominates (not necessarily strictly) $N_1$, etc. Let $K_r(A,B)$ denote the set of such possible sequences, where, unlike [2b], even $A = N_1 = N_2 = \dots = N_r$ is permissible. Sequences $N_1,\dots,N_r$ as in [2d] have been called r-dominance chains (perhaps dominant r-sequence might be clearer).

KREWERAS' DOMINANCE THEOREM (1965). With these notations, if A dominates B, then

(i) $$|K_r(A,B)| = \left\| \binom{a_i - b_j + r}{r+i-j}_+ \right\|_{k\times k}$$

with $|K_o(A,B)| = 1$ by convention.

(ii) $$|Y_r(A,B)| = \sum_{k=o}^{r} (-1)^k \binom{W(A)-W(B)+1}{k} K_{r-k}$$

for $r = 1,\dots,s$; here s is the maximum number of switchbacks possible in going from B to A through Young chains.

It would take us much too far out of the way to show how the Dominance Theorem also provides a solution to the classical problem of Simon Newcomb; however a switchback in a Young chain or tableau may be defined as follows: let $U < V < W$ be three consecutive vectors in a Young chain. A switchback occurs at V if the index of the term to be increased to go from V to W is strictly less than the index of the term to be increased to go from U to V. In the set $Y(A,B)$ $A = (3,2,2)$, $B = (2,1,0)$ all switchbacks are underlined.

The Set $Y(A,B)$

322

| | | | | | | | |
|---|---|---|---|---|---|---|---|
| 321 | 321 | 321 | 321 | 321 | 321 | $\underline{222}$ | $\underline{222}$ |
| $\underline{221}$ | 320 | 320 | $\underline{311}$ | 311 | $\underline{221}$ | 221 | 221 |
| 220 | $\underline{220}$ | 310 | 310 | $\underline{211}$ | $\underline{211}$ | $\underline{211}$ | 220 |

210

To those of you who have followed the contributions of I.R. Savage in rank-order probabilities since 1956 and Steck's more recent work since 1969, the following connections between Young chains and rank-tests in the two-sample problem must appear self-evident. Completeness forces me to refer to my monograph to appear in Mathematical Expositions (University of Toronto Press) where fullest details are available. Let us suppose we are filling a Young tableau of shape $(n,\ldots,n)$ [or going up a Young chain from $(0,\ldots,0)$ to $(n,\ldots,n)$] where all vectors involved are k-vectors and we fill in the tableau [or climb up the chain] at a uniform rate of 1 per second. Consider two independent samples $x_1,\ldots,x_n$ and $y_1,\ldots,y_k$ from populations with distribution functions $F(x)$ and $G(y)$. Rearranging the two samples in increasing order as $z_1,\ldots,z_{n+k}$, let us construct a lattice (sample) path from $(0,0)$ to $(n,k)$ by making a horizontal (vertical) step at positive integer time $i \le n+k$ if $z_i$ is an $x(y)$. It is easily seen that we obtain a (one-sided) rank test by rejecting the null hypothesis $F = G$ if the sample path lies entirely beneath the path $(n_1,\ldots,n_k)$, where $n_j$ is the distance measured parallel to the x-axis of the path from the point $(n,j-1)$ on the line $x = n$. The level of the test is given by the number of paths below or *dominated* by $(n_1,\ldots,n_k)$ divided by $\binom{n+k}{n}$. The Young chain represents the rank test and the choice of a particular $(n_1,\ldots,n_k)$ is the choice of the level of the test. From this point of view, there are as many *irreducible* tests as Young chains.

To illustrate further this correspondence between statistical tests and Young tableaux, let $A = (n,\ldots,n)$ and $B = (0,\ldots,0)$, where $A,B$ are k-vectors. The dominance theorem yields

$$[2e] \qquad |K_r(A,B)| = K_r(n,k) = \prod_{j=1}^{k} \left[\binom{n+r+j-1}{n-k+2j-1} \div \binom{n+j-1}{n-k+2j-1}\right],$$

where we have assumed without loss of generality $n \ge k$. From part (ii) of the dominance theorem we can calculate $Y_r(n,k)$. It is a remarkable fact, pointed out by Kreweras and Stanley (1972) that $|Y_r| = |Y_{r'}|$ where $r, r'$ are non-negative integers satisfying

[2f] $$r + r' = (n-1)(k-1).$$

A short table of values of $|Y(n,k)|$ is given as Table II.

As seen from Table II there should exist a natural bijection between Young tableaux with $r$ switchbacks and tableaux with $r'$ switchbacks, where $r$, $r'$ satisfy [2f]. Unfortunately such a bijection is only known in the special case where $\min(n,k) = 2$; in this case the number of switchbacks is

[2g] $$|Y_r| = \binom{n}{r}\binom{n}{r-1}/n.$$

TABLE II

Table of $Y_r(n,k)$ = Young chains on $n \times k$ rectangle with $r$ switchbacks.

<u>Note</u>: The symmetric table is extended as illustrated and $Y_0 = 1$.

| $Y_r$ / n,k | $Y_1$ | $Y_2$ | $Y_3$ | $Y_4$ | $Y_5$ | $Y_6$ | $Y_7$ |
|---|---|---|---|---|---|---|---|
| (3,3) | 10 | 20 | (10) | (1) | | | |
| (4,3) | 22 | 113 | 190 | (113) | | | |
| (5,3) | 40 | 400 | 1,456 | 2,212 | (1,456) | | |
| (4,4) | 53 | 710 | 3,548 | <u>7,700</u> | (7,700) | | |
| (6,3) | 65 | 1,095 | 7,095 | 20,760 | 29,484 | | |
| (5,4) | 105 | 2,856 | 30,422 | 151,389 | 385,029 | 523,200 | |
| (6,4) | 185 | 8,910 | 171,580 | 1,596,770 | 7,962,636 | 22,599,115 | <u>37,775,705</u> |
| (5,5) | 226 | 13,177 | 306,604 | 3,457,558 | 21,034,936 | 73,605,961 | 153,939,214 |

$Y_8 = 196,433,666$

Values where n,k are both even are symmetric with the underlined value repeating as when $n = k = 4$.

This special case has been studied by statisticians, and leads us to rank domination as follows. Consider a lattice path from the origin to the point $(n,k)$. Such a path can be uniquely described by (a) the $n$ vertical ($k$ horizontal) distances $v_1,\ldots,v_n$ ($h_1,\ldots,h_k$) of the path from the $x(y)$ axis, or equivalently by (b) its horizontal (vertical) ranks $r_1,\ldots,r_n$ ($s_1,\ldots,s_k$). Of course $R = (r_1,\ldots,r_n)$ and $S = (s_1,\ldots,s_k)$ are complements of each other w.r.t. the set $(1,2,\ldots,n+k)$; also $r_i = v_i + i$ $(i=1,\ldots,n)$, $s_j = h_j + j$ $(j=1,\ldots,k)$. Figure 1 illustrates these ideas in the case $n = 5$, $k = 4$.

Figure 1

Illustrating Lattice Duality and Rank Dominance

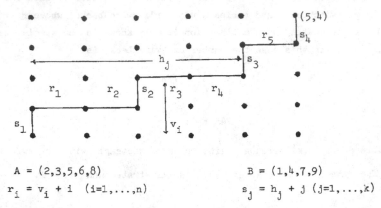

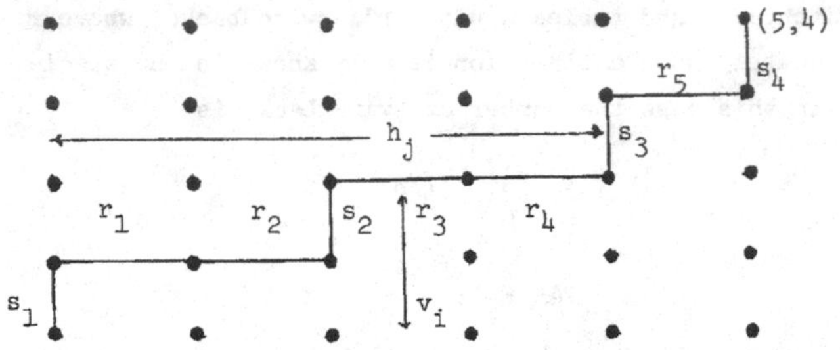


A = (2,3,5,6,8) B = (1,4,7,9)

$r_i = v_i + i \quad (i=1,\ldots,n)$ $s_j = h_j + j \quad (j=1,\ldots,k)$

It is evident from the figure that the set of all paths with vertical ranks dominating $S^\circ = (s_1,\ldots,s_k)$ is identical to the set of all paths with horizontal ranks dominated by $R^\circ = (r_1,\ldots,r_n)$ where $R^\circ$, $S^\circ$ are complementary sets w.r.t. $(1,\ldots,n+k)$. This result is known as (Young) lattice duality and Figure 2 illustrates how we explicitly obtain switchback duality by a bijection through this very natural lattice duality, in the special case k or n = 2.

We conclude our discussion of switchbacks by mentioning the very elegant computer program for switchbacks (and therefore for degrees of characters of the symmetric group--or more precisely a partitioning of these degrees) by McKay and Rohlicek. Given any partition P of n, their table (Table III) enumerates the number of Young tableau with shape P *and* having a given number of switchbacks. For example, when n = 6 and P = 4 1 1, their Table III shows there are 1,6,3 tableaux with 0,1,2 switchbacks respectively. The total 1 + 6 + 3 = 10 is the degree of the irreducible representation of S corresponding to the partition 4 1 1. This is the well-known result of Frobenius and Young established at the turn of the century and this degree is also given by the Hook theorem. With the number of switchbacks known, is it possible to obtain the results of Rasala (1977)? A typical example is, if $n \geq 9$, the first 4 minimum degrees of characters of $S_n$ are (i) 1, (ii) n-1, (iii) $\frac{1}{2}$ n(n-3), (iv) $\frac{1}{2}$ (n-1)(n-2). Similarly for $n \geq 15$, the first 7 minimum degrees are given.

Figure 2

Illustrating Switchback-Duality By Bijection

Consider the path with $n = 7$, $r = 3$ turns in Figure 2 (a)

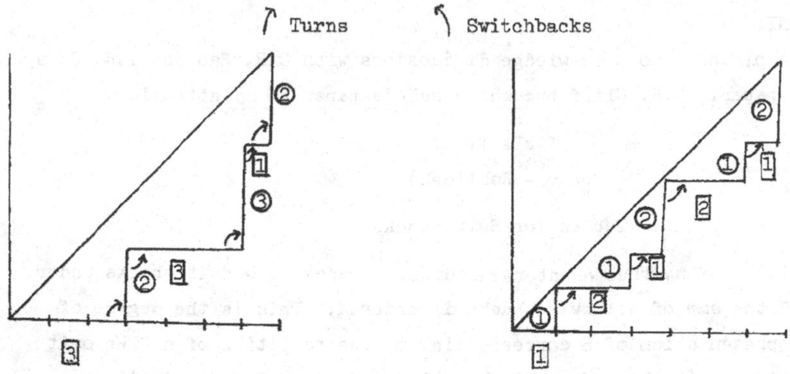


Figure 2(a) — (3,3,1) □ dominates (2,3,2) o

Figure 2(b) — Dual of Figure 2(a)

Clearly the complements of (3,6,7) □ (2,5,7) o (obtained by cumulation) w.r.t. (1,2,3,4,5,6,7) *{and adding back 7}* are (1 2 4 5 7), (1 3 4 6 7). Now note (1 3 4 6 7) o i.e. 1 2 1 2 1 dominates □ i.e. 1 1 2 1 2, as in Figure 2(b), where we have drawn the path with 5 turns. (Corresponding to 1 2 1 2 1 dominates 1 1 2 1 2.) The correspondence $r \leftrightarrow n - r + 1$ keeps the numbers $\frac{\binom{n}{r}\binom{n}{r-1}}{n}$ invariant. This proves switchback duality elegantly in the case of Young tableaux = Young chains when k or n = 2.

Problem: Can such an elegant solution (probabilistically evaluating the *mean* of the time of switchback) be given in the general case for switchback duality (first pointed out by Kreweras) on the general $k \times n$ rectangle? This would settle the "dictionary" for Young tableaux on the rectangle in a satisfactory manner.

## 3. STATISTICAL PROBLEMS

My final problems concern statistics, and will only be briefly stated in this gathering of combinatorialists. Is it possible to prove combinatorially that the Lehmann distribution (1953)

$$L^k_{m,n}(s_1,\ldots,s_n) = \frac{k^n}{\binom{m+n}{n}} \prod_{j=1}^{n} \frac{(s_j+\overline{j-1}\,k)\ldots(s_j+jk-1)}{(m+n+\overline{j-1}\,k)\ldots(m+n+jk-1)}$$

on the lattice paths in the $m \times n$ rectangle (note the change of notations) is a true probability distribution, at least in the case where k is a positive integer? This would be the starting point for many combinatorial problems, many of which stem from the work of Steck (1969, 1974).

4. ACKNOWLEDGEMENT

It is a great pleasure to acknowledge discussions with C.R. Rao and I.R. Savage in preparing this paper. G.H. Cliff brought Rasala's paper to my attention.

TABLE III

(McKay - Rohlicek)

Tables for Switchbacks

Let P = partition of n. Thus we enter the number of cases of N switchbacks under 1,2,... . Under N the *sum* of all switchbacks is entered. This is the degree of the irreducible representation of S corresponding to the partition of n. We omit the trivial partition (n itself) of n and the column of 1's under 0. Conjugate partitions are also omitted.

n = 6

| P | N | 1 | 2 | 3 | 4 | 5 | 6 |
|---|---|---|---|---|---|---|---|
| 5 1 | 5 | 4 | | | | | |
| 4 2 | 9 | 5 | 3 | | | | |
| 4 1 1 | 10 | 6 | 3 | | | | |
| 3 3 | 5 | 3 | 1 | | | | |
| 3 2 1 | 16 | 7 | 7 | 1 | | | |

n = 7

| P | N | 1 | 2 | 3 | 4 | 5 | 6 |
|---|---|---|---|---|---|---|---|
| 6 1 | 6 | 5 | | | | | |
| 5 2 | 14 | 7 | 6 | | | | |
| 5 1 1 | 15 | 8 | 6 | | | | |
| 4 3 | 14 | 6 | 6 | 1 | | | |
| 4 2 1 | 35 | 11 | 18 | 5 | | | |
| 4 1 1 1 | 20 | 9 | 9 | 1 | | | |
| 3 3 1 | 21 | 8 | 10 | 2 | | | |

REFERENCES

[1] C. Berge, *Principles of Combinatorics,* Academic Press, New York, 1971.

[2] A. Brauer, I. Gentry and K. Shaw, A new proof of a theorem of H.G. Landau on tournament matrices, *J. Comb. Th.,* 1968, 289.

[3] P. Erdös, Colloquium sur la théorie des Nombres, Bruxelles, 1955.

[4] J. Frame, G. deB. Robinson and R.M. Thrall, On the Hook graph of the symmetric group, *Can. J. Math,* 1954, 316.

[5] G.H. Hardy, E.G. Littlewood and G. Polya, *Inequalities,* Cambridge Univ. Press, Cambridge, 1952.

[6] G. Kreweras, Sur une classe de problèmes de dénombrement liés au treillis des partitions d'entiers, *Cah. BURO,* 1965, 5.

[7] G. Kreweras, Traitement simultané du "Problème de Young" et du "Problème de Simon Newcomb", *Cah. BURO,* 1967, 3.

[8] H.G. Landau, The conditions for a score structure, *Bull. Math. Biophysics,* 1953, 143.

[9] E.L. Lehmann, The power of rank tests, *Ann. Math. Stat.,* 1953, 23.

[10] W. Maurer, On most effective tournament plans with fewer matches than players, *Ann. Stat.,* 1975, 717.

[11] J. Moon, *Topics in Tournaments,* Holt, Reinhart and Winston, New York, 1968.

[12] J. Moon and N.J. Pullman, Generalised tournament matrices, *SIAM Review,* 1970, 384 .

[13] T.V. Narayana, Sur les treillis formés par les partitions d'un entier et leurs applications à la théorie des probabilités, *CR Acad. Sci,,* 1955, 1188.

[14] T.V. Narayana, *Lattice Path Combinatorics with Statistical Applications,* Univ. of Toronto Press, Toronto.

[15] T.V. Narayana and J. Zidek, Contributions to the theory of tournaments II, *Rev. (Roum.) des Math. Pures et Appl,* 1969, 1563.

[16] E.J.G. Pitman, Simple proofs of Steck's determinantal expressions for probabilities in the Kolmogorov and Smirnov tests, *Bull. Aust. Math. Soc.,*7, 1972, 227.

[17] R. Pyke, William-Jeffery Lecture, *Can. Math. Cong.,* 1973.

[18] R. Rasala, On the minimal degrees of characters of $S_n$, *J. Algebra,* 1977, 132.

[19] I.R. Savage, Contributions to the theory of rank order tests - the two sample case, *Ann. Math. Stat.,* 1956, 590.

[20] R.P. Stanley, *Ordered Structures and Partitions,* Memoirs of Amer. Math. Soc., No. 119, 1972.

[21] G.P. Steck, The Smirnov two sample tests as rank tests, *Ann. Math. Stat.,* 1969, 1449.

[22] G.P. Steck, A new formula for $P(R_i \leq b_i, 1 \leq i \leq m \mid m, n, f = G^k)$, *Ann. Prob.,* 1974, 155.

[23] L. Takacs, *Combinatorial Methods in the Theory of Stochastic Processes,* Wiley, 1967.

[24] B.E. Wynne and T.V. Narayana, Tournament configurations and weighted voting, *Cah. BURO,* to appear.

[25] A. Young, On quantative substitutional analysis, *Proc. London Math. Soc.*, 1927, 255.
[26] A. Young, *Collected Works*, Univ. of Toronto Press, Toronto, 1977.

# ON THE LINEAR INDEPENDENCE OF SETS OF $2^q$ COLUMNS OF CERTAIN (1, -1) MATRICES WITH A GROUP STRUCTURE, AND ITS CONNECTION WITH FINITE GEOMETRIES

J. N. Srivastava

Department of Statistics
Colorado State University, Fort Collins, CO 80521 U.S.A.

05B25, 05B30, 62K10

## ABSTRACT

Consider a set of m symbols (indeterminates) $F_1, \ldots, F_m$, and let G be the group of order $2^m$ generated by multiplying these symbols two, or three, or more at a time, where the multiplication is assumed commutative, and where $F_j^2 = \mu$ (the identity element of G) for all j. The elements of G can be written, in order, as $\{\mu; F_1, \ldots, F_m; F_1F_2, F_1F_3, \ldots, F_{m-1}F_m; F_1F_2F_3, \ldots; F_1F_2 \ldots F_m\}$. Consider a matrix $A(N \times 2^m)$ over the real field whose columns correspond in order to the elements of the group G. The elements of A are 1 and (-1), and are obtained as follows. The elements of A in the column corresponding to $\mu$ are all equal to 1. The next m columns of A, filled in arbitrarily, constitute an $(N \times m)$ submatrix, say A*. Finally, for all $\ell$ ($1 \leq \ell \leq m$), and all $i_1, \ldots, i_\ell$ (with $1 \leq i_1 < i_2 < \ldots < i_\ell \leq m$), the column of A corresponding to $F_{i_1}F_{i_2} \ldots F_{i_\ell}$ is obtained by taking the Schur product of the columns of A (or A*) corresponding to $F_{i_1}, F_{i_2}, \ldots, F_{i_\ell}$. The matrix A (over the real field) is said to have the property $P_t$ if and only if every set of t columns of A is linearly independent. In this paper, for all positive integers q, we obtain necessary conditions on A* such that <u>every</u> $(N \times 2^q)$ submatrix A** in A has rank $2^q$. A non-statistical introduction together with an illustrative example is provided.

## INTRODUCTION

We first execute the remark made in the last sentence above.

This subject is a part of the theory of "the design of factorial experiments of the $2^m$ type." Here, we are concerned with (statistically) planning a scientific experiment in which we are studying the effect of m factors (or variables) each at two levels, on some characteristic of the experimental material. For example, we may have an agricultural experiment with 4 (=m) factors, these being the nitrogen, phosphorus, potasium, and organic fertilizers, and the characteristic under study may be the yield of wheat. The two levels of each fertilizer (indicated by 1 and (-1) respectively) may indicate the presence and absence respectively of the fertilizer. Each row of A* then indicates a particular treatment-combination, i.e. a combination of levels of these factors. The elements of G can be interpreted as the names of certain parameters describing the effect of the various fertilizers on the yield of wheat. Thus, $\mu$ denotes the over-all average of the effects of the various treatment-combinations, $F_i$ the main effect of the ith factor, $F_iF_j$, the interaction between the ith and jth factors, $F_iF_jF_k$ the three-factor interaction between the ith, jth, and kth factors, and so on. The effect of any particular treatment-combination (which corresponds to a row of A*) is a linear function of the above parameters, the coefficients being the corresponding elements in the row of A containing this particular row of A*.

For any positive integer t, the significance of A having the property $P_{2t}$ is as follows. Suppose no random fluctuations are present, and at most t out of the $2^m$ parameters are non-zero. Also, assume that an experiment is done using the N treatment-combinations represented by the N rows of A*. Then a necessary and sufficient condition that the value of the non-zero parameters can be determined precisely is that A have property $P_{2t}$. Thus, this problem has a fundamental importance in the design of experiments, and is deeply connected with information and coding theory.

Definition 1.1 (a) Let T (N x m) be the (0,1) matrix obtained from A* by replacing (-1) by 0. Then T is called the design. (b) Let $\overline{T}$ be the matrix obtained from T by interchanging 0 and 1; we shall consider $\overline{T}$ over GF(2).

To help in clarifying ideas, we now present an example of the matrices T, A, etc. Thus, the matrix T at (1.1) below represents a design for a $2^4$ factorial experiment, the rows of T representing combinations of levels of the four factors. The matrix A corresponding to T is presented at (1.2). For convenience, the elements of the group G corresponding to each column of A is indicated at the top of the column:

$$T = \begin{bmatrix} 1&1&1&1 \\ 1&0&0&0 \\ 0&1&0&0 \\ 1&0&1&0 \\ 0&1&1&1 \\ 0&1&1&0 \end{bmatrix}, \qquad \overline{T} = \begin{bmatrix} 0&0&0&0 \\ 0&1&1&1 \\ 1&0&1&1 \\ 0&1&0&1 \\ 1&0&0&0 \\ 1&0&0&1 \end{bmatrix}, \qquad A^* = \begin{bmatrix} +&+&+&+ \\ +&-&-&- \\ -&+&-&- \\ +&-&+&- \\ -&+&+&+ \\ -&+&+&- \end{bmatrix}, \tag{1.1}$$

$$A = \begin{array}{c} \begin{matrix} \mu & F_1 & F_2 & F_3 & F_4 & F_{12} & F_{13} & F_{14} & F_{23} & F_{24} & F_{34} & F_{123} & F_{124} & F_{134} & F_{234} & F_{1234} \end{matrix} \\ \begin{bmatrix} + & + & + & + & + & + & + & + & + & + & + & + & + & + & + & + \\ + & + & - & - & - & - & - & - & + & + & + & + & + & + & - & - \\ + & - & + & - & - & - & + & + & - & - & + & + & + & - & + & - \\ + & + & - & + & - & - & + & - & - & + & - & - & + & - & + & + \\ + & - & + & + & + & - & - & - & + & + & + & - & - & - & + & - \\ + & - & + & + & - & - & - & + & + & - & - & - & + & + & - & + \end{bmatrix} \end{array} \tag{1.2}$$

where, in the above, (+) and (-) stand respectively for (+1) and (-1). Notice that

A* is constituted by the columns of A corresponding to $\{F_1,F_2,F_3,F_4\}$, and is identical with T except that 0 is changed to (-1). Also, for convenience, the elements of G at the top of the columns of A are represented by abbreviated symbols as indicated in the definition below.

Definition 1.2 If $1 \leq \ell \leq m$, and $i_1,\ldots,i_\ell$ are a set of $\ell$ distinct integers from the set $\{1,2,\ldots,m\}$, with $i_1 < i_2 < \ldots < i_\ell$, then the element $F_{i_1}F_{i_2}\ldots F_{i_\ell}$ of G will also be represented by the expressions $F_{i_1,i_2,\ldots,i_\ell}$ and $F(g_1,g_2,\ldots,g_m)$, where $g_r = 1$, if r belongs to the set $\{i_1, i_2,\ldots,i_\ell\}$, and $g_r = 0$ otherwise. The element $\mu$ will have the representation $F(0,0,\ldots,0)$. The symbols $g_r (r = 1,\ldots,m)$ which take the values 0 and 1, will be assumed to be over GF(2).

It is easy to check that if $E_1$ and $E_2$ are any two elements of G having respectively the representations $F(g_{11},\ldots,g_{m1})$ and $F(g_{12},\ldots,g_{m2})$, then their product $E_1E_2$ will have the representation $F(g_{11} + g_{12}, g_{21} + g_{22}, \ldots, g_{m1} + g_{m2})$.

As indicated in earlier papers, we shall, without loss of generality, assume that the first row of T and A* is (1,1,...,1). As a result, the first row of A will also have 1 everywhere. This convention will be used throughout the paper.

For later use, we now recall some results from Srivastava (1975).

A matrix B is said to have property $P_t$ if and only if every set of t columns of B are linearly independent. The matrix B could be over the real field or a finite field. There is a vast literature on such matrices, both in the context of factorial designs and coding theory.

Thus, we want to determine the necessary and/or sufficient condition on T such that A has $P_t$, with $t = 2^q$. Clearly, if a matrix B has property $P_t$ for some $t > 1$, then B also has property $P_{t-1}$, but not necessarily vice-versa.

Theorem 1.1. (a) The matrix A has $P_1$. (b) A necessary and sufficient condition that A has property $P_2$ is that in every (N x q) submatrix T* of T(N x m) with $1 \leq q \leq m$, there exists a row with an odd number of zeros. (c) If A has property $P_2$, then A also has $P_3$. (d) Suppose A has property $P_3$. Let $T_j(N \times m_j)$, $m \geq m_j \geq 1$, j = 1,2 be any two submatrices of T. Then a necessary and sufficient condition that A has $P_4$ is the following:

For every pair of submatrices $T_1$ and $T_2$ there exist three distinct rows (say the row number $r_1$, $r_2$, and $r_3$, where $(1 \leq r_1 < r_2 < r_3 \leq N)$) such that the following configuration arises:

| Row Number | Number of Zeros in | |
|---|---|---|
| | $T_1$ | $T_2$ |
| $r_1$ | odd | odd |
| $r_2$ | odd | even |
| $r_3$ | even | odd |

(1.3)

Of course, the values of $r_1$, $r_2$, and $r_3$ may change from one pair $(T_1,T_2)$ to another.

Theorem 1.2. (a) Let $A^0(N \times 2^m)$ be the matrix obtained from A by replacing (-1) everywhere by 0. Then A has property $P_t$ if and only if $A^0$ does. (b) Let K be a matrix with elements 0 and 1 over the real field, such that the first row of K has 1 everywhere. Let K* be the same matrix as K, except that the elements 0 and 1 belong to GF(2). Suppose K has property $P_{2t}$ (where t is a positive integer) but not $P_{2t+1}$. Then K* does not have property $P_{2t}$.

The remaining part of the section is meant to provide motivation for the more statistically minded reader, and may be skipped by other readers.

Consider a vector of observations $\underline{y}(N \times 1)$, obeying the 'Search Linear Model'

$$E(\underline{y}) = A_1\underline{\xi}_1 + A_2\underline{\xi}_2,\ V(\underline{y}) = \sigma^2 I_N, \tag{1.4}$$

where $A_1(N \times \nu_1)$, $A_2(N \times \nu_2)$ are known matrices, $\underline{\xi}_1(\nu_1 \times 1)$ is unknown, and $\underline{\xi}_2(\nu_2 \times 1)$ is partly known in the sense that it is given that there is an integer k ($\geq 0$) such that at most k elements of $\underline{\xi}_2$ are nonzero; however, we do not know exactly which k elements of $\underline{\xi}_2$ are nonzero. The problem is to search the nonzero element of $\underline{\xi}_2$ and draw inferences on these and on the elements of $\underline{\xi}_1$. The noiseless case ($\sigma^2 = 0$) is important in a basic way, since problems arising here arise also when $\sigma^2 > 0$. In the noiseless case, the necessary and sufficient condition that the above mentioned search and estimation problem can be resolved is [Srivastava (1975)] that, for <u>every</u> $(N \times 2k)$ submatrix $A_{20}$ of $A_2$, we have

$$\text{Rank}(A_1 : A_{20}) = \nu_1 + 2k. \tag{1.5}$$

Now, suppose that the observations $\underline{y}$ are the observed 'yields' corresponding to a set of N treatment combinations (say, T) in a $2^m$ factorial experiment (without block effects), and suppose $\nu_1 = 0$, $\nu_2 = 2^m$, and $k = 2^q$, where q is a non-negative integer. In this paper, we obtain necessary conditions on T so that

$$\text{Rank}(A_{20}) = 2k, \tag{1.6}$$

for <u>every</u> (N x 2k) submatrix $A_{20}$ of $A_2$ arising from such an experiment.

The above 'search linear model' now called 'search linear model with fixed effects' was introduced in Srivastava (1975), the results having been first presented at the International Symposium on Statistical Designs and Linear Models in March, 1973. Some other work in the field started by the author in 1973 is continued in Srivastava (1976), Srivastava and Ghosh (1977), Srivastava (1977), and Srivastava and Mallenby (1977). In these papers, the 'advantages,' and (in a large number of situations) the 'indispensibility' of the theory of search linear models has been explained. We shall now proceed with the main problem of this paper. In passing, it may be of interest to remark that the subject of search linear models and the related subject of 'Search Designs' ranges over a wide variety of branches of statistics and mathematics, particularly the subfields of statistical inference, multivariate distribution theory, and combinatorics. The present paper belongs to the field of search designs, and deals with statistical problems whose solutions seem to be rooted in the theory of zero-one matrices and extremal graph theory.

Consider a $2^m$ factorial experiment. Let $F_1, F_2, \ldots, F_m$ be m factors each at two levels. Treatment combinations can be denoted by the vector $(j_1, j_2, \ldots, j_m)$, where $j_r = 0, 1$ for $r = 1, 2, \ldots, m$. The 'true effect' of the treatment combination $(j_1, j_2, \ldots, j_m)$ will be denoted by $\tau(j_1, j_2, \ldots, j_m)$. Also, let $y(j_1, j_2, \ldots, j_m)$ denote the observed 'yield' corresponding to the treatment $(j_1, j_2, \ldots, j_m)$. Since we are not assuming any block effects, we shall have

$$Ey(j_1, j_2, \ldots, j_m) = \tau(j_1, j_2, \ldots, j_m). \tag{1.7}$$

The thrust of this paper is in the direction of breaking new ground regarding designs satisfying the condition (1.6). The case when block effects are present would be considered elsewhere. Let T(N x m) be a matrix whose rows correspond to treatment combinations (N in number) included in the experiment. Actually, each treatment combination can be used zero, one, or more times in the experiment. However, as will be clear later on from the nature of the material in this paper, we can assume without loss of generality that all the N treatments in T are distinct. Usually N will be much smaller than $2^m$, and in view of (1.6), would be expected to be of order 2k. Let $\underline{y}$(N x 1) be a vector containing the observations $y(j_1, j_2, \ldots, j_m)$, where the elements in $\underline{y}$ will be assumed to be arranged in the same order as the corresponding elements in T. Also, let $\underline{\tau}$(N x 1) contain the elements $\tau(j_1, j_2, \ldots, j_m)$, in the same order as the treatments in T, so that we have

$$E(\underline{y}) = \underline{\tau}, \ldots \tag{1.8}$$

As usual, the factorial effects also are $2^m$ in number, and include the general mean $\mu$, the main effects $F_i (i = 1,2,\ldots,m)$, the two-factor interactions $F_{ij} (i < j; i,j - 1,2,\ldots,m)$, the three factor interactions $F_{ijk} (i < j < k; i,j,k = 1,2,\ldots,m)$, $\ldots$, and, finally the m-factor interaction $F_{12\ldots m}$. Let $\underline{\xi}(2^m \times 1)$ be the vector of parameters given by

$$\underline{\xi}' = (\mu; F_1,\ldots,F_m; F_{12},\ldots,F_{m-1,m}; F_{123},\ldots,; \ldots; F_{12\ldots m}). \tag{1.9}$$

We now consider (1.5) again. We shall consider the case when $\sigma^2 = 0$, $\nu_1 = 0$, $\nu_2 = 2^m$, and $\underline{y}$ is as defined above. For simplicity, we will write $\underline{\xi}_2 = \underline{\xi}$, where $\underline{\xi}$ is as defined in (1.9), and $A_2 = A(N \times 2^m)$, $A_{20} = A_0(N \times 2k)$, so that the model becomes

$$\underline{y} = A\underline{\xi}. \tag{1.10}$$

The nature of the columns of A is well known, and is as mentioned in the beginning of this section.

## 2. THE PROPERTY $P_t$ WHEN t IS OF THE FORM $2^q$

From the foregoing, it is clear that given a matrix B over the real field with the property $P_t$, we want to be able to deduce (of course, under some further conditions) that B has property $P_{t+1}$. Below, if a matrix B does not have property $P_t$ then we shall say that B has property $\overline{P}_t$.

Theorem 2.1. Let B be a zero-one matrix over the real field such that the first row of B has 1 everywhere. Let B* be the same as B except that it is over GF(2). Then B has property $P_{2t+1}$ if both B and B* have property $P_{2t}$.

Proof. Suppose B does not have $P_{2t+1}$. Then, by Theorem 1.2(b), B* has $\overline{P}_{2t}$, a contradiction! This completes the proof.

Definition 2.1. Consider EG(m,2), the finite Euclidean Geometry of m dimensions based on GF(2). The design T(N x m) is said to be incident with a given u-flat $\Sigma$ of EG(m,2), if there is a point in $\Sigma$ which occurs as a point in T. (A row vector in T will sometimes be called a 'point' of T.)

Definition 2.2. A design T(N x m) is said to be a u-covering of EG(m,2) if and only if T is incident with every (m - u)-flat of EG(m,2).

Theorem 2.2. If T is u-covering of EG(m,2), with $u \geq 1$, then T is a (u - 1)-covering of EG(m,2).

Proof. Let $\Sigma$ be any $(m - u + 1)$-flat of EG(m,2), and let $\Sigma_1$ be a $(m - u)$-flat of EG(m,2) contained in $\Sigma$. Then T is incident with $\Sigma_1$ and hence with $\Sigma$. This completes the proof.

Theorem 2.3. The conditions (a) and (b) below are equivalent, and each is necessary and sufficient for A to have $P_2$:

(a) $\overline{T}$ has full rank over GF(2).

(b) $\overline{T}$ is a 1-covering of EG(m,2).

Proof. Consider a hyperplane (i.e., a $(m - 1)$-flat) $\Sigma_c$ in EG(m,2) given by the equation $x_{i_1} + \ldots + x_{i_\ell} = c$, where $c \in GF(2)$, $1 \leq \ell \leq m$, and $i_1, i_2, \ldots, i_\ell$ are $\ell$ distinct integers chosen out of the set $\{1,2,\ldots,m\}$. Then $\overline{T}$ is incident with $\Sigma_0$ in the point $(0,0,\ldots,0)$. Also, consider the $(N \times \ell)$ submatrix $\overline{T}^*$ of $\overline{T}$ obtained by taking columns number $i_1, i_2, \ldots, i_\ell$ of $\overline{T}$. From Theorem 2.1(b), it follows that there exists a row in $\overline{T}$ which has an odd number of 1's in it. Clearly, this row corresponds to a point in EG(m,2) which is in $\Sigma_1$. This proves (b). Again, if (a) does not hold, then there exists a vector $\underline{a}' = (a_1, a_2, \ldots, a_m)$ over GF(2) such that $\underline{a}$ is orthogonal to every row of $\overline{T}$. This would, however, imply that $\overline{T}$ is not incident with the $(m - 1)$-flat given by the equation $a_1x_1 + a_2x_2 + \ldots + a_mx_m = 1$. This completes the proof.

Corollary 2.1. If $T(N \times m)$ is such that A has property $P_2$, then

$$N \geq m + 1. \tag{2.1}$$

Proof. From Theorem 2.3(a), it follows that, over GF(2) the rank of $\overline{T}$ equals m. But the first row of $\overline{T}$ is the zero vector. Hence, $N - 1 \geq m$. This completes the proof.

Theorem 2.4. A necessary and sufficient condition that A has property $P_4$ is that $\overline{T}$ be a 2-covering of EG(m,2).

Proof. Let $\overline{T}_j(N \times m_j)$, $j = 1,2$, $1 \leq m_j \leq m$ be any two distinct submatrices of $\overline{T}$. By Theorem 1.1(d), it follows that a necessary and sufficient condition that A has property $P_4$ is that there exist three rows of $\overline{T}$ (say, rows number $r_1$, $r_2$ and $r_3$), such that (along with row number 1) the following configuration arises:

| Row Number | Number of 1's in | |
|---|---|---|
| | $\overline{T}_1$ | $\overline{T}_2$ |
| 1 | even | even |
| $r_1$ | odd | odd |
| $r_2$ | odd | even |
| $r_3$ | even | odd |

(2.2)

Let $i_1, i_2, \ldots, i_{m_1}$, be the columns of $\overline{T}$ included in $\overline{T}_1$, and similarly $j_1, j_2, \ldots, j_{m_2}$ be the columns corresponding to $\overline{T}_2$. Consider the (m - 2)-flat $\Sigma_{c_1,c_2}$ whose equation is given by

$$\begin{aligned} x_{i_1} + x_{i_2} + \ldots + x_{i_{m_1}} &= c_1 \\ x_{j_1} + x_{j_2} + \ldots + x_{j_{m_2}} &= c_2. \end{aligned} \tag{2.3}$$

The equations (2.3) indicate that the points of EG(m,2) corresponding respectively to rows number 1, $r_1$, $r_2$, and $r_3$ in $\overline{T}$ lie respectively in $\Sigma_{00}$, $\Sigma_{11}$, $\Sigma_{10}$, and $\Sigma_{01}$. This shows that $\overline{T}$ is a 2-covering of EG(m,2). On the other hand, we can clearly reverse the above argument which shows that if $\overline{T}$ is a 2-covering of EG(m,2), then (2.2) holds for all permissible pairs $(\overline{T}_1, \overline{T}_2)$. This completes the proof.

Definition 2.4. Let $\varepsilon_1, \varepsilon_2, \ldots, \varepsilon_n$ be a set of n factorial effects, such that $\varepsilon_r (r = 1,2,\ldots,n)$ has the representation $F(g_{1r}, g_{2r}, \ldots, g_{mr})$. Then the set of effects $(\varepsilon_1, \varepsilon_2, \ldots, \varepsilon_n)$ is said to have 'geometrical rank' (or briefly, 'rank') s if the m x n matrix G* given by

$$G^* = \begin{bmatrix} g_{11} & g_{12} & \cdots & g_{1n} \\ . & . & \cdots & . \\ . & . & \cdots & . \\ . & . & \cdots & . \\ g_{m1} & g_{m2} & & g_{mn} \end{bmatrix} \tag{2.4}$$

has rank s over GF(2). Furthermore, the effects $\varepsilon_1, \varepsilon_2, \ldots, \varepsilon_n$ will be said to be 'geometrically independent' (or, briefly, 'independent') if the set $\{\varepsilon_1, \varepsilon_2, \ldots, \varepsilon_n\}$ has geometrical rank n.

Example 1. In a $2^7$ factorial experiment the effects $F_{23}$, $F_{127}$, and $F_{137}$ are not independent, their set having rank 2. In general, let $F_{i_{11},\ldots,i_{1\ell_1}}$, $F_{i_{21},\ldots,i_{2\ell_2}}$, $\ldots, F_{i_{n1},\ldots,i_{n\ell_n}}$ be a set of n factorial effects. Then these effects do not form an independent set if each integer in the set $\{1,2,\ldots,m\}$ occurs an even number of times in the collection of $(\ell_1 + \ell_2 + \ldots + \ell_m)$ integers $\{i_{11},\ldots,i_{1\ell}; i_{21},\ldots,i_{2\ell_2}; \ldots; i_{n1},\ldots,i_{n\ell_n}\}$. This result follows by considering the matrix G* in (2.4) for the present case, and noting that the conditions of the theorem state that G* will have an even number of 1's in each row.

Theorem 2.5. A necessary condition that the matrix $A(N \times 2^m)$ have property $P_{2^q}$ $(1 \leq q \leq m)$ is that $\overline{T}$ be a q-covering of EG(m,2).

Proof. Suppose A has property $P_{2^q}$. Also, suppose that $\overline{T}$ is not a q-covering of EG(m,2), and let the (m - q)-flat $\Sigma$ given by the equations

$$\begin{aligned} b_{11}x_1 + \ldots + b_{1m}x_1 &= c_1 \\ \cdots\cdots\cdots\cdots & \\ b_{q1}x_1 + \ldots + b_{qm}x_m &= c_q \end{aligned} \tag{2.5}$$

be such that no point of $\Sigma$ corresponds to any row of $\overline{T}$. Obviously, $(c_1,c_2,\ldots,c_q)$ is not the zero vector. Consider the elements $\varepsilon_1,\varepsilon_2,\ldots,\varepsilon_q$ such that $\varepsilon_r$ has the vector representation $F(b_{r1},b_{r2},\ldots,b_{rm})$. Since $\Sigma$ is a (m - q)-flat, the effects $\varepsilon_1,\ldots,\varepsilon_q$ form an independent set, and therefore generate a subgroup of order $2^q$ whose elements are members of the set $\{\mu;\varepsilon_1,\ldots,\varepsilon_q;\varepsilon_1\varepsilon_2,\varepsilon_1\varepsilon_3,\ldots,\varepsilon_1\varepsilon_3,\ldots,\varepsilon_{q-1}\varepsilon_q;\varepsilon_1\varepsilon_2\varepsilon_3,\ldots,$ $\varepsilon_{q-2}\varepsilon_{q-1}\varepsilon_q;\ldots;\varepsilon_1\varepsilon_2\ldots\varepsilon_q\}$. Consider the $2^q$ columns of A corresponding to these $2^q$ effects, and let $A_0(N \times 2^q)$ be the submatrix of A corresponding to these columns. Since A has property $P_{2^q}$, $A_0$ must have $2^q$ distinct rows in it. Let $A^{**}(N \times q)$ be the submatrix of $A_0$ having the q-columns corresponding to the effects $\varepsilon_1,\varepsilon_2,\ldots,\varepsilon_q$. Notice that any row of A** gets completely fixed by the treatment in T corresponding to this row; indeed, if $(t_1^*,\ldots,t_m^*)$ is the treatment (where the t's are considered over the real field), then the corresponding row of A** is $(w_1^*,\ldots,w_q^*)$, where for all i, $w_i^* = \Pi_{j=1}^m (2t_j^* - 1)^{b_{ij}^*}$, where the symbols $c_i$, $b_{ij}$, $t_j$ denote the same quantities as $c_i^*$, $b_{ij}^*$, and $t_j^*$, except that the former are over GF(2) and the latter are over the reals. Let $w_i = \sum_{j=1}^m b_{ij}(1 - t_j)$, for all i. Then it is easily checked that $w_i^* = (1 - 2w_i)$, for all i, where now the $w_i$ are regarded as real numbers. Also, since the matrix A is such that the column corresponding to a product of elements of G is the product of the columns corresponding to these elements, it follows that any row of $A_0$ is completely determined by the part of this row which is in A**. Thus, the number of distinct rows in $A_0$ and A** are the same. Since, however, A** is a (1,-1) matrix, it follows that A** can have at most $2^q$ distinct rows. Now, Rank $(A_0) = 2^q$, by assumption. Hence, all the possible $2^q$ rows must occur in $A_0$ at least once. However, since $\Sigma$ given by (2.5) is not represented in $\overline{T}$, it follows that the row $(1 - 2c_1^*, 1 - 2c_2^*,\ldots,1 - 2c_q^*)$ does not occur in A**, a contradiction! This completes the proof.

The earlier results indicate that the condition in Theorem 2.5 is also sufficient when $q \leq 2$. However, for larger values of q, counterexamples against sufficiency can be easily constructed.

The above gives rise to an interesting unsolved problem. This is to find matrices $\overline{T}(N \times m)$ whose rows form a q-covering of EG(m,2), and furthermore, which minimize the value of N. Values of q in the range $1 \leq q \leq 4$ would be of greater interest. Also, it would be useful for statistical applications to consider classes of non-isomorphic q-coverings.

3. ACKNOWLEDGEMENT

This research was supported by Air Force Office of Scientific Research contract number F33615-74-1198.

4. REFERENCES

(1) J. N. Srivastava, "Designs for searching non-negligible effects," A Survey of Statistical Design and Linear Models, pp. 507-719, Edited by J. N. Srivastava, (North Holland Publishing Company, Amsterdam, 1975).

(2) J. N. Srivastava, "Optimal Search designs, or designs optimal under bias-free optimality criteria," Statistical Decision Theory and Related Topics, II, pp. 375-409, Edited by S. S. Gupta and D. S. Moore, (Purdue University Press, Lafayette, Indiana, 1977).

(3) J. N. Srivastava and S. Ghosh, "Balanced $2^m$ factorial designs of resolution V which allow search and estimation of one extra unknown effect $4 \leq m \leq 8$," Comm. Statist., A6, (1977), pp. 141-166.

(4) J. N. Srivastava and D. W. Mallenby, "Some studies on a new method of search in search linear models," (submitted for publication).

05B05, 05B20, 05B30

# THE DOEHLERT-KLEE PROBLEM: PART I, STATISTICAL BACKGROUND

Professor R.G. Stanton

Computer Science Department,
University of Manitoba,
Winnipeg, Man., Canada

## 1. INTRODUCTION

In discussing the problem of estimating an unknown function $f(x_1)$ of one independent variable $x_1$, Scheffé [14] made a plea for an "equally spaced distribution" of the levels of the variable $x_1$ at which observations are taken. The analogous problem, when one has a function $f(x_1, x_2, \ldots, x_n)$ of n independent variables, then leads one to attempt to have an equally spaced distribution of points $(x_1, x_2, \ldots, x_n)$ in n-space. What is meant by an "equally spaced distribution" of points in space is open to interpretation; Plackett [7] and Doehlert [3] have suggested criteria. Plackett advocates that the volume of the Voronoi polyhedron centred at any observation point P be constant; Doehlert introduces the criterion of a "constant distance pattern" for any P. Both of these criteria are satisfied by the designs introduced by Doehlert [3] under the title of "uniform shell designs"; we give further details in the next sections.

## 2. THE CUBOCTAHEDRON DESIGNS.

Suppose there are only two independent variables $x_1$ and $x_2$. Take a regular simplex in 2-space (an equilateral triangle); its vertices may be placed at (0, 0), (1, 0), (.5, .866). By differencing these points, we get four additional points, and thus end up with a totality of seven points, namely,

$$(0, 0), (\pm 1, 0), (\pm .5 \pm .866).$$

These form the vertices of a regular hexagon. Note that $x_1$ appears at five levels, $x_2$ appears at three levels. However, the number of levels is dependent on the co-ordinatization. If we rotate the $x_1 - x_2$ axes through $20^\circ$, then both factors appear at seven levels. This fact is illustrated in Figure 1, where the starred points represent the new levels of $x_1$.

The general case is well exemplified by what happens when one has three factors $(x_1, x_2, x_3)$. The 3-simplex is a tetrahedron with vertices (0, 0, 0), (1, 0, 0), (.5, .866, 0), and (.5, .289, .816). The complete "difference body" consists of thirteen points, namely,

$$(0, 0, 0), (\pm 1, 0, 0), (+.5, \pm .866, 0), (+.5, \pm .289, \pm .816),$$
$$(-.5, \pm .289, \mp .816), (0, \pm .577, \mp .816), (-.5, \pm .866, 0).$$

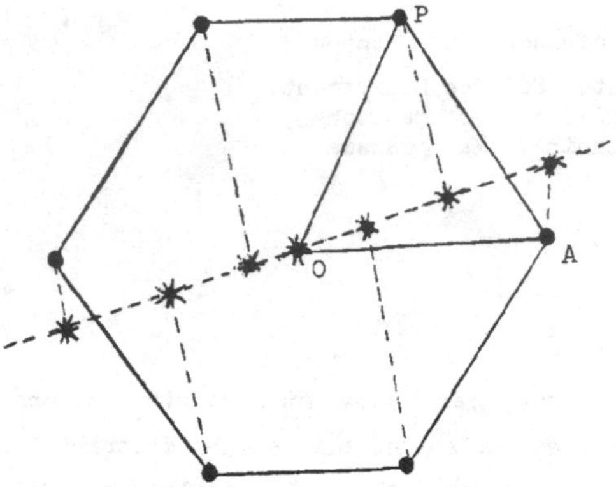


Figure 1.

(Here we have used the convention that all symbols + and - refer only to the other symbols on the same line.) The figure is a <u>cuboctahedron</u> obtained by taking a cube and joining midpoints of adjacent faces to give eight equilateral triangles (truncating the corners) and six squares (one per face).

The algorithm for d factors is exactly identical. It consists of four steps.

(1) Take the simplex in (d - 1) dimensions.

(2) Add a zero co-ordinate to increase the dimensionality.

(3) Take one extra point

$(\frac{1}{2}, \frac{1}{2\sqrt{3}}, \frac{1}{2\sqrt{6}}, \frac{1}{2\sqrt{10}}, \ldots, \frac{1}{\sqrt{2(d-1)(d-2)}}, \frac{1}{\sqrt{2d(d-1)}}, z)$, where $z = \sqrt{d+1}/\sqrt{2d}$.

(4) Difference these points.

The design properties of the collection of points are as follows.

(1) The simplex has d + 1 points (including $\theta$).

(2) The particular co-ordinatization employed has one 3-level factor; one 5-level factor; all other factors at 7 levels. We call this a 7-level design.

(3) There are $d^2 + (d+1)$ points in the design.

(4) The points are all equidistant from $\theta$; Doehlert [3] calls it a "uniform shell design", since translation produces a lattice all of whose points lie on spherical shells centred at $\theta$. Thus augmentation to any region is possible.

(5) The Voronoi polyhedron (consisting of all points closest to a given point $\theta$) is of constant volume.

(6) From any point in 3-space, the distance pattern to other points is constant (omitting $\theta$). The distances for 2 factors are

1(2), 1.732(2), 2(1).

In general, for d factors, the distance pattern is 1(2d - 2), 1.414((d - 1)(d - 2)), 1.732 (2d - 2), 2(1). This is Doehlert's "uniform distance property".

## 3. THE PROBLEM OF EXPERIMENTAL LEVELS

In practice, having many levels may be costly; it may be rendered impractical by the nature of the equipment; or, in some cases, a large number of levels may increase the experimental error.

The general problem under consideration is thus to take $C_d$, the d-dimensional cuboctahedron (or difference body of a d-dimensional simplex), and adjoin its central point $\theta$. We seek to co-ordinatize the figure, that is, choose a normal orthogonal basis, in such a manner that the set of inner products

$$I_b = \{\langle v, b\rangle : \; v \in C_d \cup \theta\}$$

has restricted cardinality.

For each b, $|I_b|$ is odd since $I_b$ includes 0 and is symmetric about 0. The result of the last section can be expressed formally as the

THEOREM. One can always choose the basis so as to have a 7-level design. Thus, if $B = \{b\}$, and

$$k = \max_{b \in B} |I_b|,$$

we can choose B so that $k = 7$.

The interesting problem is to maximize either the number of 3-level vectors in B or the number of 5-level vectors in B. Thus the original statistical problem leads us to a problem in linear algebra.

One approach is to take $C_d$ in $E^{d+1}$. Take $C_d$ as the d-simplex whose vertices are (1, 0, ..., 0, 0) up to (0, 0, ..., 0, 1). Translate one vertex to the origin $\theta$, then the simplex lies in the hyperplane $H_d$ with equation

$$H_d: \sum_{j=0}^{d} x_j = 0.$$

The process is illustrated in Figure 2, where $C_2$ appears in $E^3$. $C_3$ consists of $C_2$ with an $S_2$ "above" and "below" it. The difference body $C_d$ is thus represented in $H_d$ by the $d^2 + d$ points having two non-zero co-ordinates, the one being +1 and the other being -1.

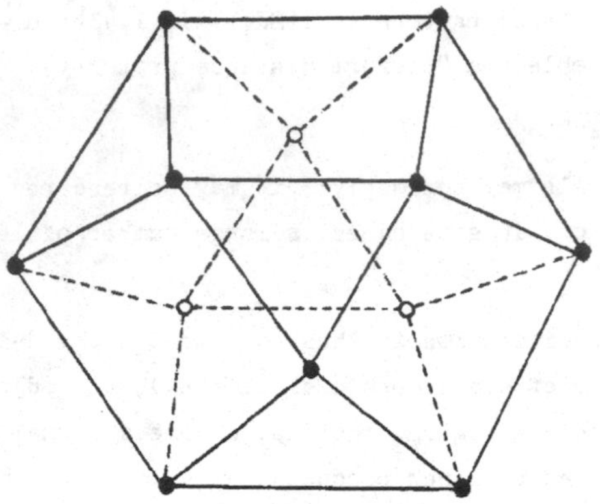

Figure 2

## 4. THREE-LEVEL AND FIVE-LEVEL DESIGNS.

Investigations of designs which maximize the number of 3-level and 5-level factors is made easy by the following two lemmas from Doehlert and Klee [4].

LEMMA 1: Three-level vectors in $H_d$ have the form: j co-ordinates of $-i$, i co-ordinates of $+j$, where $i + j = d + 1$; positive multiples of these.

LEMMA 2: Five-level vectors in $H_d$ have the form: i co-ordinates of $2k + j$, j co-ordinates of $k - i$, k co-ordinates of $-2i - j$, where $i + j + k = d + 1$; positive multiples of these.

From these two lemmas, Doehlert and Klee deduce their

MAIN THEOREM. Assume that $H(n)$, a Hadamard matrix of order n, exists for $n \equiv 0 \pmod 4$. Then the following four results hold.

(1) $H_{n-1}$ has an orthogonal basis with $n - 1$ three-level vectors.

(2) $H_n$ has an orthogonal basis with one three-level vector, $n - 1$ five-level vectors

(3) $H_{n+1}$ has an orthogonal basis with either one three-level vector and n five-level vectors; or two three-level vectors, $n - 2$ five-level vectors, one nine-level vector.

(4) $H_{n+2}$ has an orthogonal basis with one three-level vector, n five-level vectors, one seven-level vector.

Now introduce, for $\ell \geq 3$, the following quantities ($\ell$ will play the role of $d + 1$).

$T(\ell)$ = maximum m such that $H_{\ell-1}$ contains an orthogonal set with m vectors of level $\leq 3$.

$F(\ell)$ = maximum m such that $H_{\ell-1}$ contains an orthogonal set with m vectors of level $\leq 5$.

The main theorem, just quoted, shows that $F(\ell) = \ell - 1$ for all $\ell$ except $\ell \equiv 3 \pmod 4$; in the latter case, $F(\ell) \geq \ell - 2$ (whether $F(\ell) = \ell - 1$ in this case is an open question.)

In the next section, we describe the relationship of $T(\ell)$ with a purely combinatorial problem.

5. COMBINATORIAL FORMULATION OF THE DOEHLERT-KLEE PROBLEM

This section completes the changeover from Statistics, through Algebra, to Combinatorics. Doehlert and Klee [4] proved the

THEOREM. $T(\ell) \geq m$ if and only if there exist m proper subsets $X_1, \ldots, X_m$ of $\{1, 2, \ldots \ell\}$ such that

$$\frac{(\text{card } X_i)(\text{card } X_j)}{\text{card } X_i \cap X_j} = \ell$$

for all $i \neq j$.

If the only possibility is a single set $X_i = \{1, 2, \ldots, \ell\}$, we agree to set $T(\ell) = 1$ (trivially).

A particularly interesting case occurs if $H_{\ell-1}$ contains an orthogonal set of m three-level vectors which are permutation equivalent. Then we speak of $T^*(\ell)$; $T^*(\ell) \geq m$ under the same condition, with the m sets all having the same cardinality k. Thus we seek m sets of size k with

$$k^2 = \ell(\text{card } X_i \cap X_j).$$

Clearly, card $X_i \cap X_j$ is independent of i and j.

The known results until recently were:

(1) $T^*(\ell) = 1$ for $\ell$ square-free;

(2) $T^*(4a) = 4a - 1$ (subject to the Hadamard conjecture).

E. Nemeth and J.W. di Paola considered the case $\ell = 9$ in detail, and showed that $T^*(9) = 7$ (unpublished).

THE DOEHLERT-KLEE PROBLEM: PART II, A RE-INTERPRETATION

1. REFORMULATION OF THE PROBLEM

At the end of Part I, we were led to seek as many k-subsets as possible of $\{1, 2, \ldots, b\}$ such that each pair of subsets intersected in $k^2/b$ elements. Such a set of subsets produces an incidence matrix.

Now dualize this matrix, that is, interchange the roles of varieties and blocks. If, as usual, we use r to denote the replication number for a variety, $\lambda$ to denote the frequency of occurrence of a pair, then we seek "the maximum v such that we have b blocks, $r > \lambda$, and $r^2 = b\lambda$". Note that we admit null blocks (in the original problem, there might have been unused varieties). We have thus converted the problem to one of construction of $(r, \lambda)$ designs; we call these special $(r, \lambda)$ designs by the name DK designs.

It is convenient to exclude the case $b = 2r$. For, in this case, $b = 4\lambda$, $r = 2\lambda$, $\lambda$, are the parameters. The result is an SBIBD with $v = 4\lambda - 1$, $r = 2\lambda$, $\lambda$ (as a DK design, there is one empty block). This is the Hadamard case.

For $b - 2r \neq 0$, we note that $(b - r)^2 = b(b - 2r + \lambda)$, that is, the complement of a DK design is also a DK design. Hence there is no loss in generality in assuming $b - 2r > 0$ (equivalently, $r - 2\lambda > 0$).

## 2. SOME USEFUL TOOLS OF ATTACK ON THE PROBLEM

Probably the most useful principle involved in finding $T^*(b)$ is the formulation of the problem so that, given b, r, and $\lambda$, $T^*(b)$ is the maximum v such that an $(r, \lambda)$ design on v varieties and b blocks exists.

The block sizes in the design can be denoted by $k_i$; for $n = r - \lambda$, David McCarthy proved that

$$n + \lambda v - \sqrt{M} \leq 2\lambda k_i \leq n + \lambda v + \sqrt{M},$$

where $M = (n + \lambda v)(n + \lambda v - 4n\lambda)$. This is a Deza-type inequality and was very useful in obtaining early results; however, it was later replaced by the following result (see [5]), which is always stronger (here $a = r + \lambda(v - 1)$):

$$(\lambda b - r^2)k_i^2 + (vr^2 - ab + 2rn)k_i + n(ab - vr^2 - a) \geq 0.$$

Note that, in a DK design, this inequality takes the simpler form

$$k_i \leq \frac{(r-\lambda)(b-1)-\lambda v}{b-2r}. \tag{1}$$

Various results on block lengths were also given in [8] and [9], which also indicate the special role of the DK condition $r^2 - b\lambda = 0$.

I generally refer to (1) as the McCarthy-Vanstone or MV inequality. Two other very useful inequalities are easily obtained in the next two theorems.

THEOREM L. Let L be the longest block in a DK design; then

$$L \geq 1 + \frac{\lambda(v-1)}{r}.$$

PROOF. Clearly $\sum_{i=0}^{L} ib_i = rv$,

$$\sum_{i=0}^{L} \binom{i}{2} b_i = \lambda\binom{v}{2}.$$

Multiply the first equation by $\frac{L-1}{2}$, and subtract the second equation to yield

$$\sum_{i=0}^{L} \tfrac{1}{2}i(L - i)b_i = \frac{rv(L-1)}{2} - \lambda\binom{v}{2}.$$

Since the LHS $\geq 0$, we have

$$rv(L - 1) - \lambda v(v - 1) \geq 0,$$

and this reduces to the given condition on L.

Of course, by the Fisher Inequality for $(r, \lambda)$ designs, we have $b \geq v$; so we may set $v = b - d$, where d is called the "defect" of the DK design.

THEOREM D (THE DEFECT THEOREM). The defect of a DK design satisfies the inequality

$$d \geq \frac{r-\lambda}{\lambda}.$$

PROOF. By the MV inequality and Theorem L, we have

$$1 + \frac{\lambda(v-1)}{r} \leq L \leq \frac{(r-\lambda)(b-1)-\lambda v}{b-2r}.$$

Thus

$$1 + \frac{\lambda(b-d-1)}{r} \leq \frac{(r-\lambda)(b-1) - \lambda(b-1-d+1)}{b-2r},$$

$$1 + \frac{\lambda(b-1)}{r} - \frac{\lambda d}{r} \leq \frac{(r-2\lambda)(b-1)}{b-2r} + \frac{\lambda(d-1)}{b-2r}.$$

But $\frac{\lambda}{r} = \frac{r-2\lambda}{b-2r}$; hence

$$1 - \frac{\lambda d}{r} \leq \frac{\lambda(d-1)}{b-2r},$$

and we transform this result to

$$d \geq \frac{r-\lambda}{\lambda}$$

We can deduce an immediate Corollary as follows.

COROLLARY. Let $b = 9t$; one set of values for r and $\lambda$ is $r = 3t$, $\lambda = t$ (the only set if t is square free). Then $d \geq 2$ and so $T^*(9t) = 9t - 2$ whenever an SBIBD exists with parameters $(9t - 2, 3t, t)$.

This corollary handles the cases b = 9, 18, 27, 81. The remaining cases with $b \leq 100$ occur for b = 25, 45, 49, 50, 54, 63, 75, 90, 98, 99.

3. THE CASE $b = 25$.

The case $b = 25$ was handled in [6], and is especially interesting in that it provides the first DK design not a BIBD (this was unexpected). We here handle the case by somewhat different methods.

For $b = 25$, there are (up to complementarity) only the possibilities $r = 5$, $\lambda = 1$, and $r = 10$, $\lambda = 4$. If $r = 5$, $\lambda = 1$, then $d \geq 4$; thus $v \leq 21$, and a solution is provided by the projective geometry (21, 21, 5, 5, 1) with 4 empty blocks.

The question then arises as to whether we can do better with $r = 10$, $\lambda = 4$. In this case, $d \geq 2$. So we must consider the possibility that $v = 23$ or $v = 24$.

In general, we can prove [13] that a (10t, 4t) design on 25t - 2 varieties is impossible. We give the special case of this result as the

THEOREM. A (10, 4) design on 23 varieties is not possible.

PROOF. Here $k_i \leq \frac{52}{5}$, $L \geq \frac{49}{5}$, hence $L = 10$.

Take a specific variety, and let it occur in $a_i$ blocks of length i. Then

$$a_1 + a_2 + a_3 + \ldots + a_{10} = 10 ,$$

$$a_2 + 2a_2 + 3a_4 + \ldots + 9a_{10} = 88.$$

Eliminate $a_{10}$ to give $9a_1 + 8a_2 + \ldots + 2a_8 + a_9 = 2$.

Thus $a_9 = 2$, $a_{10} = 8$; or $a_8 = 1$, $a_{10} = 9$.

In either case, it follows that all blocks in the design have lengths 8, 9 or 10. Consequently,

$$b_8 + b_9 + b_{10} = 25,$$

$$8b_8 + 9b_9 + 10b_{10} = 230$$

$$28b_8 + 36b_9 + 45b_{10} = 1012.$$

Solve for $b_8$, and we get $b_8 = 67$. Since this is impossible, we must seek a (10, 4) design on 22 varieties.

Now, if $v = 22$, then $k_i \leq 11$, $L \geq 10$. We can prove [13] that a (10t, 4t) design on 25t - 3 varieties must have $L = 10t + 1$. Here we are content with the special

THEOREM. If $b = 25$, $r = 10$, $\lambda = 4$, $v = 22$, then $L = 10$ is not possible.

PROOF. It is a general result [5] that, with $a = r + \lambda(v - 1)$, we always have

$$(an - ak_i + \lambda k_i^2)(an - ak_j + \lambda k_j^2) - (\lambda k_i k_j - a\delta_{ij})^2 \geq 0,$$

where $\delta_{ij}$ is the intersection number for two blocks of lengths $k_i$ and $k_j$. If we apply this here, $a = 94$, $n = 6$; take $k_i = 10$. We find that, for $k_j = 10$,

$$24^2 - (400 - 94\,\delta_{10,10})^2 \geq 0.$$

Thus $\delta_{10,10} = 4$; similarly, $\delta_{10,k} \leq 4$ for $k < 10$.

Now consider intersections of a block of length 10 with the 24 other blocks. We obtain

$$\begin{aligned} x_0 + x_1 + x_2 + x_3 + x_4 &= 24, \\ x_1 + 2x_2 + 3x_3 + 4x_4 &= 90, \\ x_2 + 3x_3 + 6x_4 &= 135. \end{aligned}$$

Take three times the second equation and subtract twice the third; then

$$3x_1 + 4x_2 + 3x_3 = 0.$$

It follows that $x_4 = 22.5$, $x_0 = 1.5$; this contradiction proves the theorem.

We thus know that, if a design is possible, $L = 11$. The inequality used in the last theorem becomes, for blocks of lengths 11 and k,

$$14(564 - 94k + 4k^2) - (44k - 94\delta)^2 \geq 0,$$

or

$$47\delta^2 - 44k\delta + (10k^2 + 7k - 42) \leq 0.$$

Thus $47\delta$ lies in the range $22k \pm \sqrt{7(2k^2 - 47k + 282)}$. This leads to the following table for k and $\delta$.

| k | 11 | 10 | 9 | 8 | 7 | 6 | 5 | 4 | 3 | 2 | 1 | 0 |
|---|---|---|---|---|---|---|---|---|---|---|---|---|
| $\delta$ | 5 | - | 4 | 4 | 3 | 3 | 2 | 2 | 1,2 | 1 | 0,1 | 0 |

Now, for $v = b - 3$, we can use [1] to derive a stronger condition (see [13]); for b odd, this requires that

$$\begin{vmatrix} an - ak_1 + \lambda k_1^2 & -a\delta + \lambda k_1 k_2 & a - rk_1 \\ -a\delta + \lambda k_1 k_2 & an - ak_2 + \lambda k_2^2 & a - rk_2 \\ a - rk_1 & a - rk_2 & b \end{vmatrix}$$

be a perfect square. Using $k_1 = 11$, k and $\delta$ from the above table, we find the condition that

$$32\delta - 8 - 13k - (2k - 5\delta)^2$$

is a perfect square. This condition excludes the $(k, \delta)$ pairs specified by (9, 4), (7, 3), (5, 2) , (3, 1), (2, 1), (1, 1), (1, 0), (0, 0).

Again defining $x_i$ as the number of elements common to the blocks of lengths 11 and k, we see that $x_0 = x_1 = 0$. Hence the intersection equations become

$$\begin{aligned} x_2 + x_3 + x_4 + x_5 &= 24, \\ 2x_2 + 3x_3 + 4x_4 + 5x_5 &= 99, \\ x_2 + 3x_3 + 6x_4 + 10x_5 &= 165. \end{aligned}$$

Using multipliers 10, -4, 1, on these equations, we find that

$$3x_2 + x_3 = 9.$$

Also from the $(k, \delta)$ table, we see that $b_{11} = 1 + x_5$, $b_8 = x_4$, $b_6 = x_3$, $b_3 + b_4 = x_2$. But the $b_i$ satisfy the equations

$$\begin{aligned} b_3 + b_4 + b_6 + b_8 + b_{11} &= 25, \\ 3b_3 + 4b_4 + 6b_6 + 8b_8 + 11b_{11} &= 220 \\ 3b_3 + 6b_4 + 15b_6 + 28b_8 + 55b_{11} &= 924. \end{aligned}$$

In terms of the $x_i$'s, we have

$$\begin{aligned} x_2 + x_3 + x_4 + x_5 &= 24, \\ b_4 + 3x_2 + 6x_3 + 8x_4 + 11x_5 &= 209, \\ 3b_4 + 3x_2 + 15x_3 + 28x_4 + 55x_5 &= 869. \end{aligned}$$

Using multipliers 44, -9, 1, we obtain the result

$$-6b_4 + 20x_2 + 5x_3 = 44.$$

So $x_3$ is even; since we earlier had $3x_2 + x_3 = 9$, we either have $x_3 = 0$ or 6.

If $x_3 = 0$, $x_2 = 3$, it follows that $6b_4 = 16$, and this is impossible. If $x_3 = 6$, we have $x_2 = 1$, $b_4 = 1$, $x_4 = 6$, $x_5 = 11$. Thus we have the

THEOREM. If a DK (10, 4) design with 25 blocks and 22 varieties exists, then $b_4 = 1$, $b_6 = b_8 = 6$, $b_{11} = 12$.

Indeed, this design can be constructed. Actually, we shall construct the dual: a design on 25 varieties, with 22 blocks, such that 12 varieties occur 11 times, 6 varieties occur 8 times, 6 varieties occur 6 times, 1 variety occurs 4 times; all blocks have length 10; all intersections are equal to 4.

Our raw material is:

(1) the symbol $\infty$;

(2) blocks $B_1, \ldots, B_{11}$ in the SBIBD (11, 11, 5, 5, 2);

(3) any 11 blocks $C_1, \ldots, C_{11}$ from the SBIBD (13, 13, 4, 4, 1).

Define eleven blocks $\alpha_i$ by taking blocks $B_i \cup \infty \cup C_i$; these blocks have length 10

Define eleven blocks $\beta_i$ by taking blocks $\overline{B}_i \cup C_i$; these blocks likewise have length 10.

The result follows by noting that $|\alpha_i \cap \alpha_j| = |\alpha_i \cap \beta_j| = |\alpha_i \cap \beta_i| = |\beta_i \cap \beta_j|$. It is probable that this design is unique, although the question is open.

This design is the first DK design which is not a BIBD.

4. OTHER RESULTS

In [10] and [11], we give various results; a construction similar to that used to prove that T*(25) = 22 is given in greater generality. In particular, T*(49) = 46. Some lower bounds, which generally improve those of Doehlert and Klee (often being of the order of b/2 when theirs are of order $\sqrt{b}$) are given in [12]; these are recorded in the final table of the current paper.

Recently Haemers, at the Sixth British Combinatorial Conference in July of 1977, described the construction of the SBIBD (71, 71, 15, 15, 3). This result is used in [13], together with various results on (10t, 4t) designs with 25t blocks, to show that T*(75) = 71. Similar arguments show that T*(45) ≤ 40.

In the final table, we use our usual notation K(b) for T*(b). Also, we omit square-free values of b, for which K(b) = 1. We also omit values b = 4t, since K(4t) = 4t - 1, at least for all values 4t for which the Hadamard conjecture holds. With this understanding, we record the current bounds on K(b). If the symmetric designs (61, 61, 21, 21, 7) and (97, 97, 33, 33, 11) should exist, then K(63) and K(99) would be known.

| b | K(b) |
|---|---|
| 9 | 7 |
| 18 | 16 |
| 27 | 25 |
| 45 | 16 ≤ K ≤ 40 |
| 54 | 25 ≤ K ≤ 51 |
| 63 | 31 ≤ K ≤ 61 |
| 81 | 79 |
| 90 | 28 ≤ K ≤ 87 |
| 99 | 31 ≤ K ≤ 99 |
| 25 | 22 |
| 50 | 26 ≤ K ≤ 46 |
| 75 | 71 |
| 49 | 46 |
| 98 | 50 ≤ K ≤ 96 |

REFERENCES

[1] Conner, W.S., Jr., "On the Structure of Balanced Incomplete Block Designs", *Ann. Math. Stat.* 23 (1952), 52-71.

[2] Collens, R.J., "Constructing BIBD's with a Computer", *Ars Combinatoria* 2 (1976), 285-303.

[3] Doehlert, D.H., "Uniform Shell Designs", *J. Royal Stat. Soc. C,* 19 (1970), 231-239.

[4] Doehlert, D.H. and Klee, V.L., "Experimental Designs Through Level Reduction of the d-dimensional Cuboctahedron", *Discrete Math.* 2 (1972), 309-334.

[5] McCarthy, D. and Vanstone, S.A., "On the Structure of Regular Pairwise Balanced Designs", submitted to Discrete Math.

[6] McCarthy, D., Stanton, R.G. and Vanstone, S.A., "On an Extremal Class of (r, λ) Designs Related to a Problem of Doehlert and Klee", *Ars Combinatoria* 2 (1976).

[7] Plackett, R.L. and Burman, J.P., "The Design of Optimum Multifactor Experiments", *Biometrika* 33 (1946), 305-325.

[8] Stanton, R.G., "Results on (r, λ) Designs", *Cong. Num. 18 (Proc. 6th Manitoba Conf. on Numerical Maths.),* Winnipeg (1976), 411-412.

[9] Stanton, R.G., "Some Results on Block Lengths in (r, λ) Designs", *Ars Combinatoria* 2 (1976), 213-219.

[10] Stanton, R.G. and Vanstone, S.A., "Further Results on a Problem of Doehlert and Klee", to appear, Utilitas Math.

[11] Stanton, R.G. and Vanstone, S.A., "On a Problem of Doehlert and Klee", *Cong. Num. 19 (Proc. Eighth S.E. Conf. on Combinatorics, Graph Theory, and Computing)",* Baton Rouge (1977) to appear.

[12] Stanton, R.G. and Vanstone, S.A., "Some Lower Bounds on the Size of Doehlert-Klee Designs", to appear, Ars Combinatoria.

[13] Stanton, R.G. and Vanstone, S.A., "Some Theorems on DK Designs", preprint.

[14] Scheffé, H., "The Simplex-centroid Design for Experiments with Mixtures", *J. Royal Stat. Soc. B,* 25(1963), 235-263.

05C20, 20B25

# ON THE CAYLEY INDEX OF A GROUP

Mark E. Watkins
Mathematics Department,
Syracuse University,
Syracuse, New York 13210,
U.S.A.

ABSTRACT.

The set of left multiplications by the elements of a group G form a subgroup $L_G$ of the automorphism group A(X) for any Cayley graph X of G. The Cayley index c(G) of G is the minimum of the index $[A(X) : L_G]$ taken over all Cayley graphs X of G. The present article is expository, reporting what is known to date about Cayley indices of groups and where some of the results may be found. The general pattern thus far indicates that, with finitely many exceptions, c(G) is determined by whether G is (1) abelian, (2) generalized dicyclic, (3) solvable and finite but neither abelian nor generalized dicyclic, or (4) none of the above.

## 1. PRELIMINARIES.

The symbol G will always denote a group and e will denote its identity. If $H \subset G$ and $e \notin H$, we define the *Cayley graph* $X_{G,H}$ *of* G *with respect to* H to be the graph with vertex set G and edge set consisting of all 2-subsets of G of the form $\{g, gh\}$ where $g \in G$ and $h \in H$. Since $X_{G,H}$ is unaltered by the assumption $h \in H \Longleftrightarrow h^{-1} \in H$, we shall henceforth assume that H is a subset of G having this property.

If X is a graph with vertex set V, then A(X) will denote its automorphism group. If $x \in V$, then $A_x(X)$ denotes the subgroup of A(X) which fixes x, i.e., the *stabilizer* of x.

For each $g \in G$, we define $\lambda_g : G \to G$ by $\lambda_g(x) = gx$ for all $x \in G$, and let $L_G = \{\lambda_g : g \in G\}$. One easily verifies that $L_G$ is isomorphic to G and that $L_G$ is a subgroup of $A(X_{G,H})$ for any H. Moreover, $L_G$ acts on G, the vertex set of $X_{G,H}$, as a regular permutation group. G. Sabidussi [12] observed that, in fact, a necessary and sufficient condition for an arbitrary graph X to be a Cayley graph of some group G is that A(X) admit a regular subgroup isomorphic to G. We therefore define the *Cayley index* of G to be

$$c(G) = \min\{[A(X_{G,H}) : L_G] \mid H \subset G\}.$$

When $X_{G,H}$ is a Cayley graph of G for which this minimum is attained, we have $c(G) = |A_e(X_{G,H})|$ whenever c(G) is finite - and we know of no case where c(G) is larger than 16 [4 and 8].

If $c(G) = 1$, then $A(X_{G,H})$ is a regular permutation group for some $H \subset G$, and $X_{G,H}$ is called a *graphical regular representation* (GRR). We say that G is in *Class I* if it admits a GRR. *Class II* consists of those groups G such that for any subset $H \subseteq G$ which is closed with respect to inverses, there corresponds a non-identity group-automorphism $\phi$ of G such that $\phi[H] = H$. In [14] it was shown that these two classes are disjoint, and it was conjectured that every group is in one class or the other. Although many groups, both finite and infinite, have since been classified, the conjecture still stands. Observe that Class II is defined entirely in algebraic terms with no mention of graphs.

## 2. THE CAYLEY INDEX OF AN ABELIAN GROUP.

The permutation $\alpha$ on a group G given by $\alpha(g) = g^{-1}$ $(g \in G)$ certainly maps any set $H = H^{-1}$ onto itself. If G is abelian, then $\alpha$ is an automorphism of G. If, moreover, G is not of exponent 2, then it must belong to Class II. Thus in 1964, C.-Y. Chao [3] and G. Sabidussi [13] obtained simultaneously and independently the first result concerning Cayley indices, namely, that $c(G) \geq 2$ for any abelian group G not of exponent 2. Imrich [5] then showed that GRR's do not exist for the elementary abelian 2-groups of orders 4, 8, and 16 and indicated the method of construction of a GRR for the other elementary abelian 2-groups. Thus it was known for abelian groups whether or not the Cayley index exceeded 1.

It turned out [8] that the Cayley index of an abelian group is in fact at most 2 with the exception of the seven groups: $Z_2^3$, $Z_2^4$, $Z_3^2$, $Z_3^3$, $Z_4^2$, $Z_2 \times Z_4$, and $Z_2^2 \times Z_4$. (The symbol $Z_n$ denotes the cyclic group of order n.) Using a computer, D. Hetzel [4] determined that $c(Z_3^3) = 12$, the largest Cayley index for an abelian group.

## 3. GENERALIZED DICYCLIC GROUPS AND AN ANSWER TO A CONJECTURE.

A non-abelian group is *generalized dicyclic* if it is generated by an abelian group B and an element d of order 4 not in B such that $d^2 \in B$ and $d^{-1}bd = b^{-1}$ for all $b \in B$. The smallest generalized dicyclic group is the quaternion group Q, where $B = Z_4$. T. Lewin and I [14] characterized the generalized dicyclic groups as the non-abelian groups G admitting a non-identity automorphism $\beta$ such that $\beta(g) \in \{g, g^{-1}\}$ for all $g \in G$. From this it was immediate that the generalized dicyclic groups are in Class II. (By graph-theoretical means, L. Nowitz [9] had shown previously that the generalized dicyclic groups are not in Class I.) Since $\beta^2$ is the identity, c(G) is even for all generalized dicyclic groups G.

If G is generalized dicyclic but not of the form $Q \times Z_2^m$ $(m \geq 0)$ and if its abelian subgroup B of index 2 is not of the form $Z_4^2$, $Z_2 \times Z_4$, or $Z_2^2 \times Z_4$, then c(G) = 2 or 4. Things tend to become easier as groups get larger, for Imrich and I showed further [8] that under these assumptions, c(G) = 2 if $|G| > 96$. Our results combined with D. Hetzel's calculations suggest that $c(G) \leq 16$ for all generalized dicyclic groups G. The value of 16 is attained for Q and $Q \times Z_2$.

On the basis of these two infinite families of Class II groups and the discrete scattering of Class II groups I had encountered outside these families, I conjectured [16]: there exists a positive integer n such that if a non-abelian group G is in Class II, then G is generalized dicyclic or $|G| \leq n$.

At the time I knew $n \geq 32$ because of the group $Q \times Z_4$ (see [15]). Recently L. Babai [1] has proved the conjecture with n = 3057. Babai does not claim that this bound is sharp.

## 4. OTHER SOLVABLE GROUPS.

In the light of the conjecture made at the end of the previous section, it is not surprising that for non-abelian, non-generalized dicyclic groups, the general effort has been to show that, with finitely many exceptions, they all fall in Class I. This was certainly the pattern begun in [14] with certain small classes of groups such as the dihedral groups and groups of order $p^3$ for p prime.

The first strong theorem for solvable groups came in [10, 11], where it was shown that any finite non-abelian group of order coprime to 6 admits a GRR. The proof relied heavily upon the Feit-Thompson Theorem which asserts the solvability of all groups of odd order. Nowitz and I would have liked to exploit it to its fullest by replacing 6 with 2, but we were discouraged by two difficulties. First, we had shown that a normal extension of a Class I group G by $Z_n$ is also in Class I if $n \geq 5$, but we could not prove this for n = 2, 3, 4. Secondly, we had shown that the non-abelian group of order 27 and exponent 3 is in Class II. The first obstacle was reduced when it was shown [7] that for $|G| > 32$ a non-abelian normal extension by $Z_2$ or $Z_4$ is in Class I, while for $|G| > 36$ a non-abelian normal extension by $Z_3$ is in Class I. The second obstacle was completely toppled when Imrich [6] showed, again using solvability, that every non-abelian group of odd order *except* the aforementioned anomaly of order 27 admits a GRR.

The basic technique for attacking solvable groups in all of the above work was to start at some point in their normal series with a group known to be in Class I. Then, by means of our extension theorems, one could climb inductively up the normal series as far as one would like, finding only Class I groups along the way. The problem was that there remained too many "small groups" (e.g. of orders 32, 36, 48, 64, 72) to which the techniques developed in [7] do not apply. Thus there remained too many "starting points" each of which would have to be handled separately.

It was at this point that Hetzel applied a computer to determine for small groups G a Cayley graph $X_{G,H}$ with $|A_e(X_{G,H})| = c(G)$. This done, Hetzel could apply the procedures of those of us who had gone before him to obtain the following powerful result: if G is a finite, solvable group which is neither abelian nor generalized dicyclic, then $c(G) = 1$, unless G is one of ten groups, each of order at most 32, in which case $c(G) = 2$.

## 5. OTHER GROUPS.

There have been only two attacks on the problem of determining the Cayley index of groups without a finite normal series. In [16], I showed that the alternating group $A_n$ is in Class I if and only if $n \geq 5$. E. Bannai [2] generalized my method to prove the far more powerful result that every finite, non-solvable group that could be generated by two elements, one of which has order 2, admits a GRR.

The other approach is exactly the reverse of the extension process of the previous section. It has been shown [17] that every free group, with from 2 to $\aleph_0$ generators is in Class I. At this point Sabidussi's result [13] is invoked, that if $N \trianglelefteq G$ and if X is a Cayley graph of G, then the quotient graph X/N is a Cayley graph of G/N. By taking successive normal closures of relations among the generators of a free group, it was shown in [17] that every free product of at least two and at most countably many groups, each of which is at most countably generated, admits a graphical regular representation.

## REFERENCES.


[1] L. Babai, "On a conjecture of M.E. Watkins", (to appear).

[2] E. Bannai, "Graphical regular representations of non-solvable groups", (to appear).

[3] C.-Y. Chao, "On a theorem of Sabidussi", *Proc. Amer. Math. Soc.* 15 (1964), 291-294.

[4] D. Hetzel, "Über reguläre graphische Darstellung von auflösbaren Gruppen", Diplomarbeit, Technische Universität Berlin, (1977), 371 pp.

[5] W. Imrich, "Groups with transitive Abelian automorphism groups", in *Combinatorial Theory and Its Applications*, Coll. Soc. Janós Bolyai 4, Balatonfűred, Hungary, (1969), 651-656.

[6] W. Imrich, "On graphs with regular groups", *J. Combinatorial Theory* (B) 19 (1975), 174-180.

[7] W. Imrich and M.E. Watkins, "On graphical regular representations of cyclic extensions of groups", *Pacific J. Math.* 54 (1974), 149-165.

[8] W. Imrich and M.E. Watkins, "On automorphism groups of Cayley graphs", *Per. Math. Hung.* (to appear).

[9] L.A. Nowitz, "On the non-existence of graphs with transitive generalized dicyclic groups", *J. Combinatorial Theory* 4 (1968), 49-51.

[10] L.A. Nowitz and M.E. Watkins, "Graphical regular representations of non-abelian groups, I", *Canad. J. Math.* 24 (1972), 993-1008.

[11] L.A. Nowitz and M.E. Watkins, "Graphical regular representations of non-abelian groups, II", *Canad. J. Math.* 24 (1972), 1009-1018.

[12] G. Sabidussi, "On a class of fixed-point free graphs", *Proc. Amer. Math. Soc.* 9 (1958), 800-804.

[13] G. Sabidussi, "Vertex-transitive graphs", *Monatsh. Math.* 68 (1964), 426-438.

[14] M.E. Watkins, "On the action of non-abelian groups on graphs", *J. Combinatorial Theory B* 1 (1971), 95-104.

[15] M.E. Watkins, "On graphical regular representations of $C_n \times Q$", in *Graph Theory and Applications* (Ed. Y. Alavi, D.R. Lick, A.T. White), Springer-Verlag, Berlin, (1972), 305-311.

[16] M.E. Watkins, "Graphical regular representations of alternating, symmetric, and miscellaneous small groups", *Aequat. Math.* 11 (1974), 40-50.

[17] M.E. Watkins, "Graphical regular representations of free products of groups", *J. Combinatorial Theory* (B) 21 (1976), 47-56.

05B05, 05B30

# A SURVEY OF EXTREMAL $(r,\lambda)$-SYSTEMS AND CERTAIN APPLICATIONS

R.C. Mullin

University of Waterloo

The purpose of these talks is to survey some of the extremal properties of $(r,\lambda)$-systems and to explain the implications of some of these results to related configurations, namely equidistant permutation arrays (generalized Room squares). (The theory of $(r,\lambda)$-systems has recently proved fruitful in the area of Doehlert-Klee designs, but these are not discussed here since they are discussed in detail by R G Stanton elsewhere in these proceedings.) An $(r,\lambda)$-*system* is a pair $(V,F)$ where $V$ is a $v$-set of objects called varieties, and $F$ is a family of subsets (called blocks) of $V$ which satisfies the following :

(i) every variety occurs in precisely $r$ blocks;

(ii) every pair of distinct varieties occurs in precise $\lambda$ blocks.

The quantity $n = r - \lambda$ is called the *order* of the design.

We shall refer to the cardinality of a block as its *size* and for a given $(r,\lambda)$-system $D$ with blocks $B_1,B_2,\ldots,B_b$ (which need not be distinct) we let $K(D) = K = \{k_1,k_2,\ldots,k_b\}$ where $k_i = |B_i|$. If $K(D) \subseteq \{1,v\}$ then $D$ is said to be *trivial*, otherwise $D$ is *nontrivial*. Further if $K(D) \subseteq \{1,v-1,v\}$, then $D$ is said to be *near-trivial*.

For fixed $r$ and $\lambda$, it is not difficult to show that there exists a least integer $v_0(r,\lambda)$ such that for $v > v_0(r,\lambda)$ any $(r,\lambda)$-system is trivial. Lovàsz and Erdös had conjectured that $v_0(r,\lambda)$ was no worse than quadratic in $r$. This author proved that for fixed $n$ and large $\lambda$, $v_0(r,\lambda) = \lambda + 2$, and that the extremal designs are near trivial.

In 1973, M Deza [3] discovered a powerful approach which proved that any non-near-trivial system contained at most $n^2 + n + 2$ varieties. This was improved by Vanstone [27] to $n^2 + n + 1$. These results can be combined to establish the following.

THEOREM 1. $v_0(r,\lambda) \leq \max\{\lambda+2,n^2+n+1\}$.

Van Lint [14] showed that if a non-near-trivial system contained $n^2 + n + 1$ varieties, then $n$ is the order of a projective plane. Consequently in the case where $n$ is not the order of a projective plane, theorem 1 can be strengthened.

Many results have been achieved in this direction. For example a couple of the most recent (and therefore strongest) are cited below. For proofs, the reader is referred to [17].

THEOREM 2. Let D be a non-trivial $(n+1,1)$ design on $n^2 - \alpha$ varieties $(1 \le \alpha \le n)$. If $n > 2\alpha^2 + 3\alpha + 2$, then D can be embedded in a finite projective plane of order n.

THEOREM 3. Let D be an $(n+1,1)$ design on $n^2 - 2$ varieties. If $n \neq 2$, then D can be embedded in a projective plane of order n.

In the proof of his result, Deza [3] established the inequality stated below.

THEOREM 4. Let D be a $(2n,n)$ design on v varieties. If $k_i$ is the size of a block in D, then

$$k_i(v+1-k_i) \le (v+1)n .$$

This has been generalized by McCarthy and Vanstone to the following.

THEOREM 5. In an $(r,\lambda)$ design on v varieties, the size, $k_i$, of any block must satisfy the inequality

$$k_i(n+\lambda v-\lambda k_i) \le n(n+\lambda v) .$$

An alternative approach to the problem of determining the function $v_0(r,\lambda)$ is to fix the value of n. By doing so, all but a finite number of values are known in view of the results described on the previous page.

It is easy shown that

$$\begin{aligned} v_0(n+\lambda,\lambda) &= n^2 + n + 1 && \text{for } \lambda \le n^2 + n - 1 \\ &= \lambda + 2 && \text{for } \lambda > n^2 + n - 1 , \end{aligned}$$

provided that n is the order of a projective plane. Also

$$v_0(\lambda+1,\lambda) = \lambda + 2 \quad \text{for } \lambda \ge 1 .$$

Hence the first value of n for which the series is unknown is $\lambda = 6$.

It has been shown by McCarthy that

$$\begin{aligned} v_0(\lambda+6,\lambda) &= 31 && \text{for } \lambda \le 6 \\ &\quad 31 \text{ or } 32 && \text{for } 6 < \lambda < 30 , \\ &\quad \lambda + 2 && \text{for } \lambda > 30 . \end{aligned}$$

An *equidistant permutation array* $A(r,\lambda;v)$ is a set of $v$ permutations of $\{1,2,\dots,r\}$ such that any pair of distinct permutations agree in precisely $\lambda$ positions.

The relationship between equidistant permutation arrays and $(r,\lambda)$-systems is described below. An $(r,\lambda)$-system $D$ is said to be *resolvable* if the blocks can be partitioned into classes such that each variety occurs precisely once in each class. $D$ is orthogonally resolvable if it can be resolved into classes $R_1,R_2,\dots,R_r$ and also into classes $S_1,S_2,\dots,S_r$ such that $|R_i \cap S_j| \le 1$. It is known that equidistant permutation arrays $A(r,\lambda;v)$ are coexistent with orthogonally resolvable $(r,\lambda)$-systems on $v$ varieties. It is usual to let $R(r,\lambda)$ denote the maximum value of $v$ for which an $A(r,\lambda;v)$ exists. A summary of some known results on $R(r,\lambda)$ is given below.

$R(2,1) = 1$, $R(3,1) = 2$, $R(4,1) = 3$, $R(5,1) = 5$. (Bolton [1])

$R(6,1) = 10$. (Bolton [1], van Rees [20])

$R(7,1) = 13$, $R(8,1) \ge 15$. (McCarthy, Mullin [15])

$R(13,1) \ge 27$. (Schellenberg and Taylor (private communication)

$R(m,1) \ge 2n - 4$ for $n \ge 6$. (Heinrich, van Rees [11])

$R(n,1) \le n^2 - 4n - 1$ for $n \ge 6$. (Mullin, Nemeth [18])

$R(4n,n) \ge n(n-1)$ for $n$ a prime power. (Woodall [30])

$R(4n,n) \ge kn$ if $\exists k$ p.o.ℓ.s. of side $n$ (Vanstone [25])

$R(n,\lambda) \le \max\{2 + \left\lceil \frac{\lambda}{[\frac{n}{3}]} \right\rceil, n^2\}$, $n = r - \lambda$ (McCarthy and Vanstone, private communication)

$R\left(\frac{3q(q^{n-1}-1)}{(q-1)} + q, \frac{3q(q^{n-2}-1)}{(q-1)} + q\right) \ge (q-1)q^{n-1}$ (Vanstone [25])

$R(n+\lambda+2,\lambda) \ge (\lambda+1)n + 1$ if $\exists\ \lambda + 1$ p.o.ℓ.s. of order $n$ with two disjoint transversals. (Heinrich, van Rees, Wallis [12])

(For $n$ a prime power, the above becomes

$R(2n-1,n-3, \ge (n-1)^2)$.

$R(r,\lambda) = 2 + \left\lceil \frac{\lambda}{[\frac{n}{3}]} \right\rceil$, $n = r - \lambda$, $\lambda > (n^2+n)[\frac{n}{3}]$. (Vanstone [28])

BIBLIOGRAPHY


[1] D.W. Bolton, Unpublished Manuscript.

[2] V. Chvátal, On finite Δ-systems of Erdös and Rado, Acta Math. Acad. Sc. Hungaricae, 21 (1970), 341-355.

[3] M. Deza, Une proprieté extremale des plans projectifs finis dans une class de codes equidistants, Discrete Math. 6 (1973), 343-352.

[4] M. Deza, Matrices dont deux lignes quelconques coincident dans un nombre donne de positions communes, J. Combinatorial Theory Series A 20 (1976) 306-308.

[5] M. Deza, R.C. Mullin and S.A. Vanstone, Orthogonal systems, Aequationes Math. (to appear).

[6] M. Deza, R.C. Mullin and S.A. Vanstone, Room squares and equidistant permutation arrays, Ars Combinatoria 2 (1976), 235-244.

[7] D.H. Doehlest and V.L. Klee, Experimental designs through level reduction of the d-dimensional cuboctahedron, Discrete Math. 2 (1972), 309-334.

[8] J.I. Hall, Bounds for equidistant codes and partial projective planes, Discrete Math. (to appear).

[9] J.I. Hall, On two conjectures of Stanton and Mullin, J. Combinatorial Theory Series A (to appear).

[10] J.I. Hall, A.J.E.M. Janssen, A.W.J. Kolen and J.H. van Lint, equidistant codes with distance 12, Discrete Math. (1976).

[11] Katherine Heinrich and G.H.J. van Rees, Some construction for equidistant permutation arrays of index one, Utilitas Math. (to appear).

[12] Katherine Heinrich, G.H.J. van Rees and W.D. Wallis, A general construction for equidistant permutation arrays, Graph Theory and Related Topics (to appear).

[13] F. Hoffman, P.J. Schellenberg and S.A. Vanstone, A starter-adder approach to equidistant permutation arrays. Ars Combinatoria 1 (1976) 303-315.

[14] J.H. van List, A theorem on equidistant codes, Discrete Math. 9 (1973), 353-358.

[15] D. McCarthy and R.C. Mullin (Unpublished manuscript).

[16] D. McCarthy, R.G. Stanton and S.A. Vanstone, An extremal class of $(r,\lambda)$-designs related to a problem of Doehlert and Klee, Ars Combinatoria 2 (1976), 305-317.

[17] D. McCarthy and S.A. Vanstone, Embedding $(r,1)$-designs in finite projective planes, Discrete Math. (to appear).

[18] R.C. Mullin and E. Nemeth, An improved bound for equidistant permutation arrays of index one, Utilitas Math. (to appear).

[19] R.C. Mullin and S.A. Vanstone, On regular pairwise balanced designs of order 6 and index one, Utilitas Math. 8, (1975), 349-369.

[20] G.H.J. van Rees, Private communication.

[21] P.J. Schellenberg and S.A. Vanstone, Equidistant permutation arrays of index one, Proc. 6th Man. Conference on Num. Math. (1976).

[22] P.J. Schellenberg and S.A. Vanstone, Recursive constructions for equidistant permutation arrays, J. Austral. Math. Soc. (to appear).

[23] S.A. Vanstone and P.J. Schellenberg, A construction for equidistant permutation arrays of index one. J. Combinatorial Theory Series A, (to appear).

[24] S.A. Vanstone and D. McCarthy, $(r,\lambda)$-systems and finite projective planes, Utilitas Math. 11 (1977) 57-71.

[25] S.A. Vanstone, Pairwise orthogonal generalized Room squares and equidistant permutation arrays. J. Combinatorial Theory. Series A (to appear).

[26] S.A. Vanstone, Extremal $(r,\lambda)$-designs, Discrete Math. (submitted).

[27] S.A. Vanstone, A bound for $v_0(r,\lambda)$, Proc. 5th Southeastern Conference on Combinatorics, Graph Theory and Computing, Boca Raton (1974), 661-673

[28] S.A. Vanstone, The asymptotic behaviour of equidistant permutation arrays. Canad. J. Math., (submitted).

[29] P. de Witte, On the embeddability of linear spaces in projective planes of order $n$, (to appear).

[30] D. Woodall, Unpublished Manuscript.

05C30

# ON THE ENUMERATION OF CERTAIN GRACEFUL GRAPHS

C. C. Chen
Department of Mathematics
Nanyang University
Singapore

**ABSTRACT.** *A graceful graph on n vertices is said to be simple if each of its connected components has at most one cycle and the component containing the edge with end points labelled 1 and n has no cycle. Let* $s_n$ *and* $t_n$ *denote the numbers of simple graceful graphs and graceful trees on n vertices respectively. Then* $t_n \leqq s_n \leq p(A_{n-2})$ *where* $p(A_{n-2})$ *is the permanent (plus determinant) of the* $(n-2)\times(n-2)$ *matrix* $A_{n-2} = (a_{ij})$ *defined by:*

$$a_{ij} = \begin{cases} 2 & \text{if } i \leqq j \text{ and } i+j \leqq n-1 \\ 0 & \text{if } i > j \text{ and } i+j > n-1 \\ 1 & \text{otherwise.} \end{cases}$$

*More specifically, let* $c_i$ *denote the number of simple graceful graphs on n vertices with i cycles* ($i = 0,1,2,\ldots,k$, *where* $k = [(n-2)/3]$). *Then we have*

$$\sum_{i=0}^{k} 2^i c_i = p(A_{n-2}).$$

## 1. INTRODUCTION.

Let G be a graph on n vertices which are labelled by the integers in the set $N = \{1,2,\ldots,n\}$ such that each vertex of G is associated with a distinct number in N. The *weight* of an edge is defined as the absolute value of the difference between the two numbers labelled at its end points. If the weights of all edges of G are distinct then we call G a *graceful graph*. If, in addition, every connected component of G has at most one cycle and the component containing the edge with end points labelled 1 and n has no cycle, then G is called a *simple graceful graph*. Evidently, every graceful tree is simple.

Two graceful graphs $G_1$, $G_2$ on n vertices are said to be *isomorphic* if the mapping which carries each vertex in $G_1$ labelled i to the corresponding vertex in $G_2$ labelled i ($i = 1,2,\ldots,n$) is a graph-isomorphism.

We exhibit in the following all (non-isomorphic) simple graceful graphs with four and five vertices.

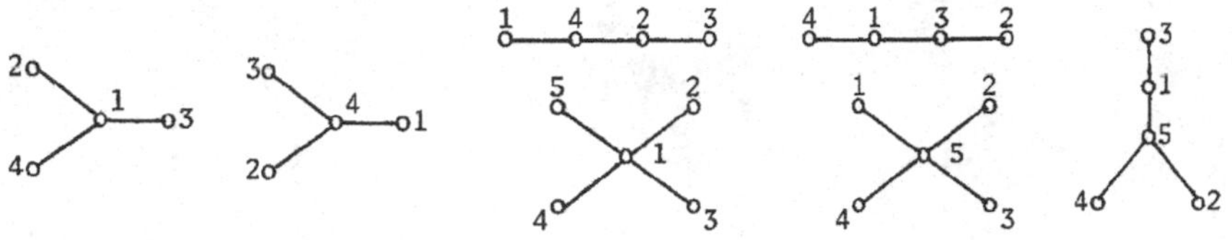

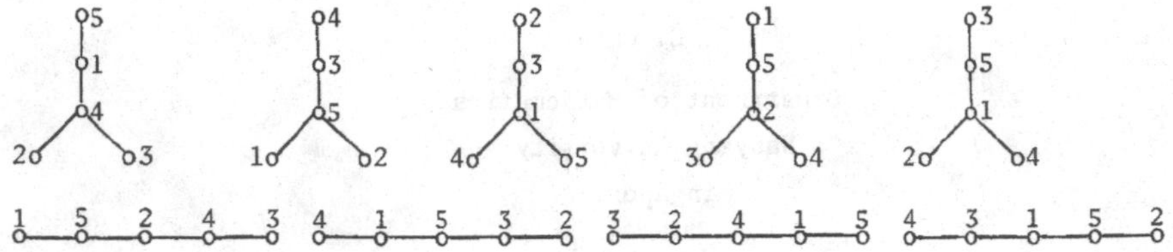


There does also exist a non-simple graceful graph without isolated vertices in which 1 and n lie in a component containing a circuit as shown below:

$G_8$:

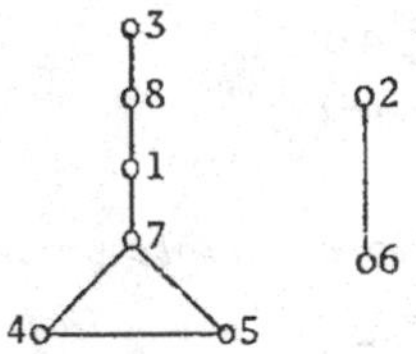


Note also that given one graceful graph G on n vertices, another one G' (called the *pair* of G) is obtained by replacing each label k by the label n-k+1. We shall show that G and G' are never isomorphic when n ≥ 3. Indeed, consider the edge e of G of weight n-2. Then we have only two possiblities:

Case 1. The end points of e are labelled n and 2.

Case 2. The end points of e are labelled n-1 and 1.

If the pair G' of G is isomorphic to G, then in case 1, the vertices labelled n-1 and 1 must be adjacent in G' and hence in G; whereas, in case 2, the vertices labelled n and 2 must be adjacent in G' and hence in G. These give rise to two edges in G of weight n-2, a contradiction.

However, in spite of this result, we shall count both a graceful graph and its pair as separate objects in this paper.

Ringel conjectured in 1963 that all trees are graceful. The term "graceful" is however due to Golomb. See references in (1) in which Cahit gave some example of graceful complete binary trees.

Note that all simple graceful graphs given above are graceful trees. In fact, it can be shown that each simple graceful graphs with less than 7 vertices is always a tree. There do however exist a non-tree simple graceful graph with 7 vertices as shown below:

$G_7$:

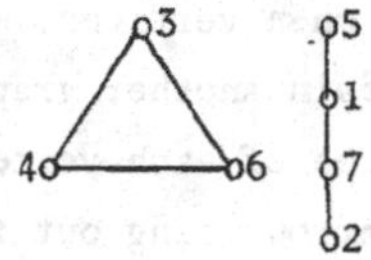


We shall denote the numbers of (non-isomorphic) graceful graphs, simple graceful graphs and graceful trees on n vertices by $g_n$, $s_n$ and $t_n$ respectively. It is easy to prove that $g_n = (n-1)!$. In this paper, we shall give an upper bound for $s_n$ and hence for $t_n$. To do this, let us define a *graceful nxn matrix* $A_n$ as the matrix $(a_{ij})$ where

$$a_{ij} = \begin{cases} 2 & \text{if } i \leq j \text{ and } i+j \leq n+1 \\ 0 & \text{if } i > j \text{ and } i+j > n+1 \\ 1 & \text{otherwise.} \end{cases}$$

The *permanent* (plus determinant) of the matrix $A_n$ is defined as :

$$p(A_n) = \sum_{\sigma \in S_n} \Pi_{i=1} a_{\sigma(i),i}$$

where $S_n$ denotes as usual the set of all permutations of $\{1,2,\ldots,n\}$.

The aim of this paper is to prove the following:

THEOREM 1. $t_n \leq s_n \leq p(A_{n-2})$ *for each* $n \geq 3$.

THEOREM 2. *Let* $c_i$ *denote the number of simple graceful graphs on n vertices with i cycles* ($i = 0,1,\ldots,k$, *where* $k = [(n-2)/3]$). *Then*

$$\sum_{i=0}^{k} 2^i c_i = p(A_{n-2}) .$$

## 2. THE PROOFS.

Let $n \geq 3$ be given. Let $B_{n-2}$ denote the $(n-2)x(n-2)$ matrix $(b_{ij})$ where each $b_{ij}$ is an ordered pair defined by $b_{ij} = \langle j+1-i,\ i+j+1 \rangle$, for $i,j = 1,2,\ldots,n-2$. A selection of components of entries of $B_{n-2}$ is said to be a *legitimate selection* of $B_{n-2}$ if:

(i) all components selected are positive and $\leq n$; and

(ii) one and only one component is selected from each column and each row of $B_{n-2}$.

It is easy to see that there are exactly $p(A_{n-2})$ legitimate selections of $B_{n-2}$. Hence to prove that $s_n \leq p(A_{n-2})$, we need only to establish a one:one mapping from the set of all simple graceful graphs on n vertices into the set of all legitimate selections of $B_{n-2}$. To do this, let G be a simple graceful graphs on n vertices. We first delete the edge with end points labelled 1 and n from G to obtain another graph G'. Next, we turn G' into a directed graph such that the out-degree of each vertex other than 1 and n is exactly one (i.e. there exists exactly one arrow going out from each vertex other than 1 and n). This is possible because each connected component of G' has at most one cycle. If there exists an arrow from a vertex v to a vertex u in G', we say that v *covers* u. Now, for each vertex v labelled i in G', we select a component of $B_{n-2}$ in the (i-1) column which is equal to the labelling of the vertex covered by v. Since G is graceful, it can easily be seen that, we do obtain a legitimate selection of $B_{n-2}$ in this manner and different simple graceful graphs give rise to different selections. This proves Theorem 1.

The reader may find the above argument easier to follow by comparing the directed graph $G_7'$ and the selection (consisting of underlined components) in the following matrix $B_5$,

$G_7'$:

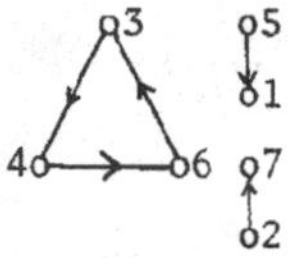


$$B_5 = \begin{pmatrix} \langle 1,3 \rangle & \langle 2,\underline{4} \rangle & \langle 3,5 \rangle & \langle 4,6 \rangle & \langle 5,7 \rangle \\ \langle 0,4 \rangle & \langle 1,5 \rangle & \langle 2,\underline{6} \rangle & \langle 3,7 \rangle & \langle 4,8 \rangle \\ \langle -1,5 \rangle & \langle 0,6 \rangle & \langle 1,7 \rangle & \langle 2,8 \rangle & \langle \underline{3},9 \rangle \\ \langle -2,6 \rangle & \langle -1,7 \rangle & \langle 0,8 \rangle & \langle \underline{1},9 \rangle & \langle 2,10 \rangle \\ \langle -3,\underline{7} \rangle & \langle -2,8 \rangle & \langle -1,9 \rangle & \langle 0,10 \rangle & \langle 1,11 \rangle \end{pmatrix}$$

Hence by Theorem 1, we have $s_7 \leq p(A_5) = 168$, where

$$A_5 = \begin{pmatrix} 2 & 2 & 2 & 2 & 2 \\ 1 & 2 & 2 & 2 & 1 \\ 1 & 1 & 2 & 1 & 1 \\ 1 & 1 & 0 & 1 & 1 \\ 1 & 0 & 0 & 0 & 1 \end{pmatrix}.$$

Finally, note that if G contains one cycle, then we may turn G' into a directed graphs in two ways such that the out-degree of each vertex other than 1 and n is one. For instance, $G'_7$ can be re-directed as follows:

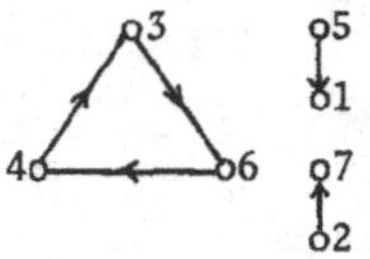


More generally, if G contains i disjoint cycles, G' can be turned into a directed graph in $2^i$ different ways to ensure that the out-degree of each vertex other than 1 and n is one. Take note of this and the fact that our process of constructing legitimate selections out of simple graceful graphs is reversible, we obtain Theorem 2 immediately.

## 3. SOME FINAL REMARKS.

Since all simple graceful graphs with less than 7 points are trees, by Theorem 2, we have $t_i = p(A_{i-2})$ for i = 3,4,5,6. From this, it is easy to compute that the number of graceful trees on 3,4,5,6 vertices are respectively 2,4,12 and 40. Again, by Theorem 2, $p(A_{n-2}) = c_o + 2c_1 + 4c_2 + \ldots + 2^k c_k$. The author believes that $t_n = c_o >> c_i$ (i = 1,2,3,...,k) for large n. Hence it is conjectured that

$$\lim_{n\to\infty} p(A_{n-2})/t_n = 1$$

## REFERENCE

(1) I. Cahit, "Are all complete binary trees graceful?" *Amer. Math. Monthly* 83(1976), 35-37.

05C25

# FIXING SUBGRAPHS OF $K_{m,n}$

Keith Chidzey
Department of Mathematics,
University of Melbourne,
Parkville, Victoria, 3052
Australia

ABSTRACT

We characterise the fixing subgraphs of complete bipartite graphs $K_{m,n}$ and study various aspects of those which are minimal. These minimal fixing subgraphs are necessarily forests. In particular, we show that any given tree with parts of unequal size is a component of a minimal fixing subgraph of some $K_{m,n}$.

## 1. INTRODUCTION

We study finite simple graphs G with vertex set V(G), edge set E(G) and automorphism group $\Gamma(G)$. We let $\underset{\sim}{S}(G)$ denote the set of spanning subgraphs of G.

Fixing subgraphs were introduced by Sheehan [1]. We use the following definition.

DEFINITION. Let $H \in \underset{\sim}{S}(G)$. If K is any spanning subgraph of G such that $K \cong H$ and if for any permutation $\alpha$ such that $H^{\alpha} = K$, then $\alpha \in \Gamma(G)$, we say that H is a *fixing subgraph* of G. We denote the set of fixing subgraphs of G by $\underset{\sim}{F}(G)$.

Remark 2 of Sheehan [1] gives us that if $H \subset K$ and $H \in \underset{\sim}{F}(G)$, then $K \in \underset{\sim}{F}(G)$.

DEFINITION. If $H \in \underset{\sim}{F}(G)$ and H contains no subgraph in $\underset{\sim}{F}(G)$, then H is a *minimal fixing subgraph* of G. We denote the set of minimal fixing subgraphs of G by $\underset{\sim}{M}(G)$.

In this paper we characterise the fixing subgraphs of complete bipartite graphs $K_{m,n}$. Further, if H is a connected minimal fixing subgraph of $K_{m,n}$, we show that H is a path of odd size and consequently $|m - n| = 1$. We then find necessary conditions for disconnected minimal fixing subgraphs of $K_{m,n}$ and observe how elements of $\underset{\sim}{F}(K_{m,n})$ and $\underset{\sim}{M}(K_{m,n})$ can be extended to elements of $\underset{\sim}{F}(K_{m+k,n+k})$ and $\underset{\sim}{M}(K_{m+k,n+k})$ respectively. Finally, given any tree with unequal parts we are also able to construct a minimal fixing subgraph of some $K_{m,n}$ containing the tree as a component.

## 2. NOTATION

The following notation will be useful in our study of $\underset{\sim}{F}(K_{m,n})$. Let $\mathbb{N}_n = \{1, 2, \ldots, n\}$, nG be the union of n copies of graph G, $A \subset \mathbb{N}_n$ denote that A is a nonempty proper subset of $\mathbb{N}_n$ and $\lceil x \rceil$ be the least integer greater than or equal to x.

For bipartite graphs let $V_r(G)$ and $V_b(G)$ (or simply $V_r$ and $V_b$) denote the two parts of G. When a colouring of V(G) is referred to, consider $V_r$ and $V_b$ to be the red vertices and blue vertices respectively. In any subgraph of G, let the vertices retain the same colour as in G. For $H \subseteq G$ let $r(H) = |V_r(H)|$ and $b(H) = |V_b(H)|$. For $K_{m,n}$ we shall arbitrarily denote $r(K_{m,n}) = m$ and $b(K_{m,n}) = n$. We let $K_{1,0}$ and $K_{0,1}$ represent an isolated red and blue vertex respectively.

If $S \subseteq V(G)$, let $\langle S\rangle$ be the maximal subgraph of G with vertex set S. If $e \in E(G)$ let $G_e$ denote the spanning subgraph of G with edge set $E(G)\backslash\{e\}$. For $v \in V(G)$ let $G_v = \langle V(G)\backslash\{v\}\rangle$ and for $H \subseteq G$ let $G_H = \langle V(G)\backslash V(H)\rangle$.

## 3. FIXING SUBGRAPHS OF $K_{m,n}$

Before determining the fixing subgraphs of $K_{m,n}$ we make an observation of the action of automorphisms of connected bipartite graphs on their parts.

LEMMA 1. Let G be a connected bipartite graph with parts $V_r$ and $V_b$, and let $\alpha \in \Gamma(G)$ be such that $v^\alpha \in V_b$ for some $v \in V_r$. Then $V_r^\alpha = V_b$, $V_b^\alpha = V_r$ and $|V_r| = |V_b|$.

We now characterise the fixing subgraphs of $K_{m,n}$ in terms of the sizes of the parts of their components.

THEOREM 1. If $H \in \underset{\sim}{S}(K_{m,n})$ has components $H_i$, $i \in \mathbb{N}_k$, then $H \in \underset{\sim}{F}(K_{m,n})$ if and only if for each $A \subset \mathbb{N}_k$, $\sum_{i\in A} [r(H_i) - b(H_i)] \neq 0$.

Proof: Let H be as given.

($\Rightarrow$) Assume $H \in \underset{\sim}{F}(K_{m,n})$ and suppose there exists $A \subset \mathbb{N}_k$ such that $\sum_{i\in A} [r(H_i) - b(H_i)] = 0$. Let $t = \sum_{i\in A} r(H_i) = \sum_{i\in A} b(H_i)$. Now

$$H = \bigcup_{i\in A} H_i \cup \bigcup_{i\in \mathbb{N}_k\backslash A} H_i \quad \text{and} \quad \bigcup_{i\in A} H_i \subseteq K_{t,t} \subseteq K_{m,n}.$$

Thus

$$H \subseteq K_{t,t} \cup \bigcup_{i\in \mathbb{N}_k\backslash A} H_i \subseteq K_{m,n}.$$

By Remark 2 of Sheehan [1],

$$K_{t,t} \cup \bigcup_{i\in \mathbb{N}_k\backslash A} H_i \in \underset{\sim}{F}(K_{m,n})$$

since $H \in \underset{\sim}{F}(K_{m,n})$. Let $\alpha$ be the automorphism of

$$K_{t,t} \cup \bigcup_{i\in \mathbb{N}_k\backslash A} H_i$$

which interchanges the t red vertices of $K_{t,t}$ with its t blue vertices and which fixes the vertices of $\bigcup_{i\in \mathbb{N}_k\backslash A} H_i$. Clearly $\alpha \notin \Gamma(K_{m,n})$ by Lemma 1. Therefore

$$K_{t,t} \cup \bigcup_{i\in \mathbb{N}_k\backslash A} H_i \notin \underset{\sim}{F}(K_{m,n})$$

in contradiction to the opposite conclusion above. Hence there is no $A \subset \mathbb{N}_k$ such that $\sum_{i\in A} [r(H_i) - b(H_i)] = 0$.

($\Leftarrow$) Assume $\sum_{i\in A} [r(H_i) - b(H_i)] \neq 0$ for each $A \subset \mathbb{N}_k$. Suppose $M \cong H$ where $M \in \underset{\sim}{S}(K_{m,n})$. Let $\beta$ be an isomorphism such that $H^\beta = M$.

Suppose $\beta$ maps red (blue) vertices onto red (blue) vertices only. Then $\beta \in \Gamma(K_{m,n})$ since $\Gamma(K_{m,n})$ contains all permutations of vertices which preserve its parts.

Suppose $\beta$ interchanges all the red vertices with all the blue vertices and thus $m = n$. Then $\beta \in \Gamma(K_{m,n})$ since $\Gamma(K_{m,n})$ contains all such permutations.

Suppose there exists $B \subset \mathbb{N}_k$ such that $V_r(H_i)^\beta \subseteq V_r(K_{m,n})$, $V_b(H_i)^\beta \subseteq V_r(K_{m,n})$ for $i \in B$, and $V_r(H_i)^\beta \subseteq V_b(K_{m,n})$, $V_b(H_i)^\beta \subseteq V_r(K_{m,n})$ for $i \in \mathbb{N}_k\backslash B$. Then

$$r(\bigcup_{i\in B} H_i) = r(\bigcup_{i\in B} H_i^\beta) \tag{i}$$

$$r(\bigcup_{i\in \mathbb{N}_k\backslash B} H_i) = b(\bigcup_{i\in \mathbb{N}_k\backslash B} H_i^\beta) \tag{ii}$$

As $r(H) = r(\bigcup_{i\in B} H_i) + r(\bigcup_{i\in \mathbb{N}_k\backslash B} H_i)$, (i) and (ii) give

$$r(H) = r(\bigcup_{i\in B} H_i^\beta) + b(\bigcup_{i\in \mathbb{N}_k\backslash B} H_i^\beta). \tag{iii}$$

Further

$$r(M) = r(\bigcup_{i\in B} H_i^\beta) + r(\bigcup_{i\in \mathbb{N}_k\backslash B} H_i^\beta) \tag{iv}$$

and as $r(M) = r(K_{m,n}) = r(H)$, (iii) and (iv) give

$$r(\bigcup_{i\in \mathbb{N}_k\backslash B} H_i^\beta) - b(\bigcup_{i\in \mathbb{N}_k\backslash B} H_i^\beta) = 0 \tag{v}$$

and thus we deduce that $\sum_{i\in \mathbb{N}_k\backslash B} [r(H_i) - b(H_i)] = 0$, contrary to assumption.

Therefore $\beta \in \Gamma(K_{m,n})$ and hence $H \in \underset{\sim}{F}(K_{m,n})$. □

COROLLARY 1.1. Any minimal fixing subgraph of $K_{m,n}$ is a forest.

COROLLARY 1.2. Given $F \in \underset{\sim}{M}(K_{m,n})$ with components $F_i$, $i \in \mathbb{N}_k$, and $r(F_i) \geq b(F_i)$ for each $i \in \mathbb{N}_k$, then $r(F_i) = b(F_i) + 1$ for each $i \in \mathbb{N}_k$.

## 4. MINIMAL FIXING SUBGRAPHS OF $K_{m,n}$

We are now in a position to characterise the connected minimal fixing subgraphs of $K_{m,n}$ showing that they are all paths of odd size.

THEOREM 2. H is a connected minimal fixing subgraph of $K_{m,n}$ if and only if $H = P_{m+n}$ and $|m - n| = 1$.

Proof: ($\Rightarrow$) Let H be a connected minimal fixing subgraph of $K_{m,n}$. By Corollary 1.1, H is a tree.

Assume $|m - n| \neq 1$. Let e be an edge incident with an endvertex v. Then $H_e$ has two components, $H_v$ and $\langle v \rangle$, neither having parts of equal size and so by Theorem 1, $H_e \in \underset{\sim}{F}(K_{m,n})$. This contradicts the fact that $H \in \underset{\sim}{M}(K_{m,n})$, so $|m - n| = 1$.

Suppose H has both red and blue endvertices. Let e be an edge incident with an endvertex v with $|r(H_v) - b(H_v)| = 2$. As $|r(\langle v \rangle) - b(\langle v \rangle)| = 1$, by Theorem 1 we have the contradiction $H_e \in \underset{\sim}{F}(K_{m,n})$. Hence the endvertices of H are all of the same colour. Also as H is a tree and $|m - n| = 1$, we deduce that H is a path of size $m + n$.

($\Leftarrow$) Let $H = P_{m+n}$ and $|m - n| = 1$. By Theorem 1, $H \in \underset{\sim}{F}(K_{m,n})$. Further, the removal of any edge results in a graph with two components, one of which has the same number of red and blue vertices. By Theorem 1, $H_e \notin \underset{\sim}{F}(K_{m,n})$ and therefore $H \in \underset{\sim}{M}(K_{m,n})$. □

The next result gives a necessary condition for disconnected elements of $\underset{\sim}{M}(K_{m,n})$. We then look specifically at two and three component elements of $\underset{\sim}{M}(K_{m,n})$.

THEOREM 3. Let F be a forest with components $F_i$, $i \in \mathbb{N}_k$, $k > 1$. If $F \in \underset{\sim}{M}(K_{m,n})$ then $|\sum_{i\in A} [r(F_i) - b(F_i)]| = 1$ for some $A \subset \mathbb{N}_k$.

Proof: Let F be as given and assume without loss of generality that $r(F) \geq b(F)$.

(1) Assume $r(F_i) > b(F_i)$ for each $i \in \mathbb{N}_k$. Then $r(F_i) - b(F_i) = 1$ for each $i \in \mathbb{N}_k$ by Corollary 1.2.

(2) Assume $r(F_i) \not> b(F_i)$ for some $i \in \mathbb{N}_k$. By Theorem 1, $|\sum_{i\in A} [r(F_i) - b(F_i)]| \neq 0$ for each $A \subset \mathbb{N}_k$. Now suppose $|\sum_{i\in A} [r(F_i) - b(F_i)]| \geq 2$ for each $A \subset \mathbb{N}_k$. Let $j \in \mathbb{N}_k$ such that $r(F_j) < b(F_j)$. Let e be a pendant edge incident with a blue endvertex of $F_j$. Then $F_{j_e} = T_j \cup K_{0,1}$ where $T_j$ is a tree with $r(T_j) - b(T_j) = r(F_j) - b(F_j) + 1$. Put $T_i = F_i$ for each $i \in \mathbb{N}_k \backslash \{j\}$ and put $T_{k+1} = K_{0,1}$. Consequently $F_e = \bigcup_{i\in\mathbb{N}_{k+1}} T_i$ and $r(T_i) - b(T_i) \neq 0$ for each $i \in \mathbb{N}_{k+1}$.

Let $B \subset \mathbb{N}_{k+1}$ such that $|B| \geq 2$.

(a) Suppose $j, k+1 \in B$. Then $B \backslash \{k+1\} \subset \mathbb{N}_k$ and by hypothesis

$$|\sum_{i \in B} [r(T_i) - b(T_i)]| = |\sum_{i \in \{j,k+1\}} [r(T_i) - b(T_i)] + \sum_{i \in B \backslash \{j,k+1\}} [r(T_i) - b(T_i)]|$$

$$= |r(F_j) - b(F_j) + \sum_{i \in B \backslash \{j,k+1\}} [r(F_i) - b(F_i)]|$$

$$= |\sum_{i \in B \backslash \{k+1\}} [r(F_i) - b(F_i)]|$$

$$\geq 2 .$$

Similarly it can be shown that $|\sum_{i \in B} r(T_i) - b(T_i)| \neq 0$ for (b) $j \in B$, $k+1 \notin B$, (c) $j \notin B$, $k+1 \in B$, and (d) $j, k+1 \notin B$. Therefore by Theorem 1, $F_e \in \underset{\sim}{F}(K_{m,n})$ contrary to $F$ being minimal. Hence we deduce that $|\sum_{i \in A} [r(F_i) - b(F_i)]| = 1$ for some $A \subset \mathbb{N}_k$. □

It is convenient to use the following notation in illustrating the differences in the number of vertices in the parts of components of the bipartite graphs we study. In particular this notation will be useful for obtaining necessary conditions for two and three component elements of $\underset{\sim}{M}(K_{m,n})$.

NOTATION. Given a graph $G$ with components $G_i$, $i \in \mathbb{N}_k$, let $s(G)$ denote the unordered collection of possibly repeated integers $[r(G_i) - b(G_i) : i \in \mathbb{N}_k]$. Also let $-[n_1, n_2, \ldots, n_k] = [-n_1, -n_2, \ldots, -n_k]$.

REMARK 1. By consideration of Theorem 1 it is apparent that if $F \in \underset{\sim}{S}(K_{m,n})$, $H \in \underset{\sim}{S}(K_{m,n})$ and $s(F) = s(H)$ then $F \in \underset{\sim}{F}(K_{m,n})$ if and only if $H \in \underset{\sim}{F}(K_{m,n})$. Further if $F \in \underset{\sim}{S}(K_{m,n})$, $H \in \underset{\sim}{S}(K_{m',n'})$ and $s(F) = s(H)$ then $F \in \underset{\sim}{F}(K_{m,n})$ if and only if $H \in \underset{\sim}{F}(K_{m',n'})$.

It can be shown that if $F \in \underset{\sim}{S}(K_{m,n})$ and $H$ is a union of $F$, $P_{2k}$ and an edge e which is incident with an endvertex of both $F$ and $P_{2k}$, $k \geq 1$, then $F \in \underset{\sim}{F}(K_{m,n})$ if and only if $H \in \underset{\sim}{F}(K_{m+k,n+k})$, and $F \in \underset{\sim}{M}(K_{m,n})$ if and only if $H \in \underset{\sim}{M}(K_{m+k,n+k})$, i.e. any (minimal) fixing subgraph of $K_{m,n}$ can be extended to a (minimal) fixing subgraph of $K_{m+k,n+k}$ for any $k \geq 1$.

Now we obtain some corollaries to Theorem 3.

COROLLARY 3.1. Let $F$ be a two component minimal fixing subgraph of $K_{m,n}$. Then $s(F) = \pm[1, 1]$ or $\pm[1, -k]$ for some $k \in \mathbb{N}$.

Proof: Follows immediately from Corollary 1.2 and Theorem 3.

COROLLARY 3.2. Let $F$ be a three component minimal fixing subgraph of $K_{m,n}$. Then s(F) is one of the following; $\pm[1, 1, 1]$, $\pm[1, 1, -k] : k \geq 2$, $\pm[1, k, -2] : k \geq 3$, $\pm[1, -k, k+1] : k \geq 2$, $\pm[2, k, -(k+1)] : k \geq 2$, $\pm[k, k, -(k+1)] : k \geq 1$.

We omit the proof which shows for the other possible s(F) not mentioned that there exists an edge which can be removed from any such F yielding a smaller fixing subgraph.

REMARK 2. Each integer triple mentioned in Corollary 3.2 is a realisable s(F) for some minimal fixing subgraph F of some complete bipartite graph. Such a subgraph can be simply constructed by the union of k-stars. More specifically, given a triple $[i, j, -\ell]$ allowed by Corollary 3.2 such that $i,j,\ell > 0$, then $K_{i+1,1} \cup K_{j+1,1} \cup K_{1,\ell+1} \in \underset{\sim}{M}(K_{i+j+3,\ell+3})$.

It is convenient to use the following notation in the next result.

NOTATION. Let T be a tree and let $e \in E(T)$. Then $T_e$ is a forest with two components which are trees and we denote them by $T_{e_1}$ and $T_{e_2}$. Let $r_{e_i} = r(T_{e_i})$ and $b_{e_i} = b(T_{e_i})$, $i \in \mathbb{N}_2$.

Finally, we construct a minimal fixing subgraph of some $K_{m,n}$ containing any given tree with unequal parts as a component. This result shows that although components of elements of $\underset{\sim}{M}(K_{m,n})$ are trees, the variety of such trees is rather extensive.

THEOREM 4. The tree T is a component of a minimal fixing subgraph of some $K_{m,n}$ if and only if $r(T) \neq b(T)$.

Proof: ($\Rightarrow$) The necessity follows from Theorems 1 and 2.

($\Leftarrow$) Let T be a tree such that $r(T) \neq b(T)$.

We consider several cases of $r(T) - b(T)$. In each case we construct a particular fixing subgraph, with T as a component, of some $K_{m,n}$ and prove that it is minimal by showing that the removal of any edge does not result in another fixing subgraph of $K_{m,n}$. We only need consider $r(T) > b(T)$ as the choice of colours is arbitrary.

*Case 1.* $r(T) - b(T) = 1$.

Let $F = T \cup p_1 K_{1,0}$ where $p_1 = \max_{e \in E(T)} (b_{e_i} - r_{e_i}) : i \in \mathbb{N}_2$. As each component of F has one more red than blue vertex, by Theorem 1, $F \in \underset{\sim}{F}(K_{m,n})$ where $m = r(F) = r(T) + p_1$ and $n = b(F) = b(T)$.

Let $e \in E(T)$. Then as $r(T) - b(T) = 1$, $r_{e_1} - b_{e_1} + r_{e_2} - b_{e_2} = 1$. Therefore there exists $j \in \mathbb{N}_2$ such that $r_{e_j} - b_{e_j} \leq 0$. Now, if $r_{e_j} - b_{e_j} = 0$ then by Theorem 1, $F_e \notin \underset{\sim}{F}(K_{m,n})$. If $r_{e_j} - b_{e_j} < 0$, then $X = T_{e_j} \cup (b_{e_j} - r_{e_j})K_{1,0}$ is a union of components of $F_e$ with $r(X) = b(X)$ and so by Theorem 1, $F_e \notin \underset{\sim}{F}(K_{m,n})$. Hence $F \in \underset{\sim}{M}(K_{m,n})$

*Case 2.* $r(T) - b(T) = 2$.

Let $F = T \cup K_{1,4} \cup K_{5,1} \cup p_2K_{1,6} \cup K_{1,8}$ where $p_2 = \left[\frac{1}{5} \max_{e \in E(T)} (r_{e_i} - b_{e_i}) : i \in \mathbb{N}_2\right]$. Now T and $K_{5,1}$ are the only components of F with more red than blue vertices. By applying Theorem 1, it is readily verified that $F \in \underset{\sim}{F}(K_{m,n})$, where $m = r(T) + p_2 + 7$ and $n = b(T) + 6p_2 + 13$.

By finding a nonempty proper subset of components whose union has parts of equal size we show that $F_e \notin \underset{\sim}{F}(K_{m,n})$ for each e in F.

(a) $e \in E(K_{5,1})$. As $(K_{5,1})_e = K_{4,1} \cup K_{1,0}$, $F_e$ has $K_{4,1}$ and $K_{1,4}$ as components whose union has parts both of size 5.

(b) $e \in E(K_{1,t})$, $t = 6, 8$. $(K_{1,5})_e$ has components $K_{1,t-1}$ and $K_{0,1}$, so $K_{1,4} \cup K_{5,1} \cup K_{0,1}$ is a union of components of $F_e$ with both parts of size 6.

(c) $e \in E(K_{1,4})$. $(K_{1,4})_e$ has components $K_{1,3}$ and $K_{0,1}$, so $T \cup K_{1,3}$ is a union of components of $F_e$ with both parts of size $r(T) + 1$.

(d) $e \in E(T)$. As $r(T) - b(T) = 2$,

$$r_{e_1} - b_{e_1} + r_{e_2} - b_{e_2} = 2.$$

Hence there exists $j \in \mathbb{N}_2$ such that $r_{e_j} - b_{e_j} \geq 1$ and $\ell \neq j$ such that $r_{e_\ell} - b_{e_\ell} \leq 1$. Let $r_{e_j} - b_{e_j} = s + 5t$ where $0 < s < 5$ and $t \geq 0$. Then

$$\begin{aligned} p_2 &= \left[\frac{1}{5} \max_{f \in E(T)} (r_{f_i} - b_{f_i}) : i \in \mathbb{N}_2\right] \\ &\geq \left[\frac{1}{5}(r_{e_j} - b_{e_j})\right] \\ &= \left[\frac{1}{5}(s - 5t)\right] \\ &= 1 + t\,. \end{aligned}$$

(i) Suppose $r_{e_j} - b_{e_j} = 1 + 5t$, $t \geq 0$. Then $T_{e_j} \cup K_{5,1} \cup (1 + t)K_{1,6}$ is a union of components of $F_e$ with both parts of size $r_{e_j} + 6 + t$.

(ii) Suppose $r_{e_j} - b_{e_j} = 2 + 5t$, $t \geq 0$. If $t = 0$ then $T_{e_\ell}$ has $r_{e_\ell} = b_{e_\ell}$. If $t > 0$ then $T_{e_j} \cup K_{1,8} \cup (t - 1)K_{1,6}$ is a union of components of $F_e$ with both parts of size $r_{e_j} + t$.

(iii) Suppose $r_{e_j} - b_{e_j} = 3 + 5t$, $t \geq 0$. Then $T_{e_j} \cup K_{5,1} \cup K_{1,8} \cup t\,K_{1,6}$ is a union of components of $F_e$ with both parts of size $r_{e_i} + 6 + t$.

(iv) Suppose $r_{e_j} - b_{e_j} = 4 + 5t$, $t \geq 0$. Then $T_{e_j} \cup K_{1,4} \cup K_{5,1} \cup (1+t)K_{1,6}$ is a union of components of $F_e$ with both parts of size $r_{e_j} + 7 + t$.

(v) Suppose $r_{e_j} - b_{e_j} = 5t$, $t \geq 1$. Then $T_{e_j} \cup t\,K_{1,6}$ is a union of components of $F_e$ with parts both of size $r_{e_j} + t$.

We have now shown for any $e \in E(F)$ that $F_e$ contains a nonempty proper subset of components whose union has parts of equal size, so by Theorem 1, $F_e \notin \underset{\sim}{F}(K_{m,n})$ and thus $F \in \underset{\sim}{M}(K_{m,n})$.

*Case 3.* $r(T) - b(T) = k \geq 3$.

Let $F = T \cup K_{1,k+2} \cup K_{k+3,1} \cup K_{1,k+4} \cup K_{1,k+5} \cup \cdots \cup K_{1,2k+1} \cup p_k K_{1,2k+2}$ where $p_k = \left[\frac{1}{2k+1} \max_{e \in E(T)} (r_{e_i} - b_{e_i}) : i \in \mathbb{N}_2\right]$. As T and $K_{k+3,1}$ are the only components of F with more red than blue vertices, at least one of them must be contained in any nonempty proper subset of components of F whose union Y has parts of equal size, if such a subset exists.

Suppose such a Y with $r(Y) = b(Y)$ exists.

(A) Assume T but not $K_{k+3,1}$ is a component of Y. The remaining components of Y can only be of the form $K_{1,s}$ with $s \geq k+2$, and as $b(K_{1,s}) - r(K_{1,s}) = s - 1 \geq k + 1$ it follows that $b(Y) - r(Y) \geq 1$, contrary to assumption.

(B) Assume $K_{k+3,1}$ but not T is a component of Y. If $K_{1,k+2}$ is a component of Y then as $r(K_{k+3,1} \cup K_{1,k+2}) - b(K_{k+3,1} \cup K_{1,k+2}) = 1$, the inclusion of another component of F which can only be of the form $K_{1,s}$, $s \geq k+4$, in Y will give $b(Y) - r(Y) \geq k+2 > 0$ contrary to assumption. If $K_{1,k+2}$ is not a component of B, then $K_{1,s}$, $s \geq k+4$, are the only other possible components of Y. This gives $b(Y) - r(Y) \geq k+4 - (k+2) = 2$ contrary to assumption.

(C) Assume both T and $K_{k+3,1}$ are components of Y. Now $r(T \cup K_{k+3,1}) - b(T \cup K_{k+3,1}) = 2k+2$. As no single component R of F has $b(R) - r(R) = 2k+2$, Y has at least two more components. But the two components with the least excess of blue over red vertices are $K_{1,k+2}$ and $K_{1,k+4}$ for which $b(K_{1,k+2} \cup K_{1,k+4}) - r(K_{1,k+2} \cup K_{1,k+4}) = 2k+4$ and the inclusion of them in Y gives $b(Y) - r(Y) \geq 2$, contrary to assumption.

Hence, by Theorem 1, $F \in \underset{\sim}{F}(K_{m,n})$ where $m = r(F) = r(T) + 2k + 2 + p_k$ and $n = b(F) = b(T) + \frac{1}{2}(k-1)(3k+4) + (2k+2)p_k$.

It remains to show that $F \in \underset{\sim}{M}(K_{m,n})$. This is equivalent to showing for any $e \in E(F)$ that $F_e$ has a nonempty proper subset of components whose union has parts of equal size. We consider edges from particular components of F separately.

(a) $e \in E(K_{1,s})$, $s \geq k+4$. $(K_{1,s})_e = K_{1,s-1} \cup K_{0,1}$, so $K_{1,k+2} \cup K_{k+3,1} \cup K_{0,1}$ is a union of components of $F_e$ with both parts of size $k+4$.

(b) $e \in E(K_{k+3,1})$. $(K_{k+3,1})_e = K_{k+2,1} \cup K_{0,1}$, so $K_{k+2,1} \cup K_{1,k+2}$ is a union of two components of $F_e$ with both parts of size $k+3$.

(c) $e \in E(K_{1,k+2})$. $(K_{1,k+2})_e = K_{1,k+1} \cup K_{0,1}$, so T and $K_{1,k+1}$ is a union of two components of $F_e$ with both parts of size $r(T)+1$.

(d) $e \in E(T)$. $r_{e_1} - b_{e_1} + r_{e_2} - b_{e_2} = r(T) - b(T) = k$, so there exist $j, \ell \in \mathbb{N}_2$ such that $r_{e_j} - b_{e_j} \geq \frac{1}{2}k$ and $r_{e_\ell} - b_{e_\ell} \leq \frac{1}{2}k$, $j \neq \ell$. Now $r_{e_j} - b_{e_j} = s + t(2k+1)$ where $0 < s \leq 2k+1$, $t \geq 0$. Then

$$p_k = \left[\frac{1}{2k+1} \max_{f \in E(T)} (r_{f_i} - b_{f_i}) : i \in \mathbb{N}_2\right]$$

$$\geq \left[\frac{1}{2k+1} (r_{e_j} - b_{e_j})\right]$$

$$= \left[\frac{1}{2k+1} (s + t(2k+1))\right]$$

$$= 1 + t.$$

(i) Suppose $r_{e_j} - b_{e_j} = q + t(2k+1)$ where $\frac{1}{2}k \leq q \leq k-1$, $t = 0$ or $1 \leq q \leq k-1$, $1 \leq t < p_k$. Then $T_{e_j} \cup K_{k+3,1} \cup K_{1,k+3+q} \cup t\,K_{1,2k+2}$ is a union of components of $F_e$ with parts both of size $r_{e_j} + k + 4 + t$.

(ii) Suppose $r_{e_j} - b_{e_j} = k + t(2k+1)$ where $0 \leq t \leq p_k$. Then $T_{e_\ell} \cup t\,K_{1,2k+2}$ is a union of components of $F_e$ with both parts of size $r_{e_\ell} + t$.

(iii) Suppose $r_{e_j} - b_{e_j} = q + t(2k+1)$ where $q = k+1$ or $k+3 \leq q \leq 2k+1$, and $0 \leq t < p_k$. Then $T_{e_j} \cup K_{1,q+1} \cup t\,K_{1,2k+2}$ is a union of components of $F_e$ with both parts of size $r_{e_j} + 1 + t$.

(iv) Suppose $r_{e_j} - b_{e_j} = k + 2 + t(2k+1)$ where $0 \leq t < p_k$. Then $T_{e_j} \cup K_{1,k+2} \cup K_{k+3,1} \cup K_{1,k+4} \cup t\,K_{1,2k+2}$ is a union of components of $F_e$ with both parts of size $r_{e_j} + k + 5 + t$.

For any $e \in E(F)$ we have found a nonempty proper subset of components of $F_e$, whose union has parts of equal size. By Theorem 1, we deduce that $F_e \notin \underset{\sim}{F}(K_{m,n})$ for each $e \in E(F)$ and hence $F \in \underset{\sim}{M}(K_{m,n})$. □

REFERENCE

[1] J. Sheehan, "Fixing subgraphs", *J. Comb. Th.*, (*B*), 12 (1972), 226-244.

# HADAMARD EQUIVALENCE

Joan Cooper, James Milas and W.D. Wallis
Department of Mathematics,
University of Newcastle,
New South Wales, 2308,
AUSTRALIA

05B20

ABSTRACT

We introduce a new measure, the *profile*, of an Hadamard matrix, which seems to be useful as an indicator of Hadamard equivalence. Some results on the profile are given, and its usefulness is indicated in the case of matrices of order 36.

## INTRODUCTION

Two Hadamard matrices H and K are called *equivalent* (or *Hadamard equivalent*, or *H-equivalent*) if one can be obtained from the other by a sequence of row negations, row permutations, column negations and column permutations.

The discussion of Hadamard equivalence is quite difficult, principally because of the lack of a good canonical form. The exact results which have been discovered are as follows: Hadamard matrices of orders less than 16 are unique up to equivalence; there are precisely five equivalence classes at order 16; there are precisely three equivalence classes at order 20. (See [4], [5].) Various lower bounds have been found for other orders (see, for example, [10]). Given two Hadamard matrices of the same order, it can be quite difficult to tell whether or not they are equivalent. We have attempted to use more coarse equivalence relations to study Hadamard equivalence, namely *integral equivalence*, which we discuss in the next section, and *weight* which we saw in [11] to be a bad discriminator between equivalence classes.

Our aim in this paper is to introduce a new test which seems to discriminate between inequivalent Hadamard matrices rather better than the previously known tests (other than the impossibly long technique of searching through all permissible negations and permutations), and illustrate it with a discussion of Hadamard matrices of order 36.

We assume a general knowledge of the properties of Hadamard matrices; see [12].

## INTEGRAL EQUIVALENCE

We define A and B as being *integrally equivalent* if we can obtain A from B by a sequence of the following row operations:

add an integer multiple of one row to another;
negate a row;
permute the rows;

or the corresponding column operations. It will be observed that Hadamard equivalence implies integral equivalence, so that - for example - any lower bound on the number of integral equivalence classes of Hadamard matrices of a given order is also a lower bound on the number of Hadamard equivalence classes.

The most important properties of integral equivalence are summarized in the following Theorem (see, for example,[7]).

THEOREM 1. Integer matrices A and B are integrally equivalent if and only if there are square integer matrices P and Q, each with determinant $\pm 1$, which satisfy

$$B = PAQ.$$

Given a square matrix A with non-zero determinant, there is a unique diagonal matrix D,

$$D = \text{diag}(d_1, d_2, \ldots, d_n),$$

integrally equivalent to A, with the properties that each $d_i$ is positive and divides $d_{i+1}$.

This matrix D is called the Smith normal form of A.

Several papers have been written discussing the behaviour of Hadamard matrices under integral equivalence. Two main results, limiting the variability of the Smith normal form, have been discovered ([13], [8], [9]):

THEOREM 2. Let A be an Hadamard matrix of order 4n. Suppose D is the Smith normal form of A, where

$$D = \text{diag}(d_1, d_2, \ldots, d_{4n}).$$

Then $d_1 = 1$ and $d_{4n} = 4n$, and

$$\text{(i)} \quad d_1 d_{4n} = d_2 d_{4n-1} = \ldots = d_i d_{4n-i+1} = 4n;$$

$$\text{(ii)} \quad d_2 = d_3 = \ldots = d_{\alpha+1} = 2, \text{ where } \alpha \geq [\log_2(4n-1)] + 1.$$

For certain orders, the number $\alpha$ of invariants equal to 2 determines the integral equivalence class of an Hadamard matrix. For example, if an Hadamard matrix of order 36 has $\alpha$ invariants equal to 2, then it must have $\alpha$ equal to 18 and $34-2\alpha$ equal to 6. In this case we refer to $\alpha$ as the number of the *Smith class* of the matrix.

There are four possible integral equivalence classes of Hadamard matrices of order 16, and eleven classes of order 32. We have shown [8, 10] that all classes are represented. Over the past six years we have tried to find representatives of the classes for order 36; however, only five of the twelve possible classes have as yet been found.

## THE PROFILE

Suppose H is an Hadamard matrix of order 4n with typical entry $h_{ij}$. We write

$p_{ijk\ell}$ to mean the absolute value of the generalised inner product of rows i, j, k and $\ell$:

$$p_{ijk\ell} = \left| \sum_{x=1}^{4n} h_{ix}\, h_{jx}\, h_{kx}\, h_{\ell x} \right|.$$

REMARK. $p_{ijk\ell} \equiv 4n \pmod 8$.

PROOF. For convenience simply write p instead of $p_{i,j,k,\ell}$. First observe that p is not affected by negations or permutations of the columns of H. So there is no loss of generality in assuming that columns have been negated and arranged such that rows i, j, k and $\ell$ look like:

| a times | b times | c times | d times | e times | f times | g times | h times | |
|---|---|---|---|---|---|---|---|---|
| 1 1 ··· | 1 1 ··· | 1 1 ··· | 1 1 ··· | 1 1 ··· | 1 1 ··· | 1 1 ··· | 1 1 ··· | (i) |
| 1 1 ··· | 1 1 ··· | 1 1 ··· | 1 1 ··· | - - ··· | - - ··· | - - ··· | - - ··· | (j) |
| 1 1 ··· | 1 1 ··· | - - ··· | - - ··· | 1 1 ··· | 1 1 ··· | - - ··· | - - ··· | (k) |
| 1 1 ··· | - - ··· | 1 1 ··· | - - ··· | 1 1 ··· | - - ··· | 1 1 ··· | - - ··· | ($\ell$) |

;

and

$$\pm p = a - b - c + d - e + f + g - h.$$

Taking the inner product of row $\ell$ with rows i, j and k, we obtain respectively

$$a + c + e + g = b + d + f + h,$$
$$a + c + f + h = b + d + e + g,$$
$$a + d + e + h = b + c + f + g.$$

Adding these three equations we have

$$3a - 3b + c - d + e - f - g + h = 0;$$

and if we add the left hand expression to the equation for p we get

$$\pm p = 4a - 4b.$$

Now considering the inner products of rows i, j and k, we see that

$$a + b = c + d = e + f = g + h = n;$$

hence $\pm p = 4n - 8b$, and $p \equiv 4n \pmod 8$.

We shall write $\pi(m)$ for the number of sets $\{i,j,k,\ell\}$ of four distinct rows such that $p_{ijk\ell} = m$. From the definition and from the above remark, $\pi(m) = 0$ unless $m \geq 0$ and $m \equiv 4n \pmod 8$. We call $\pi(m)$ the *profile* (or 4-*profile*) of H.

THEOREM 3. Equivalent Hadamard matrices have the same profile.

PROOF. It is clear that p is unaltered by the column equivalence operations. Row negation does not have any effect, as p is an absolute value. Row permutation has the effect of renaming $p_{ijk\ell}$ as $p_{qrst}$ for some q, r, s, t, but it leaves unchanged the *totality* of all values p.

As we stated above, the profile seems to distinguish between inequivalent

Hadamard matrices quite well. This will be illustrated in the later sections.

The next theorem enables us to say that, if there are t Hadamard matrices of order 4n which all have different profiles, then there will be at least t different profiles at order 8n.

THEOREM 4. Suppose H is an Hadamard matrix of order 4n with profile $\pi$. Then the Hadamard matrix

$$G = \begin{bmatrix} H & H \\ H & -H \end{bmatrix}$$

of order 8n has profile $\sigma$, where

$$\begin{aligned} \sigma(8n) &= 8\pi(4n) + \binom{4n}{2}, \\ \sigma(m) &= 8\pi(\tfrac{1}{2}m) \text{ if } m \neq 0 \text{ or } 8n, \\ \sigma(0) &= 8\pi(0) + \binom{8n}{4} - 8\binom{4n}{4} - \binom{4n}{2} \\ &= 8\pi(0) + 8n(2n-1)(4n-1)(4n+3)/3. \end{aligned}$$

PROOF. It will be convenient to write $\hat{i}$ for $i - n$, and so on. We let $p_{ijk\ell}$ and $s_{ijk\ell}$ be the absolute values of generalised inner products of rows i, j, k, $\ell$ of H and G respectively, so that $\pi$ and $\sigma$ count the number of times p and s attain various values. We observe that

(A) *if* $i < j < k < \ell \leq 4n$, *then* $s_{ijk\ell} = 2p_{ijk\ell}$;

(B) *if* $4n < i < j < k < \ell$, *then* $s_{ijk\ell} = 2p_{\hat{i}\hat{j}\hat{k}\hat{\ell}}$.

Moreover, if $i < j < k \leq 4n < \ell$, then

$$\begin{aligned} s_{ijk\ell} &= \sum_{x=1}^{8n} g_{ix}g_{jx}g_{kx}g_{\ell x} \\ &= \sum_{x=1}^{4n} g_{ix}g_{jx}g_{kx}g_{\ell x} + \sum_{x=4n+1}^{8n} g_{ix}g_{jx}g_{kx}g_{\ell x} \\ &= \sum_{x=1}^{4n} h_{ix}h_{jx}h_{kx}h_{\hat{\ell}x} + \sum_{x=1}^{4n} h_{ix}h_{jx}h_{kx}(-h_{\hat{\ell}x}) \\ &= 0, \end{aligned}$$

and similarly in the case $i \leq 4n < j < k < \ell$. So:

(C) *if* $i < j < k \leq 4n < \ell$, *then* $s_{ijk\ell} = 0$;

(D) *if* $i \leq 4n < j < k < \ell$, *then* $s_{ijk\ell} = 0$.

Finally suppose $i < j \leq 4n < k < \ell$. Clearly $s_{ijk\ell} = 2p_{ij\hat{k}\hat{\ell}}$. Suppose i, j, $\hat{k}$, $\hat{\ell}$ are all distinct; as we range through each of the sets $\{i,j,\hat{k},\hat{\ell}\}$ with all elements distinct and $i < j$, $\hat{k} < \hat{\ell}$, we obtain the sets $\{i,j,\hat{k},\hat{\ell}\}$ with $i < j < k < \ell$ six times each. If $i = \hat{k}$,

$$p_{ijk\hat{k}\hat{\ell}} = \sum_{x=1}^{4n} h_{ix}h_{jx}h_{ix}h_{\hat{\ell}x}$$

$$= \sum_{x=1}^{4n} h_{jx}h_{\hat{\ell}x},$$

which is 4n when $j = \hat{\ell}$ and 0 when $j \neq \hat{\ell}$. Similarly p will be zero when $j = \hat{\ell}$ but $i \neq k$. So:

(E) *If* $i < j \leq 4n < k < \ell$, $s_{ijk\ell}$ *attains each non-zero value* m *on* $6\pi(\frac{1}{2}m)$ *occasions, except that* 8n *occurs* $6\pi(4n) + \binom{4n}{2}$ *times.*

Combining (A) - (E) we get the result.

MATRICES OF SMALL ORDERS

We found the profiles of representatives of all equivalence classes of Hadamard matrices up to order 20.

The (unique) matrices of orders 4, 8 and 12 have profiles

$\pi(4) = 1$

$\pi(0) = 56, \quad \pi(8) = 14$

$\pi(4) = 495, \quad \pi(12) = 0$

respectively.

The five equivalence classes at order 16 gave four distinct profiles: referring to the names given to the classes in [4], we have

| | | | |
|---|---|---|---|
| class I : | $\pi(0) = 1680$, | $\pi(8) = 0$, | $\pi(16) = 140$, |
| class II : | $\pi(0) = 1488$, | $\pi(8) = 256$, | $\pi(16) = 76$, |
| class III : | $\pi(0) = 1392$, | $\pi(8) = 484$, | $\pi(16) = 44$, |
| class IV : | $\pi(0) = 1344$, | $\pi(8) = 448$, | $\pi(16) = 28$, |
| class V : | $\pi(0) = 1344$, | $\pi(8) = 448$, | $\pi(16) = 28$. |

Observe that the matrices of class V are the transposes of the matrices of class IV.

The three classes at order 20 all gave the same profile:

$$\pi(4) = 4560, \ \pi(12) = 285, \ \pi(20) = 0.$$

CONSTRUCTIONS - ORDER 36

Many of the known constructions for Hadamard matrices give rise to matrices of order 36. We constructed Hadamard matrices using all the methods available to us, and calculated their Smith classes $\alpha$ and profiles $\pi$. With a view to brevity and lack of tedium we give only a reference to each construction, which the reader may consult at his leisure. When we write $\pi = (a,b,c,d,e)$, we mean $\pi(4) = a, \pi(12) = b, \cdots, \pi(36) = e$.

CONSTRUCTION 1. [12,p343]. The eighty Steiner triple systems, or (15,35,7,3,1) designs, were used to construct eighty Hadamard matrices. All eighty Hadamard matrices were of Smith class 12. However seventy-seven different profiles were found, so at least seventy-seven Hadamard equivalence classes were represented. They are listed in Table 1 (the systems are numbered as in [1]). Notice that systems 39 and 41 give the same profile, as do 48 and 58, and also 66 and 78.

| | Profile | | Profile |
|---|---|---|---|
| 1 | π=(52920,5040,0,945,0) | 41 | π=(48849,9750,292,14,0) |
| 2 | π=(50936,7152,448,369,0) | 42 | π=(48945,9600,350,10,0) |
| 3 | π=(50232,7920,576,177,0) | 43 | π=(48839,9762,292,12,0) |
| 4 | π=(49800,8520,440,145,0) | 44 | π=(48821,9786,288,10,0) |
| 5 | π=(49920,8376,440,169,0) | 45 | π=(48864,9726,302,13,0) |
| 6 | π=(49400,9096,280,129,0) | 46 | π=(48840,9756,300,9,0) |
| 7 | π=(49680,8712,360,153,0) | 47 | π=(48833,9768,294,10,0) |
| 8 | π=(49464,8928,432,81,0) | 48 | π=(48847,9744,306,8,0) |
| 9 | π=(49246,9222,378,59,0) | 49 | π=(48826,9774,298,7,0) |
| 10 | π=(49282,9174,386,63,0) | 50 | π=(48843,9756,294,12,0) |
| 11 | π=(49058,9486,314,47,0) | 51 | π=(48850,9744,300,11,0) |
| 12 | π=(49223,9246,384,52,0) | 52 | π=(48806,9810,278,11,0) |
| 13 | π=(49372,9048,416,69,0) | 53 | π=(48827,9780,286,12,0) |
| 14 | π=(49484,8880,472,69,0) | 54 | π=(48902,9666,326,11,0) |
| 15 | π=(49116,9420,308,61,0) | 55 | π=(48866,9720,308,11,0) |
| 16 | π=(49952,8232,616,105,0) | 56 | π=(48829,9744,292,10,0) |
| 17 | π=(49244,9240,352,69,0) | 57 | π=(48844,9744,312,5,0) |
| 18 | π=(49144,9372,332,57,0) | 58 | π=(48847,9744,306,8,0) |
| 19 | π=(48968,9624,264,49,0) | 59 | π=(48908,9660,324,13,0) |
| 20 | π=(49019,9534,312,40,0) | 60 | π=(48872,9708,316,9,0) |
| 21 | π=(48869,9750,252,34,0) | 61 | π=(48923,9660,294,28,0) |
| 22 | π=(48890,9714,270,31,0) | 62 | π=(48830,9768,300,7,0) |
| 23 | π=(48983,9558,344,20,0) | 63 | π=(48824,9786,282,13,0) |
| 24 | π=(48970,9576,340,19,0) | 64 | π=(48881,9696,318,10,0) |
| 25 | π=(49013,9516,354,22,0) | 65 | π=(48856,9732,308,9,0) |
| 26 | π=(49072,9432,376,25,0) | 66 | π=(48853,9738,304,10,0) |
| 27 | π=(48881,9708,298,18,0) | 67 | π=(48844,9750,302,9,0) |
| 28 | π=(48932,9630,326,17,0) | 68 | π=(48819,9786,292,8,0) |
| 29 | π=(48998,9534,354,19,0) | 69 | π=(48832,9768,296,9,0) |
| 30 | π=(48911,9660,318,16,0) | 70 | π=(48838,9762,294,11,0) |
| 31 | π=(49017,9510,356,22,0) | 71 | π=(48814,9792,292,7,0) |
| 32 | π=(48898,9678,314,15,0) | 72 | π=(48834,9762,302,7,0) |
| 33 | π=(48895,9678,320,12,0) | 73 | π=(48877,9702,316,10,0) |
| 34 | π=(48901,9672,318,14,0) | 74 | π=(48795,9822,280,8,0) |
| 35 | π=(48830,9786,270,19,0) | 75 | π=(48902,9660,336,7,0) |
| 36 | π=(48797,9822,276,10,0) | 76 | π=(48785,9840,270,10,0) |
| 37 | π=(48897,9666,336,6,0) | 77 | π=(48827,9768,306,4,0) |
| 38 | π=(48856,9744,288,17,0) | 78 | π=(48853,9738,304,10,0) |
| 39 | π=(48849,9750,292,14,0) | 79 | π=(48933,9630,324,18,0) |
| 40 | π=(48836,9768,288,13,0) | 80 | π=(48675,9990,240,0,0) |

TABLE 1. Profiles of matrices from Steiner triple systems

CONSTRUCTION 2. [3]. Seventeen Hadamard matrices were constructed from the seventeen non-isomorphic latin squares of order 6, fifteen were integrally equivalent of Smith class 13 and two were in class 11. Eleven different profiles arose; they can be found in table 2. (The squares are numbered as in [2].)

CONSTRUCTION 3. [12,p339]. All Hadamard matrices constructed from the conference matrix of order 18 (and permutations of the conference matrix) were integrally equivalent ($\alpha$=17) and they also all had the same profile which was

$$\alpha = 17, \quad \pi = (48960,9792,0,153,0)$$

CONSTRUCTION 4. [12,p280]. Using the six (36,15,6) designs listed in [6], we obtained three Hadamard matrices in Smith class 11 with identical profiles (from the designs attributed to Menon, Wallis and Blackwelder in [6]), one in class 15 (attributed to Spence), and two in class 13 with different profiles (the design KT attributed to Takeuchi, and the design KS attributed to Kingsley and Stanton). The profiles were

$$\alpha = 11, \quad \pi = (48240,10656,0,9,0)$$
$$\alpha = 15, \quad \pi = (49752,8388,756,9,0)$$
$$\alpha = 13, \quad \pi = (48195,10710,0,0,0) \quad \text{(design KT)}$$
$$\alpha = 13, \quad \pi = (52920,5040,0,945,0) \quad \text{(design KS).}$$

Notice that the design KS yields a matrix with the same profile as that derived from Steiner triple system 1, but the two are definitely inequivalent because they belong to different Smith classes; also the matrix from design KT has the same Smith class and profile as the matrix from Latin square XIV.

CONSTRUCTION 5. [12,p325]. The Williamson array gave an Hadamard matrix in class 17 with profile

$$\pi = (48973,9543,389,0,0).$$

CONSTRUCTION 6. [12,p329]. The Goethals-Seidel matrix gave an Hadamard matrix in class 17 with profile

$$\pi = (49025,9486,380,14,0)$$

CONSTRUCTION 7. [12,p279]. The Hadamard matrix constructed from the difference set (35,17,8) was of class 12, with profile

$$\pi = (48195,10710,0,0,0)$$

(Notice that this is the same as the matrices from block design KT and Latin square XIV, but they are Hadamard inequivalent as they are in different Smith classes.)

CONSTRUCTION 8. [12,p346]. This construction makes use of a symmetric conference matrix of order 6. Its integral equivalence class is 13 and its profile is:

$$\pi = (48675,9990,240,0,0).$$

This matrix has the same profile as the matrix from Steiner triple system 80, but a different Smith class.

CONSTRUCTION 9. [12,p341]. The cross product of a matrix of order 5 and one of order 7 was used in this construction. Its integral equivalence class was 12 and the profile was

$$\pi = (48195,10710,0,0,0),$$

so the matrix has the same profile and class as that in Construction 7.

CONSTRUCTION 10. [12,p393]. This construction makes use of four supplementary difference sets on $Z_9$. The matrices constructed fall into two integral equivalence classes, 15 and 17. Several supplementary difference sets could be used to give the same class and profile. Table 3 lists the classes and profiles obtained and gives suitable supplementary difference sets. Two matrices arose which were indistinguishable from matrices earlier.

CONSTRUCTION 11. [14]. This construction uses four supplementary difference sets in $Z_8$. The sets and profiles are listed in Table 4.

From all the constructions we have obtained only five integral equivalence classes: classes 11, 12, 13, 15 and 17. The profiles give interesting results, and we have constructed 107 different profiles. These enable us to say that there are at least 110 Hadamard equivalence classes at order 36.

| | Class | Profile |
|---|---|---|
| α = 13: | I,II | π = (48396,10416,88,5,0) |
| | III | π = (48383,10434,84,4,0) |
| | IV | π = (48476,10308,108,13,0) |
| | V,VI | π = (48412,10404,76,13,0) |
| | VII | π = (48584,10176,112,33,0) |
| | VIII,IX | π = (48516,10260,108,21,0) |
| | XI,XII | π = (48312,10584,36,9,0) |
| | XIV | π = (48195,10710,0,0,0) |
| | XV,XVI | π = (48540,10260,60,45,0) |
| | XVII | π = (48600,10224,0,81,0) |
| α = 11: | X,XIII | π = (48240,10656,0,0,0) |

TABLE 2. Profiles of matrices from Latin squares.

| | 4-(9;3;3) sets | Profile | |
|---|---|---|---|
| α = 15 | 013,013,015,025 | π = (49079,9399,417,10,0) | |
| | 012,024,036,048 | π = (48870,9738,270,27,0) | |
| | 4-(9;4,4,3,2;4) sets | | |
| α = 17 | 0123,0146,025,04 | π = (48825,9774,300,6,0) | |
| | 0124,0146,014,02 | π = (48969,9558,372,6,0) | |
| | 0124,0146,013,04 | π = (49025,9486,380,14,0) | (*) |
| | 0124,0146,025,01 | π = (48983,9540,374,8,0) | |
| | 0124,0134,025,04 | π = (48933,9618,344,10,0) | |
| | 0124,0125,025,03 | π = (49082,9399,411,13,0) | |
| | 0124,0137,014,04 | π = (49171,9276,438,20,0) | |
| | 0124,0137,015,03 | π = (48973,9543,389,0,0) | (**) |
| | 0124,0136,015,02 | π = (48865,9708,330,2,0) | |
| | 0134,0137,015,02 | π = (48950,9544,350,11,0) | |
| | 0134,0134,024,04 | π = (48936,9638,308,25,0) | |

TABLE 3. Construction using four supplementary difference sets modulo 9
Matrix (*) has the same class and profile as that of construction 6;
matrix (**) has the same class and profile as that of construction 5.

| | Difference Sets | Profile |
|---|---|---|
| α = 17 | 0125,0135,0135,012 | π = (49000,9552,320,33,0) |
| | 0125,0135,0135,014 | π = (49012,9504,376,13,0) |
| | 0123,0135,0135,014 | π = (48992,9528,376,9,0) |
| α = 15 | 0123,0136,0145,024 | π = (50080,8060,640,121,0) |

TABLE 4. Construction 11.

REFERENCES

[1] F.C. Bussemaker and J.J. Seidel, Symmetric Hadamard matrices of order 36. *Report 70-WSK-02*, Technological University, Eindhoven, 1970.

[2] R.A. Fisher & F. Yates, *Statistical Tables For Biological, Agricultural and Medical Research*, 2nd Ed., Oliver & Boyd, 1943.

[3] J-M. Goethals and J.J. Seidel, Strongly regular graphs derived from combinatorial designs. *Canadian J. Math.* 22 (1970), 597-614.

[4] Marshall Hall Jr., Hadamard matrices of order 16, *JPL Research Summary No. 36-10*, 1 (1961), 21-26.

[5] Marshall Hall Jr., Hadamard matrices of order 20, *JPL Technical Report No. 32-76*, 1 (1965).

[6] R.A. Kingsley and R.G. Stanton, A survey of certain balanced incomplete block designs, *Proc. 3rd S-E Conf. Combinatorics, Graph Theory and Computing*, UMPI (1972), 305-310.

[7] Morris Newman, *Integral Matrices*, Academic Press, New York (1972).

[8] W.D. Wallis, Integral equivalence of Hadamard matrices, *Israel J. Math.* 10 (1971), 359-368.

[9] W.D. Wallis, Some notes on integral equivalence of combinatorial matrices, *Israel J. Math.* 10 (1971), 457-464.

[10] W.D. Wallis, On the number of inequivalent Hadamard matrices, *Proc. 2nd Manitoba Conference Numerical Math.*, UMPI (1972), 383-401.

[11] W.D. Wallis, On the weights of Hadamard matrices, *Ars Combinatoria* (to appear)

[12] W.D. Wallis, Anne Penfold Street and Jennifer Seberry Wallis, *Combinatorics: Room Squares, Sum-Free Sets, Hadamard Matrices*, Springer-Verlag, Berlin (1972).

[13] W.D. Wallis and Jennifer Wallis, Equivalence of Hadamard matrices, *Israel J. Math.* 7 (1969), 122-128.

[14] Albert Leon Whiteman, Hadamard matrices of order 4(2p+1), *J. Number Theory* 8 (1976), 1-11.

05B30, 05C99

# A NOTE ON EQUIDISTANT PERMUTATION ARRAYS

R.B. Eggleton and A. Hartman
Department of Mathematics,
University of Newcastle,
Newcastle, New South Wales, 2308,
Australia

ABSTRACT

Two permutations on $n$ elements are at (Hamming) distance $\mu$ if they disagree in exactly $\mu$ places. An equidistant permutation array is a collection of permutations on $n$ elements, every pair of which is at distance $\mu$. A permutation graph $G(n,\mu)$ is a graph with vertex set comprising all permutations on $n$ elements, and edges between each pair of permutations at distance $\mu$. These graphs enable the relevant permutation structure to be visualised; in particular, the cliques correspond to maximal equidistant permutation arrays. We obtain various structural theorems for these graphs, and conjecture several properties for their cliques.

## 1. INTRODUCTION

Let $S_n$ denote the set of all permutations on $\{1,2,\ldots,n\}$. Two permutations in $S_n$ are at (Hamming) distance $\mu$ from each other if they differ in precisely $\mu$ places. An $(n,k,\lambda)$ *equidistant permutation array* (E.P.A.) is a set of $k$ distinct permutations from $S_n$, each pair of which is at distance $\mu = n-\lambda$. We refer to $n$, $k$, $\mu$ as the *degree*, *size* and *distance*, respectively, of the array.

Although E.P.A.'s appeared in various settings around the turn of the century, their current study was initiated by Bolton [1], and has been taken up by a number of others, including Deza, Mullin and Vanstone [3] and Heinrich, van Rees and Wallis [4]. The central problem in the study is the determination of the maximum size $K(n,\mu)$ of E.P.A.'s of degree $n$ and distance $\mu$. (In Bolton's notation this maximum size is denoted by $R(r,\lambda)$, where his $r$ corresponds to our $n$; we prefer a notation which involves $\mu$, since this appears to be a more natural parameter.) In particular, any two permutations in $S_n$ differ by at least one transposition, so are at distance $\mu \geq 2$. Hence $K(n,0) = K(n,1) = 1$. Also an E.P.A. of degree $n$ and distance $\mu = n$ corresponds to the rows of a Latin rectangle, which can be extended to a Latin square but no further, so $K(n,n) = n$.

Much of the work until now has approached E.P.A.'s from the direction of generalised Room squares, or equivalently, orthogonally resolvable $(r,\lambda)$-systems; it was shown in [3] that there is a one-to-one correspondence between these structures and E.P.A.'s. Our approach is to give a graphic setting to the problem. Structural properties of the graphs introduced correspond to properties of E.P.A.'s and other configurations of permutations in which some but not all pairs are at distance $\mu$. The visual nature of the graphic setting, in our opinion, is advantageous. A maximal E.P.A. corresponds to a clique (that is, a maximal complete subgraph) in this setting. We study here some of the main structural features of the graphs in question, and their relationship to E.P.A.'s.

[Note. It has come to our attention that Dénes [2] has recently applied Turán's theorem to a problem in coding theory where he uses a graph like our $G(n,\mu)$, but two vertices are adjacent just if their Hamming distance is at least $\mu$.]

## 2. PERMUTATION GRAPHS

For any natural numbers $n$, $\mu$ with $\mu \leq n$, we define the *permutation graph* $G(n,\mu)$ of *degree* $n$ and *distance* $\mu$ to be the graph with vertex set $S_n$, comprising all permutations of $\{1,2,\ldots,n\}$, and adjacency corresponding to (Hamming) distance $\mu$ between permutations. Thus $G(n,0)$ comprises $n!$ disconnected vertices each carrying a loop, and $G(n,1)$ comprises $n!$ disconnected vertices without loops. We are interested in the structure of $G(n,\mu)$ for $\mu \geq 2$. In Diagram 1 two permutation graphs of degree 3 are shown, with cycle notation for the vertices. In general, we denote the identity of $S_n$ by $e$, with composition from left to right. For any $a \in S_n$, by $\pi_a: S_n \to S_n$ we denote the map $x \mapsto xa$; our maps are written on the right of the elements on which they act. Also $S_n(\mu)$ will denote all permutations in $S_n$ which derange exactly $\mu$ elements, that is, all permutations at distance $\mu$ from $e$.

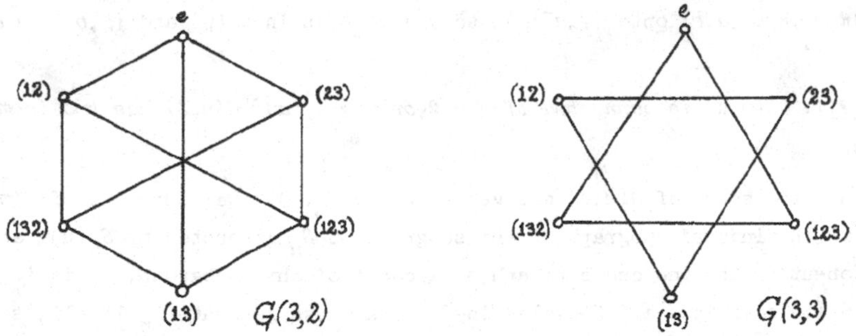


DIAGRAM 1. Permutation graphs of degree 3 and distances 2 and 3.

THEOREM 1. *$G(n,\mu)$ is vertex transitive.*

Proof. Let $a,b \in S_n$ be at distance $\mu$. For any $c \in S_n$ and any $i \in \{1,2,\ldots,n\}$ we have $iac = ibc$ just if $ia = ib$, so $ac$ and $bc$ are also at distance $\mu$. Thus $\pi_c$ is an automorphism of $G(n,\mu)$. Hence for any $u,v \in S_n$ the map $\pi_{u^{-1}v} = \pi_u^{-1}\pi_v$ is an automorphism of $G(n,\mu)$ which maps $u$ to $v$, so $G(n,\mu)$ is vertex transitive. □

COROLLARY 1.1. *$G(n,\mu)$ is regular of degree $\binom{n}{\mu}D(\mu)$, where $D(\mu)$ is the number of derangements of $\mu$ elements, given by $D(\mu) = \mu!\left(1 - \frac{1}{1!} + \frac{1}{2!} - \ldots + \frac{(-1)^\mu}{\mu!}\right)$.*

Proof. Vertex transitivity implies regularity. A permutation $a \in S_n$ is at distance $\mu$ from $e$ if it deranges exactly $\mu$ elements in $\{1,2,\ldots,n\}$. For each of the $\binom{n}{\mu}$ choices of these $\mu$ places of discord, there are $D(\mu)$ permutations which move each one. □

If we think of E.P.A.'s as generalised Latin rectangles, the problem of counting the number of possible second rows for an E.P.A. with standard first row is the analogue of the Derangement Problem. Corollary 1.1 answers the generalised problem.

Note that the involution $\theta: S_n \to S_n$ defined by $a \mapsto a^{-1}$ is an automorphism of $G(n,\mu)$, since $a$ and $a^{-1}$ move exactly the same elements. Using this automorphism, we get a result stronger than Theorem 1:

THEOREM 2. *$G(n,\mu)$ is the Cayley graph of the symmetric group $S_n$, with edges generated by the permutations in $S_n(\mu)$.*

Proof. If $c \in S_n(\mu)$ then $\pi_c$ generates edges of $G(n,\mu)$. For if $a$ is any vertex of $G(n,\mu)$, the automorphism $\theta\pi_a^{-1}\theta$ maps the edge $\{e,c\}$ onto $\{a,ac\} = \{a,a\pi_c\}$, so the latter is an edge of $G(n,\mu)$. Conversely, if $\{a,b\}$ is any edge of $G(n,\mu)$, the automorphism $\theta\pi_a\theta$ maps $\{a,b\}$ onto $\{e,a^{-1}b\}$, so $d = a^{-1}b$ is in $S_n(\mu)$ and $\{a,b\} = \{a,a\pi_d\}$. □

COROLLARY 2.1. *$G(n,\mu)$ is connected if $\mu = 2$ or $\mu \geq 4$, and $G(n,3)$ has two isomorphic components.*

Proof. Since the edges of $G(n,\mu)$ are generated by $S_n(\mu)$, the component of $G(n,\mu)$ containing $e$ is the induced subgraph of the subgroup of $S_n$ generated by $S_n(\mu)$, and any other component is the induced subgraph of a coset of this subgroup, so is isomorphic. Evidently $S_n(2)$ comprises all 2-cycles in $S_n$, and these generate $S_n$ itself, so $G(n,2)$ is connected. Also $S_n(3)$ comprises all 3-cycles in $S_n$, and these generate the alternating group $A_n$. Since $[S_n : A_n] = 2$, $G(n,3)$ has two isomorphic components. When $\mu \geq 4$, $S_n(\mu)$ contains $a = (1324\ldots\mu)$ and $b = (1\mu\ldots54)(23)$, where $ia = i+1$ if $4 \leq i < \mu$, and $ib = i-1$ if $4 < i \leq \mu$. Then $ab = (12)$, so clearly $S_n(\mu)$ generates all 2-cycles and so all of $S_n$. Then $G(n,\mu)$ is connected. □

## 3. STABILISERS AND EDGE TRANSITIVITY

It is helpful now to consider the subgraphs of $G(n,\mu)$ which are neighbourhoods: the *closed neighbourhood* $G'_a(n,\mu)$ of $a$ is the induced subgraph of the vertex $a$ and all vertices adjacent to $a$, and the *open neighbourhood* $G^0_a(n,\mu)$ of $a$ is the induced subgraph of all vertices adjacent to $a$. When $a = e$, the notation is simplified by omitting the subscript $e$. The vertex set of $G^0(n,\mu)$ is just $S_n(\mu)$. Diagram 2 shows two neighbourhoods of $e$ in permutation graphs of degree 4.

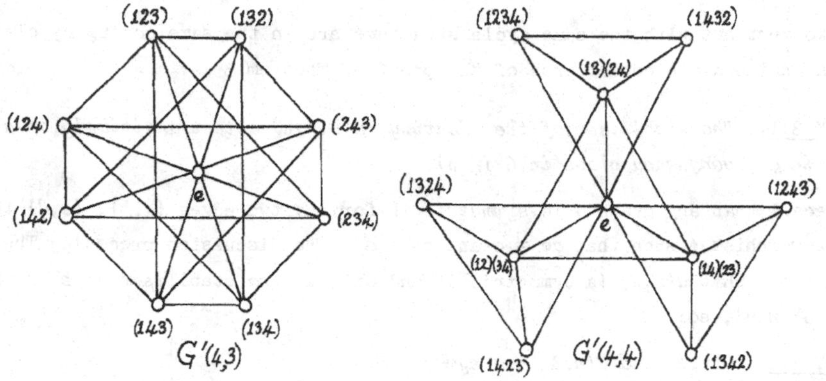


DIAGRAM 2. Two closed neighbourhoods of $e$ in permutation graphs of degree 4.

Now consider whether $G(n,\mu)$ is edge transitive. Because $G(n,\mu)$ is vertex transitive, it is edge transitive if the stabiliser of the identity (the automorphism subgroup fixing $e$) acts transitively on $S_n(\mu)$, the vertex set of $G^0(n,\mu)$. For suppose this condition holds, and let $\{a,b\}$, $\{c,d\}$ be any two edges of $G(n,\mu)$. Since the automorphism $\pi_a^{-1}$ maps $a$ to $e$ and $b$ to $ba^{-1}$, it follows that $ba^{-1} \in S_n(\mu)$; similarly $dc^{-1} \in S_n(\mu)$. By hypothesis, there is an automorphism $\psi$ which fixes $e$ and maps $ba^{-1}$ to $dc^{-1}$. Then $\pi_a^{-1}\psi\pi_c$ is an automorphism with the required property that it maps $a$ to $c$ and $b$ to $d$.

Conversely, if $G(n,\mu)$ is edge transitive, the stabiliser of the identity does act transitively on $G^0(n,\mu)$. For suppose $a,b \in S_n(\mu)$; edge transitivity ensures that either there is an automorphism which fixes $e$ and maps $a$ to $b$, or else there is an automorphism $\phi$ which maps $a$ to $e$ and $e$ to $b$. But in the latter case the automorphism $\theta\pi_a\phi$ fixes $e$ and maps $a$ to $b$. Thus, $G(n,\mu)$ is edge transitive if and only if the stabiliser of $e$ is transitive on $G^0(n,\mu)$.

THEOREM 3. *$G(n,2)$ and $G(n,3)$ are edge transitive.*

<u>Proof</u>. Let $a,b \in S_n(\mu)$. We show that if $\mu = 2$ or $3$, there is an automorphism in the stabiliser of $e$ which maps $a$ to $b$, so the stabiliser is transitive on $G^0(n,\mu)$, and $G(n,\mu)$ is edge transitive by the preceding remarks.

Note that for any $\mu$ and any $c \in S_n$, the automorphism $(\theta\pi_c)^2$ maps $a$ to $d = c^{-1}ac$ which, as is well known, has the same cycle structure as $a$. Indeed, each $i \in \{1,2,\dots,n\}$ in the cycle notation for $a$ is replaced by $ic$ in the cycle notation for $d$, since $icd = iac$. In particular, if $\mu = 2$ or $3$ then necessarily $a$, $b$ are both $\mu$-cycles, so there exist suitable permutations $c$ such that $a(\theta\pi_c)^2 = b$. □

Two vertices with the same cycle structure are in the same conjugacy class of $S_n$. Thus an immediate consequence of the proof of Theorem 3 is:

<u>COROLLARY 3.1</u>. *The stabiliser of the identity in $G(n,\mu)$ acts transitively on all vertices in any conjugacy class in $G^0(n,\mu)$.*

Recall that any graph $G$ is *symmetric* if for any two edges $\{a,b\}$, $\{c,d\}$ there is an automorphism $\phi$ such that $a\phi = c$ and $b\phi = d$. The discussion preceding Theorem 3 actually shows that $G(n,\mu)$ is symmetric if and only if the stabiliser of $e$ is transitive on $G^0(n,\mu)$, so:

<u>COROLLARY 3.2</u>. *$G(n,2)$ and $G(n,3)$ are symmetric.*

## 4. VALENCES IN OPEN NEIGHBOURHOODS

We now study the valence of vertices in $G^0(n,\mu)$. In this connection it is helpful to study the *permutation digraphs* associated with the edges. A digraph is a permutation digraph if its vertices are either isolated or are incident with just one in-arc and one out-arc. (For simplicity we do not put loops on the isolated vertices.) The digraph associated with some $d \in S_n(\mu)$, briefly referred to as the digraph $d$, has vertex set $\{1,2,\dots,n\}$ and exactly $\mu$ arcs, there being an arc from $i$ to $id$ if $id \neq i$. The vertices with incident arcs are said to be *moved* by $d$; the isolated vertices are *fixed* by $d$.

If $\{a,b\}$ is an edge in $G^0(n,\mu)$, and $\nu$ is the number of elements fixed by $a$ but moved by $b$, the *spread* of $\{a,b\}$ is defined to be $\nu$. Since $d = a^{-1}b \in S_n(\mu)$, by Theorem 2, the edges of spread $\nu$ incident with $a$ in $G^0(n,\mu)$ correspond to the permutation digraphs $d$ which move $\nu$ vertices fixed by $a$, have $\nu$ arcs which oppose arcs of $a$ and so fix $\nu$ vertices which are moved by $a$. Thus $a$ and $b$ agree on precisely $\nu$ elements; moreover, for $a$ to move enough vertices to accommodate the action of $d$ we must have $\mu \geq 2\nu$, so $\nu \leq \lfloor \frac{1}{2}\mu \rfloor$.

For any given $a \in S_n(\mu)$ systematic determination of the digraphs $d$ for which $\{a,ad\}$ is an edge in $G^0(n,\mu)$ is not difficult when $\mu$ is small. We did this for $\mu \leq 5$, and show unlabelled representatives for all cases with $\mu = 4$ in Diagram 3. By computing the weight of each digraph, that is, the number of labellings for a given

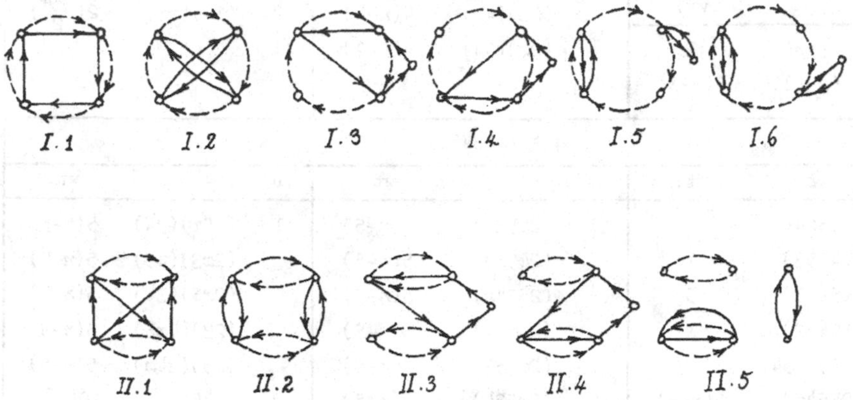


DIAGRAM 3. Unlabelled permutation digraphs $d$ (solid arcs) corresponding to edges $\{a,ad\}$ in $G^0(n,4)$ for each permutation digraph $a$ (broken arcs) which moves 4 vertices.

$a \in S_n(\mu)$, we can determine the contribution to the valence of $a$ in $G^0(n,\mu)$ made by edges of each possible spread. In Table 1 this information is summarized, by conjugacy class representatives for $a$ and $b = ad$. This calculation establishes the following theorem.

THEOREM 4. *In $G^0(n,2)$ each vertex has valence 0; in $G^0(n,3)$ each vertex has valence $3n-8$; in $G^0(n,4)$ each 4-cycle has valence $16n-62$ and each pair of 2-cycles has valence $n^2+7n-40$; in $G^0(n,5)$ each 5-cycle has valence $\frac{1}{2}(15n^2-5n-324)$ and each product of a 3-cycle and a 2-cycle has valence $6n^2+18n-228$.*

COROLLARY 4.1. *$G(n,4)$ and $G(n,5)$ are not edge transitive.*

Proof. $G^0(n,4)$ and $G^0(n,5)$ each have two different classes of vertex, and these never have equal valence, so the stabilisers of the identity in $G(n,4)$ and $G(n,5)$ are never transitive on the vertices of the open neighbourhoods of $e$. □

$a = (12)$

| $\nu$ | $b$ | wt. |
|---|---|---|
| 0 | - | - |
| 1 | - | - |

$a = (123)$

| $\nu$ | $b$ | wt. |
|---|---|---|
| 0 | $(132)$ | 1 |
| 1 | $(23x)$ | $3(n-3)$ |

$a = (1234)$

| $\nu$ | $b$ | wt. |
|---|---|---|
| 0 | $(1432)$ | 1 |
| | $(13)(24)$ | 1 |
| 1 | $(23x4)$ | $4(n-4)$ |
| | $(2x34)$ | $4(n-4)$ |
| | $(23)(4x)$ | $4(n-4)$ |
| | $(2x)(34)$ | $4(n-4)$ |
| 2 | - | - |

$a = (12)(34)$

| $\nu$ | $b$ | wt. |
|---|---|---|
| 0 | $(1324)$ | 2 |
| | $(13)(24)$ | 2 |
| 1 | $(234x)$ | $8(n-4)$ |
| | $(2x34)$ | $8(n-4)$ |
| 2 | $(34)(xy)$ | $2\binom{n-4}{2}$ |

$a = (12345)$

| $\nu$ | $b$ | wt. | $\nu$ | $b$ | wt. | $\nu$ | $b$ | wt. |
|---|---|---|---|---|---|---|---|---|
| 0 | $(13524)$ | 1 | 1 | $(2453x)$ | $5(n-5)$ | 1 | $(25x)(34)$ | $5(n-5)$ |
| | $(14253)$ | 1 | | $(245x3)$ | $5(n-5)$ | | $(2x3)(45)$ | $5(n-5)$ |
| | $(15243)$ | 5 | | $(2534x)$ | $5(n-5)$ | | $(2x5)(34)$ | $5(n-5)$ |
| | $(15432)$ | 1 | | $(25x34)$ | $5(n-5)$ | | $(23)(4x5)$ | $5(n-5)$ |
| | $(13)(254)$ | 5 | | $(2x453)$ | $5(n-5)$ | | $(25)(34x)$ | $5(n-5)$ |
| 1 | $(2354x)$ | $5(n-5)$ | | $(2x534)$ | $5(n-5)$ | 2 | $(345xy)$ | $10\binom{n-5}{2}$ |
| | $(235x4)$ | $5(n-5)$ | | $(235)(4x)$ | $5(n-5)$ | | $(345)(xy)$ | $5\binom{n-5}{2}$ |
| | $(23x54)$ | $5(n-5)$ | | $(245)(3x)$ | $5(n-5)$ | | | |

$a = (123)(45)$

| $\nu$ | $b$ | wt. | $\nu$ | $b$ | wt. | $\nu$ | $b$ | wt. |
|---|---|---|---|---|---|---|---|---|
| 0 | $(14325)$ | 6 | 1 | $(245x3)$ | $6(n-5)$ | 1 | $(245)(3x)$ | $6(n-5)$ |
| | $(14)(253)$ | 6 | | $(2x345)$ | $6(n-5)$ | | $(14)(23x)$ | $6(n-5)$ |
| 1 | $(1423x)$ | $6(n-5)$ | | $(2x453)$ | $6(n-5)$ | | $(1x)(234)$ | $6(n-5)$ |
| | $(1x234)$ | $6(n-5)$ | | $(14x)(23)$ | $6(n-5)$ | | $(2x)(345)$ | $6(n-5)$ |
| | $(234x5)$ | $6(n-5)$ | | $(1x4)(23)$ | $6(n-5)$ | 2 | $(123xy)$ | $6\binom{n-5}{2}$ |
| | $(2453x)$ | $6(n-5)$ | | $(234)(5x)$ | $6(n-5)$ | | $(3xy)(45)$ | $6\binom{n-5}{2}$ |

TABLE 1. Neighbours $a,b \in S_n(\mu)$ for $\mu \leq 5$, listed by spread $\nu$.

Note that the valence in $G^0(n,\mu)$ of any $a \in S_n(\mu)$ is the number of E.P.A.'s of degree $n$, distance $\mu$ and size 3, with first two rows $e$ and $a$. This is a generalisation of the Ménage Problem. (See Riordan [6,Chap.8].)

By adapting the techniques used for Theorem 4, we can show:

THEOREM 5. *Let $v(a)$ be the valence of $a \in S_n(\mu)$ in $G^0(n,\mu)$, with $\mu \geq 3$. Then $v(a) = O(n^\alpha)$, where $\alpha = \lfloor \frac{1}{2}\mu \rfloor$ except when $\mu$ is even and $a$ has no set of cycles of total length $\frac{1}{2}\mu$, in which case $\alpha = \frac{1}{2}\mu - 1$. In particular, let $a$ be a $\mu$-cycle and $a'$ the product of a $(\mu-2)$-cycle and a 2-cycle, with $\mu$ odd. Then*

$$\lim_{n\to\infty} \frac{v(a')}{v(a)} = \begin{cases} \frac{3}{5} & \text{if } \mu = 5, \\ 2 - \frac{4}{\mu} & \text{if } \mu \geq 7. \end{cases}$$

Proof. First note that the number of edges of spread $\nu$ incident with any given $a \in S_n(\mu)$ in $G^0(n,\mu)$ is $O(n^\nu)$, since the number of suitable digraphs $d$ depends only on $\mu$, $\nu$ and the conjugacy class of $a$, while the number of labellings of the $\nu$ vertices fixed by $a$ but moved by $d$ is at least $\binom{n-\mu}{\nu}$ and at most $\nu!\binom{n-\mu}{\nu}$, both of which are $O(n^\nu)$.

Now, for fixed $a \in S_n(\mu)$ consider the possible structure of any suitable digraph $d$ giving an edge of maximum spread $\lfloor \frac{1}{2}\mu \rfloor$. Such a digraph moves $\lfloor \frac{1}{2}\mu \rfloor$ vertices fixed by $a$, and opposes $\lfloor \frac{1}{2}\mu \rfloor$ arcs of $a$. There are two cases, depending on the parity of $\mu$.

Case 1: $\mu = 2m$. All $2m$ arcs of $d$ are accounted for, so the arcs opposing arcs of $a$ must form one or more circuits, which oppose circuits of $a$. Thus, a suitable digraph $d$ exists just if $a$ has one or more circuits of total length $m$. (It is easy to check that for $m \geq 2$ there is a suitable $d$ giving edges of spread $m-1$.)

Case 2: $\mu = 2m+1$. All arcs of $d$ but one are accounted for; the remaining one is either between two vertices moved by $a$ (without opposing an arc of $a$) or else from a vertex moved by $a$ to one fixed by $a$. For $m \geq 1$, such digraphs always exist. In any circuit containing some arcs which oppose arcs of $a$ and some which do not, all the opposing arcs must be consecutive, and there can be at most one such circuit. In particular, if $a$ has a $(2m+1)$-circuit, there are just two ways the opposing arcs of $d$ can occur in this circuit before labelling: (A) with a chord, or (B) connected to one or more vertices fixed by $a$ (Diagram 4). Each has $2m+1$ labellings for the vertices moved by $a$. In either case, if $m \geq 2$ there is a corresponding $d$ for the digraph $a'$ which has a $(2m-1)$-circuit and a 2-circuit: there are $2m-1$ labellings for the vertices moved by $a'$. Also in either case, if $m \geq 3$ there is a corresponding $d'$ for $a'$ having a 2-circuit coinciding with that of $a'$, and $m-2$ opposing arcs in the $(2m-1)$-circuit: there are again $2m-1$ labellings for $a'$. Thus if $m = 2$ the contribution to the valence of $a'$ by edges of spread $m$ falls short of that for $a$ by the factor $(2m-1)/(2m+1) = 3/5$, while if $m \geq 3$ the contribution to the valence of $a'$ by edges of spread $m$ exceeds that for $a$ by the factor $(4m-2)/(2m+1) = 2-4/(2m+1)$. In combination, these results establish the theorem. □

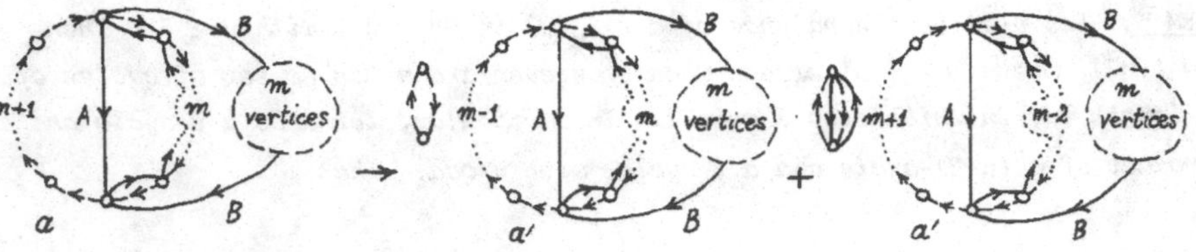


DIAGRAM 4. Suitable digraphs $d$ (solid arcs) for $a'$ (broken arcs) corresponding to each suitable digraph $d$ for $a$.

By the same reasoning used to derive the Corollary to Theorem 4, we can obtain from Theorem 5:

COROLLARY 5.1. *For any $\mu \geq 6$ and all sufficiently large $n$, $G(n,\mu)$ is not edge transitive.*

5. CONSTANT SPREAD SUBGRAPHS

It is helpful to study the structure of $G^0(n,\mu)$ in terms of *constant spread subgraphs* $G^0(n,\mu,\nu)$ of spread $\nu$; here $G^0(n,\mu,\nu)$ is the (possibly empty) subgraph of $G^0(n,\mu)$ comprising all edges of spread $\nu$ and their incident vertices.

THEOREM 6. *$G^0(n,\mu)$ is the union of its $\lfloor \frac{1}{2}\mu \rfloor + 1$ edge disjoint constant spread subgraphs $G^0(n,\mu,\nu)$, and each $G^0(n,\mu,\nu)$ comprises $\binom{n}{\mu+\nu}$ edge disjoint copies of $G^0(\mu+\nu,\mu,\nu)$.*

Proof. It suffices to note that the vertices of any edge of spread $\nu$ in $G^0(n,\mu)$ move a subset of exactly $\mu+\nu$ elements of $\{1,2,\ldots,n\}$, and the subgraph induced by all the vertices which move $\mu$ of these elements is isomorphic to $G^0(\mu+\nu,\mu,\nu)$. □

COROLLARY 6.1. *Let $v(a)$ and $v_\nu(a)$ be the valences of $a \in S_n(\mu)$ in $G^0(n,\mu)$ and $G^0(n,\mu,\nu)$ respectively. Then*

$$v(a) = \sum_{\nu \leq \lfloor \frac{1}{2}\mu \rfloor} \binom{n-\mu}{\nu} v_\nu(a).$$

Proof. Each copy of $G^0(\mu+\nu,\mu,\nu)$ incident in $G^0(n,\mu)$ with $a$ determines a subset of exactly $\nu$ elements of $\{1,2,\ldots,n\}$ which are moved by its vertices other than $a$, but are among the $n-\mu$ elements fixed by $a$, since the edges have spread $\nu$. The copies of $G^0(\mu+\nu,\mu,\nu)$ are edge disjoint, and the valence of $a$ in each is $v_\nu(a)$, so the result follows. □

Thus if $a = (123)$ then $v_0 = 1$, $v_1 = 3$. This can be deduced from Diagrams 1 and 2, or from Table 1. Similarly $v_0 = 2$, $v_1 = 16$, $v_2 = 0$ when $a = (1234)$; $v_0 = 4$, $v_1 = 16$, $v_2 = 2$ when $a = (12)(34)$; $v_0 = 13$, $v_1 = 80$, $v_2 = 15$ when $a = (12345)$; $v_0 = 12$, $v_1 = 84$, $v_2 = 12$ when $a = (123)(45)$.

COROLLARY 6.2. *$G^0(n,3)$ is the edge disjoint union of $\binom{n}{4}$ copies of the 3-cube $Q_3$ and $\binom{n}{3}$ vertex disjoint copies of the complete graph $K_2$. Alternatively, $G^0(n,3)$ is the edge disjoint union of $2\binom{n}{2}$ copies of the complete graph $K_{n-2}$ and $\binom{n}{3}$ vertex disjoint copies of $K_2$.*

Proof. $G^0(3,3,0) = K_2$ from Diagram 1, and $G^0(4,3,1) = Q_3$ from Diagram 2. Also Table 1 shows that the edges of spread 1 in $G^0(n,3)$ have the form $\{a,b\}$ represented by $a = (123)$ and $b = (23x)$. Thus for any fixed $i,j \in \{1,2,\ldots,n\}$, the $n$-2 vertices of the form $(ijx)$ induce a complete subgraph in $G^0(n,3,1)$. The corollary now follows, as there are $2\binom{n}{2}$ ordered choices for $i$, $j$. □

In the next corollary we use $P_2$ to denote the path of two adjacent edges.

COROLLARY 6.3. *Any path $P_2$ in $G^0(n,3)$ has at least one edge of spread 1, and any triangle $K_3$ comprises only edges of spread 1.*

Proof. Since edges of spread 0 are vertex disjoint in $G^0(n,3)$, it contains no $P_2$ comprising only edges of spread 0. Also it contains no $K_3$ with an edge of spread 0, for among the vertices $(12x)$, $(1x3)$, $(23x)$ which are adjacent to $(123)$ by edges of spread 1, there is none adjacent to $(132)$. □

COROLLARY 6.4. *The maximum clique size in $G(n,3)$ is $K(n,3) = n-1$ if $n \geq 4$, and when $n = 3$ it is $K(3,3) = 3$.*

Proof. The proof of Corollary 6.2 shows that any edge of spread 1 belongs to a unique subgraph $K_{n-2}$ in which all edges have spread 1. In particular, $G^0(n,3)$ has a $K_3$ just if $n \geq 5$. Any maximum clique in $G(n,3)$ can be assumed to contain $e$, by vertex transitivity, and then the rest of its vertices form a maximum clique in $G^0(n,3)$, which is a $K_2$ if $n = 3$ or $K_{n-2}$ if $n \geq 4$. □

Corollary 6.3 may be regarded as showing that certain configurations do not occur in $G^0(n,3)$, namely $P_2$ with edge spreads 0,0 and $K_3$ with edge spreads 0,1,1. Analogous results can be obtained for $G^0(n,\mu)$ with $\mu \geq 4$. For example, in $G^0(n,4)$ there can be no $K_3$ with edge spreads of any of the following types: 0,0,1; 0,0,2; 0,1,2; 0,2,2; 1,2,2. (The other five types do occur.) If $\mu \geq 4$, it is obvious that $G^0(n,\mu)$ has no $K_3$ with edge spreads $0,0,\nu$ for $\nu > 0$. Similar results can be obtained for larger configurations.

## 6. CONJECTURES

The interest in maximum clique size of $G(n,\mu)$ arises, of course, because it corresponds to maximum E.P.A. size for degree $n$ and distance $\mu$. Part of the value of the study of permutation graphs $G(n,\mu)$ in this connection is that it has drawn our attention to the usefulness of the spread parameter $\nu$. Define a clique of *constant spread* $\nu$ to be a clique containing $e$ and having all its other vertices in $G^0(n,\mu,\nu)$. Let $K(n,\mu,\nu)$ denote the maximum size of any clique of constant spread $\nu$. We conjecture for fixed $\mu$, $\nu$ that $K(n,\mu,\nu)$ is bounded if $\nu < \frac{1}{3}\mu$. Further, we conjecture for fixed $\mu$ that there is a constant $c$ (depending on $\mu$ only) such that every maximum clique of $G(n,\mu)$ is a clique of constant spread if $n > c\mu$. This would imply that $K(n,\mu) = K(n,\mu,\nu)$, for some $\nu$, if $n > c\mu$.

A theorem of Mullin [4] concerning near-trivial $(r,\lambda)$-systems implies that there is a $k_0 \leqslant \max\{n-\mu+2, \mu^2+\mu+1\}$ for which any E.P.A. of degree $n > \mu^2+2\mu-1$ and size $k \geqslant k_0$ corresponds to a clique of constant spread. A recent result of Vanstone [7] shows that for $\mu \geqslant 3$,

$$K(n,\mu) = 2 + \left\lfloor \frac{n-\mu}{\lceil \frac{1}{3}\mu \rceil} \right\rfloor \quad \text{if} \quad n > \lceil \tfrac{1}{3}\mu \rceil (\mu^2+\mu) + \mu,$$

so under this condition the maximum cliques are cliques of constant spread. We can construct E.P.A.'s of degree $n$, distance $\mu \geqslant 3$ and size $k = 2 + \lfloor (n-\mu)/\nu \rfloor$, for each $\nu$ satisfying $\lceil \frac{1}{3}\mu \rceil \leqslant \nu \leqslant \lfloor \frac{1}{2}\mu \rfloor$, and these correspond to complete subgraphs of constant spread $\nu$. These facts lend strength to the conjectures stated.

## REFERENCES


(1) D.W. Bolton, "Problem", *Combinatorics*, ed. D.Y.A. Welsh and D.R. Woodall, Math. Inst. Oxford (1972). pp.351-352.

(2) J. Dénes, "Latin squares and codes", to appear in Proc. Internat. Conf. on Information Theory, Paris, July 1977.

(3) M. Deza, R.C. Mullin and S.A. Vanstone, "Room squares and equidistant permutation arrays", *Ars Combinatoria* 2 (1976), 235-244.

(4) Katherine Heinrich, G.H.J. van Rees and W.D. Wallis, "A general construction for equidistant permutation arrays", to appear in *Graph Theory and Related Topics*, Proc. of Conf. on Graph Theory, Waterloo, July 1977.

(5) R.C. Mullin, "An asymptotic property of $(r,\lambda)$-systems", *Utilitas Math.* 3 (1973), 139-152.

(6) John Riordan, *An Introduction to Combinatorial Analysis*, Wiley (1958).

(7) S.A. Vanstone, "The asymptotic behaviour of equidistant permutation arrays", submitted to *Canad. J. Math.*

# THE COMBINATORICS OF ALGEBRAIC GRAPH THEORY IN THEORETICAL PHYSICS

I.G. Enting

Research School of Physical Sciences,
The Australian National University,
P.O. Box 4, Canberra, 2600 Australia

82A05, 05C15, 68A20, 05C10

## ABSTRACT

For many years physicists have been using graphs to represent the combinatorial properties of complicated algebraic expressions by defining sets of graphs whose generating functions are the physical quantities under investigation. In statistical mechanics there has recently been a trend towards the use of transformations which reduce the complexity of the graphical enumeration at the expense of an increase in algebraic complexity. A combinatorial analysis of algorithms used in actual computations shows why such a 'trade-off' is desirable. Series expansions for the limit of chromatic polynomials and for lattice models in statistical mechanics are considered as examples.

## 1 INTRODUCTION

Graph-theoretical characterisations are used extensively in theoretical physics. The way in which they arise is that algebraic expressions of considerable complexity are analysed so as to separate the purely algebraic aspects from the combinatorial aspects of the problem. The combinatorial problem arising from the algebraic structure is then mapped onto a graph-theory problem, typically a graph enumeration problem. Once the combinatorial problem has been solved the results are fed back into the original problem. In the simplest cases where the remaining algebraic complexity is minimal reconstructing the original problem consists merely of constructing the generating function for the appropriate class of graphs (14).

One of the simplest examples of this procedure is the high-temperature expansion of the Ising model (16). This is a model defined on a graph G with vertex set V and edge set E with a binary variable $\sigma_i = \pm 1$ associated with each vertex $i \in V$. We wish to evaluate

$$Z = c\sum_{\sigma_1=\pm 1} \sum_{\sigma_2=\pm 1} \cdots \sum_{\sigma_N=\pm 1} \prod_{(i,j)\in E} (1 + v\sigma_i\sigma_j) \tag{1}$$

where $N = |V|$. The product expands into $2^{|E|}$ terms each of which can be mapped onto a distinct subgraph of G. An edge (i,j) occurs in a given subgraph if the corresponding term in equation 1 includes the contribution $v\sigma_i\sigma_j$ from the factor associated with that edge. In any term the power of any $\sigma_i$ will be equal to the degree of vertex i in the associated subgraph. Using the property

$$\sum_{\sigma_i=\pm 1} \sigma_i^{\,m} = 0 \ (\text{m odd}), \quad = 2 \ (\text{m even}) \tag{2}$$

we have

$$Z = 2^N \sum_{g} v^{e(g)} \tag{3}$$

where the sum is over all subgraphs of G which have only vertices of even degree. (The Ising model is used to model various aspects of the temperature dependent properties of binary alloys, magnets and liquid-gas systems, particularly phase-transition phenomena.)

The generalisations of this problem to two-variable cases are known in various contexts as Whitney polynomials, Tutte polynomials, Potts model and random cluster models. There is a considerable literature concerning the various graphical formulations (3, 4, 5, 11, 13, 18, 27, 28, 29). Expression 3 generalises to

$$Z = \sum_{g} q^{c(g)} x^{e(g)} \tag{4}$$

where the sum is over all edge subgraphs of G (all the graphs have the same vertex set), and c(g) and e(g) are the number of connected components and number of edges in g. A special case of (4) is the chromatic polynomial since putting x = -1 in the right-hand side of equation 4 gives Birkhoff's expression for the chromatic polynomial C(q), (7, 21). In the guise of Potts model this system is also of interest as a model of phase transition phenomena. It generalises the Ising model by replacing the binary variables $\sigma_i$ by q-valued variables (23,13).

## 2 SERIES AND TRANSFORMATIONS

In the theory of phase transitions we are always interested in the limit of the graph G becoming arbitrarily large because phase transitions only appear in such limiting cases. Such limits are usually defined using a sequence of finite subgraphs $G_n$ of some infinite regular graph, G. For a few special sets of values of q, x, Z(q,x) as defined by equation 4 can be obtained for various two-dimensional lattices. However in the general case we have to rely on series expansions for Z in powers of some appropriate combination of variables. Such series expansions can be obtained in terms of graph enumeration problems as indicated above and low-order coefficients in the series can be found by constructing all the graphs involved at each order and simply counting them. One important result which both simplifies the enumeration and which helps in constructing the large-size limit is what is known as the logarithmic transformation (5,29). If we expand $\ell nZ$ instead of Z then in many cases we can find expansions which involve only connected subgraphs. As long as edge effects decrease to zero as the size of $G_n$ increases, the number of embeddings of a subgraph g will tend to become proportional to $V(G_n)$ the number of vertices of $G_n$ and so in these circumstances, $\left[V(G_n)\right]^{-1} \ell nZ(G_n)$ should tend to a limit.

The logarithmic transformation is an example of the procedure of changing both the graphical and algebraic problems in such a way that the graphical problem is simplified at the expense of increased complication in the algebra and the overall problem is left invariant. In obtaining series expansions in statistical mechanics there has been a strong trend towards the use of such transformations. (References 12, 22, 25 and 30 exemplify this trend.) Before discussing the reasons for this trend in terms of combinatorial properties, it must be pointed out that some of the motivation for using algebraic techniques comes from the nature of the computing tools available. It seems to be easier to use present-day general-purpose programming languages to implement algebraic manipulations than to implement graph manipulations and, compared to algebraic manipulation languages, special purpose graph-manipulation languages are few and somewhat restricted (24). The positive side of this observation is that a graphical formulation contains more richness and complexity than is managable or necessary in the series expansion context.

From the computing point of view a major problem with graphical expressions is that they are generally defined recursively. It has been remarked that 'if computers had existed in the middle ages, programmers would have been burnt at the stake .... for heresy ..... one of the main heresies would have been a belief (or disbelief) in recursion.'(2). To see the problems consider the recursive definition of the Fibonacci numbers ($f_n = f_{n-1} + f_{n-2}$). A computer procedure using this directly would evaluate $f_2$ 13 separate times in the course of evaluating $f_8$. Obviously what we have to do is keep track of which calculations we have already done and preserve the results. In graph theoretical problems where typically we have to work through all subgraphs of a given graph we have to be able to tell if a given subgraph is equivalent to one that has already been considered. It is necessary to construct a graph directory based on a canonical form of the graphs, used as the basis for a canonical ordering (20).

## 3 RECTANGULAR GRAPH EXPANSIONS

Two extreme examples of series expansion techniques which have reduced combinatorial complexity at the expense of algebraic complexity are the techniques for obtaining series for the limit of Whitney polynomials in terms of rectangular graphs (12, 22). The graphs are denoted $G_{nm}$ as indicated in the figure. The square lattice is the limit as $m \to \infty$ of $G_{mm}$. If we denote the chromatic polynomial on one of these graphs by $C(q,G_{mm})$ then we can consider the limit

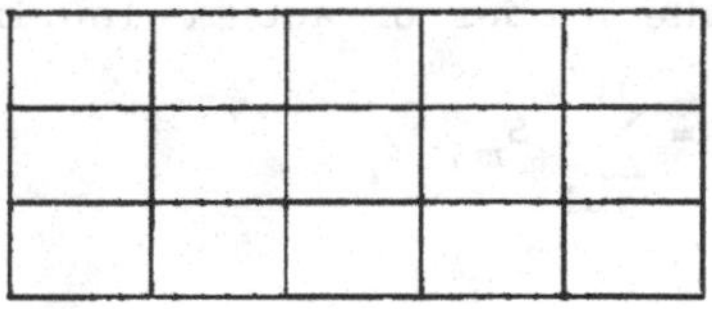

$G_{46}$

$$C(q) = \lim_{m\to\infty} \left[C(q,G_{mm})\right]^{1/m^2}$$

$$= (q-1)^2 q^{-1} \left[1 + \sum_{n=1}^{\infty} a_n (q-1)^{-n}\right]$$

$$= (q-1)^2 \, q^{-1} \left[ 1 + (q-1)^{-3} + (q-1)^{-7} + 3(q-1)^{-8} \ldots\ldots \right] \quad (5)$$

Expansions for $C(q)$ have been obtained by Nagle (21) and by Kim and Enting (19).

A simple prescription for obtaining coefficients is given by the approximation (12)

$$C(q) \approx C(q,G_{mm}) \, C(q,G_{m-1,m-1}) / C(q,G_{m,m-1})^2 \quad (6)$$

which gives all coefficients through to $a_{2m}$ correctly. Kim and Enting obtained coefficients through to $a_{18}$ by taking products of $C(q,G_{mn})$ for $m + n \leq 11$. If these expressions are compared to Nagle's expressions which involve subgraphs with no vertices of degree 1 and no bridges, it is clear that an incredible reduction in graphical complexity has been achieved. The increase in algebraic complexity has not been very great because we can calculate chromatic polynomials for the $G_{mn}$ by using transfer matrix techniques. Biggs and Meredith (6) describe one way of doing this. We can put

$$C(q,G_{mn}) = \underset{\sim}{a}(q) \, V_m(q)^n \, \underset{\sim}{e} \quad . \quad (7)$$

The size $S_m$, of the matrices $V_m(q)$ is the number of distinct proper colourings the m sites in a column, distinct implying that colourings obtained from one another by permutations of colours are the same. If $S_{mj}$ is the number of such colourings which use j colours then

$$S_m = \sum_{j=1}^{m} S_{mj} \quad (8)$$

$$S_{mj} = S_{m-1,j-1} + (j-1) S_{m-1,j} \quad . \quad (9)$$

It is simple to show that

$$S_m \leq (m-1)! \quad (10)$$

and using $S_{mm} = 1$,

$$S_{mj} \geq (j-1)^{m-j} \quad . \quad (11)$$

Equation 11 shows that $S_m$ will, for large m increase faster than any exponential $y^m$ no matter how large a value of y is chosen. The othe] case which is of particular interest is the case of Whitney polynomials with q in equation 4 fixed at integer values. For these cases the transfer matrices for $G_{mn}$ are of dimension $q^m$.

## 4 COMPARISON OF GROWTH RATES

There are a number of problems which make it difficult to accurately compare the efficiencies of alternative computational techniques in a meaningful way. Some of these are:

(i) the number of elementary operations involved in a process will depend on exactly which operations are defined as elementary, whether for example you regard multiplication as elementary or whether you allow for the fact that for large numbers the amount of time taken to perform a multiplication increases with the size of a number. Such changes in definition will multiply the number of operations by a factor which is usually bounded above by a polynomial in n (1) and so much work on computational complexity focuses attention on whether the problem as a whole is bounded by a polynomial in n, where n is some measure of the size of the problem (1, 10, 17).

(ii) The efficiency of a technique will depend on the extent to which existing information can be used. If extensive tabulations of graphical data are available then the appropriate counts can be plugged into the algebraic expressions. For large n however, all we really have is the possibility of doing several problems at once and so the amount of calculation has to be compared to the amount of calculation needed to solve the individual problems by algebraic techniques.

(iii) For graphical formulations it is difficult to determine the average properties of classes of graphs. For general graphs there is a conjecture that 'almost all' graphs are k-connected (15) but it is by no means as certain that the same 'almost-all' property applies to the class of graphs embeddable on particular lattices. The degree of connectedness will largely determine the depth of recursion needed in recursive definitions of graph weights.

If we want to compare alternative procedures which grow with n as $P_1(n)\exp(a_1n)$ and $P_2(n)\exp(a_2n)$ where $P_1$ and $P_2$ are polynomials then for large n the more efficient process will be that with the smaller of $a_1$, $a_2$ .

For the fixed q case the matrices used by de Neef and Enting grow as $q^{n/4}$ when n series coefficients are to be calculated. The number of operations involved in multiplying such matrices will grow as $q^{3n/4}$ if direct multiplication techniques are used and as $q^{2.81n/4}$ if more sophisticated techniques are used (1). In contrast, in the graphical approach even the number of closed polygons grows as $2.6385^n$, (8, 26) and since $2.6385^{4/2.81} = 3.979$, for q = 3 the rectangular graph expansion would be more efficient than direct graph counting even if we only had to count polygons (self-avoiding rings). An analysis of other graphs involved would presumably indicate that the algebraic technique is superior for q = 4. To show that the algebraic techniques are superior for larger values of q might well require an analysis of the recursive aspects of the graphical formalisms.

Both the algebraic and graphical expressions for series expansions for the limit of chromatic polynomials exhibit a roughly n! growth. A comparison of their efficiencies would require a much more detailed analysis and it may well be that in this case the only significant advantage of algebraic techniques is that they are easier to program.

REFERENCES


(1) A.V. Aho, J.E. Hopcroft and J.D. Ullman. The design and analysis of computer algorithms. (1974) Addison-Wesley.

(2) D.W. Barron, 'Recursive techniques in programming', 1969 Macdonald.

(3) R.J. Baxter, 'Potts model at the critical temperature', J. Phys. C 6 (1973) L445-8.

(4) R.J. Baxter, S.B. Kelland and F.Y. Wu, 'Equivalence of the Potts model or Whitney polynomial with an ice-type model', J. Phys. A 9 (1976) 397-406.

(5) N.L. Biggs, Algebraic graph theory. (1974) Cambridge U.P.

(6) N.L. Biggs and G.H.J. Meredith, 'Approximations for chromatic polynomials', J. Comb. Theory B 20 (1976) 5-19.

(7) G.D. Birkhoff, 'A determinantal formula for the number of ways of colouring graphs', Ann. of Math. (2) 14 (1912) 42-6.

(8) C. Domb, 'Self avoiding walks in Ising and Heisenberg models', J. Phys. C 3 (1970) 256-84.

(9) C. Domb, 'Configurational studies of the Potts models', J. Phys. A 7 (1974) 1335-48.

(10) J. Edmonds, 'Paths trees and flowers', Canad. J. Math. 17 (1965) 449-67.

(11) I.G. Enting, 'Series expansions for the Potts model : high-field expansions', J. Phys. A 7 (1974) 1617-26.

(12) I.G. Enting and R.J. Baxter, 'A special series expansion technique for the square lattice', J. Phys. A 10 (1977) L117-9.

(13) C.M. Fortuin and P.W. Kasteleyn, 'On the random-cluster model, I, Introduction and relation to other models', Physica 57 (1972) 536-64.

(14) F. Harary (ed.), 'Graph Theory and Theoretical Physics', (1967) Academic.

(15) F. Harary and E.M. Palmer, 'Graphical Enumeration', (1973) Academic.

(16) E. Ising, 'Beitrag zur Theorie des Ferromagnetismus', Z. Phys. 31 (1925) 253-8.

(17) R.M. Karp, 'The fast approximate solutions of hard combinatorial problems', Proceedings of the sixth Southeastern conference on combinatorics, graph theory and computing. (1975) 15-21 Utilitas Mathematica Publishing.

(18) T. Kihara, Y. Midzuno and T. Shizume, 'Statistics of two-dimensional lattices with many components', J. Phys. Soc. Jap. 9 (1954) 681-7.

(19) D. Kim and I.G. Enting, 'The limit of chromatic polynomials', J. Comb. Theory B (1978) to be published.

(20) J.F. Nagle, 'On ordering and identifying undirected linear graphs', J. Math. Phys. 7 (1966) 1588-92.

(21) J.F. Nagle, 'A new subgraph expansion for obtaining colouring polynomials for graphs', J. Comb. Theory B 10 (1971) 42-59.

(22) T. de Neef and I.G. Enting, 'Series expansions from the finite lattice method', J. Phys. A 10 (1977) 801-5.

(23) R.B. Potts, 'Some generalised order-disorder transformations', Proc. Camb. Phil. Soc. 48 (1952) 106-9.

(24) J.E. Sammet, 'Roster of programming languages for 1974-5', C.A.C.M 19 (1976) 655-69.

(25) M.F. Sykes, J.W. Essam and D.S. Gaunt, 'Derivation of low-temperature expansions for the Ising model of a ferromagnet and an antiferromagnet', J. Math. Phys. 6 (1965) 283-298.

(26) M.F. Sykes, D.S. McKenzie, M.G. Watts and J.L. Martin, 'The number of self-avoiding rings on a lattice', J. Phys. A 5 (1972) 661-6.

(27) H.N.V. Temperley, 'Transformation of graph-theoretical problems into one another', Proceedings of the fifth British Combinatorial Conference 1975. 585-9. Utilitas Mathematica Publishing.

(28) W.T. Tutte, 'On dichromatic polynomials', J. Comb. Theory 2 (1967) 301-20.

(29) H. Whitney, 'The colouring of graphs', Ann. of Math. 33 (1932) 688-718.

(30) M. Wortis, 'Linked cluster expansions', Chapter 3 of Phase Transitions and Critical Phenomena Vol. 3. (1974) eds. C. Domb and M.S. Green. Academic.

05C25

# GRAPHS, GROUPS AND POLYTOPES

C.D. Godsil
Department of Mathematics,
University of Melbourne,
Parkville, Victoria, 3052
Australia

ABSTRACT

With each eigenspace of the adjacency matrix A of a graph X there is an associated convex polytope. Any automorphism of X induces an orthogonal transformation of this polytope onto itself. These observations are used to obtain information on the relation between the automorphism group of X and the multiplicities of the eigenvalues of A. This approach yields new results on this topic as well as improvements of previously known ones.

## 1. PRELIMINARIES

Any undefined graph theoretic terminology is based on [4].

Throughout this paper X is a graph with vertex set {1, 2, ..., n}. N(i) denotes the subgraph of X induced by the vertices adjacent to i; it does not contain i. X has adjacency matrix $A = (a_{ij})$ where $a_{ij} = 1$ or 0 according as vertices i and j are adjacent or not. The set of distinct eigenvalues of A will be denoted by $\sigma(A)$ and the multiplicity of the eigenvalue $\lambda$ by $m(\lambda)$. Since X is a graph, A is symmetric and as a result the elements of $\sigma(A)$ are real. We recall from [1 : Proposition 3.1] that if X is regular of degree d then $d \in \sigma(A)$ and for any $\lambda \in \sigma(A)$ we have $|\lambda| \leq d$.

$O_m(\lambda)$ will denote the group of orthogonal $m \times m$ matrices over the extension field $\mathbb{Q}(\lambda)$. $Z_n$ denotes the cyclic group of order n and $D_{2n}$ the dihedral group of order 2n. Aut(X) denotes the automorphism group of the graph X. By [1 : Proposition 15.2] we can identify Aut(X) with the group of $n \times n$ permutation matrices which commute with A.

## 2. WEIGHT VECTORS AND POLYTOPES

Let $\{z_i : i = 1, 2, \ldots, m(\lambda)\}$ be an orthonormal basis of the eigenspace of A, associated with the eigenvalue $\lambda$. Let $Z_\lambda$ be the $n \times m(\lambda)$ matrix with $i^{th}$ column $z_i$. Since the vectors $z_i$ are orthonormal $Z_\lambda^T Z_\lambda = I_{m(\lambda)}$ (where $I_k$ is the $k \times k$ identity matrix) and $Z_\lambda Z_\lambda^T$ is the orthogonal idempotent associated with $\lambda$ in the spectral decomposition of A. We set $A_\lambda = Z_\lambda Z_\lambda^T$ and thus have

$$A = \sum_{\lambda \in \sigma(A)} \lambda A_\lambda \qquad (1)$$

where $A_\lambda^2 = A_\lambda$ and $A_\lambda A_\mu = A_\mu A_\lambda = 0$ for $\lambda$, $\mu$ in $\sigma(A)$ with $\lambda \neq \mu$. We denote the $i^{th}$ row of $Z_\lambda$ by $\underset{\sim}{w}_\lambda(i)$ and call it the *weight vector on* $\lambda$ of the vertex i of X. Since $AZ_\lambda = \lambda Z_\lambda$ we find that

$$\lambda \underset{\sim}{w}_\lambda(i) = \sum_{j\in N(i)} \underset{\sim}{w}_\lambda(j) . \tag{2}$$

The following result indicates some of the significance of weight vectors. Since none of our later results depend on it, we only give an outline of the proof.

2.1 LEMMA. Let $W_{ij}^r$ be the number of walks of length r from vertex i to vertex j in X and let $<\underset{\sim}{x}, \underset{\sim}{y}>$ denote the inner product of the vectors $\underset{\sim}{x}$ and $\underset{\sim}{y}$. Then

$$W_{ij}^r = \sum_{\lambda\in\sigma(A)} \lambda^r <\underset{\sim}{w}_\lambda(i), \underset{\sim}{w}_\lambda(j)> .$$

Hence, for vertices i, j, k and $\ell$ in X we have $W_{ij}^r = W_{k\ell}^r$ for all $r = 0, 1, 2, \ldots$ if and only if

$$<\underset{\sim}{w}_\lambda(i), \underset{\sim}{w}_\lambda(j)> = <\underset{\sim}{w}_\lambda(k), \underset{\sim}{w}_\lambda(\ell)>$$

for all $\lambda \in \sigma(A)$.

Proof. By [1: Lemma 2.5], $W_{ij}^r$ is equal to the $i$-$j^{th}$ entry of $A^r$. From (1) above we have

$$A^r = \sum_{\lambda\in\sigma(A)} \lambda^r A_\lambda .$$

The expression given for $W_{ij}^r$ follows from this, on observing that the $i$-$j^{th}$ entry of $A_\lambda$ is $<\underset{\sim}{w}_\lambda(i), \underset{\sim}{w}_\lambda(j)>$. The remaining claim follows by straightforward algebra. □

We will refer to $<\underset{\sim}{w}_\lambda(i), \underset{\sim}{w}_\lambda(i)>$ as the *weight* of i on $\lambda$ and to $<\underset{\sim}{w}_\lambda(i), \underset{\sim}{w}_\lambda(j)>$ as the *angle* between i and j on $\lambda$. Since the $W_{ii}^r$ $(r = 0, 1, 2, \ldots)$ determine the weights of the vertex i on the distinct eigenvalues of A, it follows that vertices lying in the same orbit under the action of Aut(X) on X have equal weights on each eigenvalue.

Although we will not prove it here, it is worth noting that $W_{ii}^r$ and $W_{jj}^r$ will be equal for all $r = 0, 1, 2, \ldots$, if and only if the subgraphs obtained from X by deleting the vertices i and j in turn are cospectral i.e. if their adjacency matrices have the same characteristic polynomial.

The convex hull of the set of weight vectors for X on $\lambda$ is a convex polytope which we will denote by $C_\lambda(X)$. (For the definition and a detailed discussion of the properties of convex polytopes, see [3]). The symmetry group of a convex polytope C is the group of orthogonal transformations which map the set of vertices of C onto itself.

2.2 THEOREM. For each $\lambda$ in $\sigma(A)$ there exists a homomorphism $\phi_\lambda$ mapping $G = \mathrm{Aut}(X)$ into the symmetry group of $C_\lambda(X)$. Any two vertices lying in a given orbit of $N_\lambda = \ker \phi_\lambda$ have the same weight vector on $\lambda$. The subgroup $\cap\{N_\lambda : \lambda \in \sigma(A)\}$ is trivial.

Proof. Suppose $P \in G$. We define $\phi_\lambda(P)$ by

$$\phi_\lambda(P) = Z_\lambda^T P Z_\lambda \ . \tag{3}$$

Since P commutes with A it commutes with every polynomial in A and therefore also with $A_\lambda = Z_\lambda Z_\lambda^T$. Hence

$$\begin{aligned} Z_\lambda \cdot \phi_\lambda(P) &= Z_\lambda Z_\lambda^T P Z_\lambda \\ &= P Z_\lambda Z_\lambda^T Z_\lambda \\ &= P Z_\lambda \quad (\text{since } Z_\lambda^T Z_\lambda = I_{m(\lambda)}) \ . \end{aligned} \tag{4}$$

Thus if Q is another element of G

$$\begin{aligned} Z_\lambda \cdot \phi_\lambda(PQ) = PQZ_\lambda &= PZ_\lambda \cdot \phi_\lambda(Q) \\ &= Z_\lambda \cdot \phi_\lambda(P)\phi_\lambda(Q) \end{aligned}$$

and as $Z_\lambda^T Z_\lambda = I_{m(\lambda)}$ we find that

$$\phi_\lambda(PQ) = \phi_\lambda(P)\phi_\lambda(Q)$$

and so $\phi_\lambda$ is a homomorphism.

We now show that $\phi_\lambda(P)$ is an orthogonal matrix. It then follows that $\phi_\lambda$ is a homomorphism of G into $O_m(\lambda)$, where $m = m(\lambda)$. We have

$$\phi_\lambda(P)^T = (Z_\lambda^T P Z_\lambda)^T = Z_\lambda^T P^T Z_\lambda$$
$$= Z_\lambda^T P^{-1} Z_\lambda \qquad \text{(since P is orthogonal)}$$
$$= \phi_\lambda(P^{-1})$$
$$= \phi_\lambda(P)^{-1} \qquad \text{(since } \phi_\lambda \text{ is a homomorphism)} .$$

Thus $\phi_\lambda(P)$ is indeed orthogonal.

From (4) above we see that

$$\underset{\sim}{w}_\lambda(P(i)) = \underset{\sim}{w}_\lambda(i).\phi_\lambda(P) ,$$

where $P(i)$ denotes the image of the vertex $i$ under the action of the automorphism of $X$ represented by the matrix $P$. Hence for all $i$ in $X$, $\underset{\sim}{w}_\lambda(i).\phi_\lambda(P)$ lies in $C_\lambda(X)$ and so we conclude that $\phi_\lambda$ is a homomorphism of $G$ into the symmetry group of $C_\lambda(X)$, as claimed.

If $P \in N_\lambda = \ker \phi_\lambda$ then $PZ_\lambda = Z_\lambda$. This implies that $\underset{\sim}{w}_\lambda(P(i))$ and $\underset{\sim}{w}_\lambda(i)$ are equal for all $i$ in $X$ and in consequence we see that any two vertices in a given orbit of $N_\lambda$ have equal weight vectors on $\lambda$.

Finally, if $P \in \cap\{N_\lambda : \lambda \in \sigma(A)\}$ then $PZ_\lambda = Z_\lambda$ for each $\lambda$ in $\sigma(A)$ and so $P\underset{\sim}{z}_j = \underset{\sim}{z}_j$ for each eigenvector $\underset{\sim}{z}_j$ of $A$. As $A$ is symmetric, its eigenvectors span $\mathbb{R}^n$ and hence we must have $P = I_n$. □

2.2 is our main tool. In section 3 we obtain a number of results as corollaries of it. However we first derive some properties of the homomorphism $\phi_\lambda$.

2.3 LEMMA. Let $X$ be a connected transitive graph of degree $d$. Then $N_\lambda$ acts transitively on $X$ if and only if $\lambda = d$.

Proof. Let $\underset{\sim}{j}$ denote the vector in $\mathbb{R}^n$ with each entry equal to $\frac{1}{\sqrt{n}}$, where $n = |X|$. Then $A\underset{\sim}{j} = d\underset{\sim}{j}$ and by [1 : Proposition 3.1], $m(d) = 1$. Hence $C_d(X)$ is the interval $[0, \frac{1}{\sqrt{n}}]$ of the real line. As, for each $P \in G = \mathrm{Aut}(X)$, $\phi_\lambda(P)$ fixes the origin in $\mathbb{R}^{m(\lambda)}$, it follows that the image of $G$ in the symmetry group of $C_d(X)$ is the identity and so $\phi_\lambda$ is trivial. Hence $N_\lambda = G$ and therefore acts transitively on the vertices of $X$.

Conversely, assume $N_\lambda$ acts transitively on $X$. Then by 2.2, the weight vectors on $\lambda$ are all equal. It follows now from (2) above that

$$\lambda \underset{\sim}{w}_\lambda(i) = \sum_{j \in N(i)} \underset{\sim}{w}_\lambda(j) = d\underset{\sim}{w}_\lambda(i)$$

and therefore $\lambda = d$. □

2.4 LEMMA. Let $B_\lambda(i) = \{j \in X : \underset{\sim}{w}_\lambda(j) = \underset{\sim}{w}_\lambda(i)\}$. Then $B_\lambda(i)$ is a block for $G = Aut(X)$ The permutation group induced by the action of G on the set $\{B_\lambda(i) : i \in X\}$ is isomorphic to the group of permutations of the vertices of $C_\lambda(X)$ induced by $\phi_\lambda(G)$.

Proof. Let $j \in B_\lambda(i)$ and let $P \in G$. Then

$$\underset{\sim}{w}_\lambda(P(j)) = \underset{\sim}{w}_\lambda(j).\phi_\lambda(P) = \underset{\sim}{w}_\lambda(i).\phi_\lambda(P) = \underset{\sim}{w}_\lambda(P(i))$$

and so $P(j) \in B_\lambda(P(i))$. Since any two sets $B_\lambda(i)$ and $B_\lambda(j)$ are either equal or disjoint, it follows at once that $B_\lambda(i)$ is a block for G in its action on the vertices of X. The remaining claim follows directly from 2.2. □

2.4 affords another proof of our claim made following 2.1 that vertices in the same orbit under the action of Aut(X) on X have equal weights on any given eigenvalue $\lambda$.

2.5 LEMMA. Assume $G = Aut(X)$ acts primitively on the vertices of X. Then either $B_\lambda(i) = i$, or $B_\lambda(i) = X$ and $\lambda = d$.

Proof. If G is primitive then we must have either $B_\lambda(i) = i$ or $B_\lambda(i) = X$, since a primitive group has only trivial blocks. If $B_\lambda(i) = X$ then the argument employed in the proof of 2.3 shows that $\lambda = d$. □

## 3. APPLICATIONS OF 2.2

It follows immediately from 2.2 that $G = Aut(X)$ is the subdirect product of the quotient groups $G/N_\lambda$ $(\lambda \in \sigma(A))$. (The subdirect product is defined in [5 : I§9.11], although there it is referred to as "direkte Produkt mit ... vereinigter Faktorgruppe"). The following result generalizes Theorem 15.4 of [1], due independently to Mowshowitz [6] and to Petersdorf and Sachs. (They describe the case $M = 1$).

3.1 THEOREM. If, for some integer M, $m(\lambda) \leq M$ for all $\lambda$ in X, Aut(X) is a subdirect product of subgroups of $O_M(\lambda)$. In particular, if $M = 1$ then Aut(X) is isomorphic to $Z_2^m$ for some integer m, and if $M = 2$ then Aut(X) is isomorphic to a subdirect product of dihedral groups and cyclic groups.

Proof. The first claim follows from our earlier remarks. The remaining claims follow from the observation that $O_1(\lambda) \cong Z_2$ and that the finite subgroups of $O_2(\lambda)$ are dihedral or cyclic. □

Let X be a graph with automorphism group G. We call the subset S of X a *basis* for G on X if $G_S = \langle 1\rangle$ and $G_T \neq \langle 1\rangle$ for any proper subset T of S. (If $Y \subseteq X$, $G_Y$ denotes the subgroup $\langle P \in G : P(y) = y\ \forall y \subset Y\rangle$ of G). If G admits no basis S' with $|S'| < |S|$, we call S *minimal*.

3.2 THEOREM. Let X be a vertex transitive graph and let $\lambda$ be an eigenvalue of X such that the weight vectors $\underset{\sim}{w}_\lambda(i)$ are all distinct. Then G = Aut(X) has a basis S with $|S| \leq m(\lambda)$. Moreover if G has a minimal basis S' with $|S'| = m(\lambda)$, then G has a non-trivial normal subgroup H which is isomorphic to a real Coxeter group.

Proof. By 2.2, our hypotheses imply that $N_\lambda = \langle 1\rangle$ and so $\phi_\lambda$ is faithful. For any subset Y of X, let $w_\lambda(Y) = \{\underset{\sim}{w}_\lambda(i) : i \in Y\}$ and let $V_\lambda(Y)$ denote the vector space over $\mathbb{Q}(\lambda)$ spanned by the elements of $w_\lambda(Y)$.

Let S be a subset of X such that $w_\lambda(S)$ is a basis for $V_\lambda(X)$. Then if $P \in G_S$, for each $i \in S$ we have

$$\underset{\sim}{w}_\lambda(i) = \underset{\sim}{w}_\lambda(P(i)) = \underset{\sim}{w}_\lambda(i).\phi_\lambda(P) .$$

Hence $\phi_\lambda(P)$ fixes each vector in the basis $w_\lambda(S)$ of $V_\lambda(X)$ and $\phi_\lambda(P) = 1$. Since $\phi_\lambda$ is faithful this implies $P = 1$. Therefore $G_S = \langle 1\rangle$ and accordingly S must contain a basis for G on X.

Assume now that G has a minimal basis S' with $|S'| = m(\lambda)$. Let T be a minimal proper subset of S'. By our choice of S', $G_T = \langle 1\rangle$ and therefore contains a non-identity element P, say. $\phi_\lambda(P)$ fixes each element of $V_\lambda(T)$, which is a hyperplane in $V_\lambda(S') = V_\lambda(X)$. As $\phi_\lambda(P)$ is orthogonal, it follows that it induces a reflection in the real vector space $V_\lambda(X)$, fixing the hyperplane $V_\lambda(T)$.

Let H be the normal subgroup of G generated by all the conjugates of P. Then $\phi_\lambda(H)$ is a group generated by real reflections in $V_\lambda(X)$ i.e. a real Coxeter group. Since $\phi_\lambda$ is faithful, H is our required subgroup. □

Groups generated by real reflections are classified in [2]. We note that if Aut(X) acts primitively on the vertices of X and X has degree d, then by 2.5 every eigenvalue $\lambda \in \sigma(A)\backslash d$ satisfies the hypothesis of 3.2. Furthermore H will act transitively on the vertices of X in this case, since a normal subgroup of a primitive group is always transitive (see [8 : Proposition 7.1]).

3.3 THEOREM. Let X be a connected graph with G = Aut(X) acting transitively on both its vertices and its edges. Let $\lambda$ be an eigenvalue of X such that $m(\lambda) = 2$ and all weight vectors on $\lambda$ are distinct. Then $X \cong C_n$, the cycle on $n = |X|$ vertices.

Proof. As in 3.3, let $V_\lambda(X)$ denote the space spanned by the vectors $\underset{\sim}{w}_\lambda(i)$ over $\mathbb{Q}(\lambda)$. Since $m(\lambda) = 2$, $C_\lambda(X)$ is a polygon and by 2.4, $\phi_\lambda(G)$ acts transitively on its vertices As $G = \mathrm{Aut}(X)$ acts transitively on the edges of X the angles $<\underset{\sim}{w}_\lambda(i), \underset{\sim}{w}_\lambda(j)>$ are equal for any adjacent vertices i and j in X. By hypothesis, distinct vertices of X have distinct weight vectors on $\lambda$, and so the number of vertices x in X such that

$$<\underset{\sim}{w}_\lambda(i), \underset{\sim}{w}_\lambda(j)> = <\underset{\sim}{w}_\lambda(i), \underset{\sim}{w}_\lambda(x)> ,$$

is at most two. Hence either X has degree $d = 1$ and is consequently isomorphic to $K_2$, or it has degree $d = 2$ and is isomorphic to $C_n$. As $K_2$ has only isolated eigenvalues, the result follows. □

3.5 THEOREM. Let X be a graph with $G = \mathrm{Aut}(X)$ acting transitively on both its vertices and its edges. Let $\lambda$ be an eigenvalue such that all the weight vectors of X on $\lambda$ are distinct. Then if X contains a clique on c vertices, $m(\lambda) \geq c-1$ and if $m(\lambda) = c-1$, $\lambda = -d/(c-1)$.

Proof. Let the degree of X be d. Then $d \in \sigma(A)$ and by our assumption on $\lambda$ we have $\lambda \neq d$. From (2) in section 2 we find that

$$\lambda\underset{\sim}{w}_\lambda(i) = \sum_{j \in N(i)} \underset{\sim}{w}_\lambda(j) .$$

Taking inner products of both sides of this expression with $\underset{\sim}{w}_\lambda(i)$ and observing that, since G acts transitively on the edges of X the angles $<\underset{\sim}{w}_\lambda(i), \underset{\sim}{w}_\lambda(j)>$ $(j \in N(i))$ are all equal, we obtain

$$\lambda<\underset{\sim}{w}_\lambda(i), \underset{\sim}{w}_\lambda(i)> = d<\underset{\sim}{w}_\lambda(i), \underset{\sim}{w}_\lambda(j)> . \tag{5}$$

Since G acts transitively on the vertices of X, the weights of X on $\lambda$ are all equal and we may therefore rewrite (5) as

$$\lambda = d \cos \alpha \tag{6}$$

for suitable $\alpha$.

Thus to a clique on c vertices in X there corresponds a set of c points on the unit sphere in $V_\lambda(X)$ with the angular distance between any pair of points equal to $\cos \alpha$. The maximum number of points in such a configuration occurs when they lie at the vertices of a regular k-simplex, in which case we have $\cos \alpha = -1/k$. Since a k-simplex has k+1 vertices we thus have $c \leq m(\lambda) + 1$, with equality implying that cos $\alpha = 1/m(\lambda)$. By (6) this implies $\lambda = d/(c-1)$. □

We remark that $K_m$ provides an example of a graph satisfying the hypotheses of 3.3 and for which the bound given is obtained. The cartesian product $K_2 \times K_m$ provides a second example.

REFERENCES

[1] Biggs, N. *Algebraic Graph Theory,* C.U.P., London, (1974).

[2] Coxeter, H. and Moser, W. *Generators and relations for discrete groups,* 2nd. ed. Springer, Berlin (1965).

[3] Grünbaum, B. *Convex Polytopes,* Wiley, New York (1967).

[4] Harary, F. *"Graph Theory",* Addison-Wesley, Reading, Ma., (1969).

[5] Huppert, B. *"Endliche Gruppen I",* Springer, Berlin, (1967).

[6] Mowshowitz, A. The group of a graph whose adjacency matrix has all distinct eigenvalues, in *Proof Techniques in Graph Theory,* Academic, New York (1969), 109-110.

[7] Petersdorf, M. and Sachs, H. Spektrum und Automorphismengruppe eines Graphen, in *Combinatorial theory and its applications, III,* North-Holland, Amsterdam (1970), 891-907.

[8] Wielandt, H. *"Finite Permutation Groups",* Academic Press, New York, (1964).

05C20

# DECOMPOSITIONS OF COMPLETE SYMMETRIC DIGRAPHS INTO THE FOUR ORIENTED QUADRILATERALS

Frank Harary
Department of Mathematics,
University of Michigan,
Ann Arbor, Michigan 48109,
UNITED STATES OF AMERICA

W.D. Wallis and Katherine Heinrich
Department of Mathematics,
University of Newcastle,
New South Wales, 2308,
AUSTRALIA

ABSTRACT

We provide necessary and sufficient conditions for the existence of a decomposition of the set of arcs of a complete symmetric digraph into each of the four oriented quadrilaterals.

## 1. INTRODUCTION

In recent years there has been increasing interest in the decomposition of the edge sets of various species of graphs into isomorphic copies of some prescribed subgraph. Under the names of G-designs and isomorphic factorisations, general surveys of the subject were given in [1] and [4]. One example is when the prescribed subgraph is complete and the species at hand is some multiple of a complete graph; as noted in [5], such a decomposition is precisely a balanced incomplete block design.

Our present object is to settle the case when the species is a complete symmetric digraph and the prescribed subgraph is any oriented quadrilateral. For this purpose we require appropriate notation; all concepts not defined here can be found in [2].

A graph G has point set V and edge set E; a digraph has point set V and arc (directed edge) set X. The *digraph* DG *of the graph* G has the same point set as G, and for each edge {u,v} of G there are two arcs (u,v) and (v,u) in DG. The digraph $DK_p$ of the complete graph $K_p$ is called the *complete symmetric digraph*. By ${}_2K_p$ is meant the multigraph obtained from $K_p$ by doubling each edge.

The undirected cycle of order n is written $C_n$. An *oriented cycle* or *semicycle* is an orientation of $C_n$, obtained by assigning a direction to each edge. It is well known that there are just two oriented triangles, the cyclic triple and the transitive triple. We shall later display the four oriented quadrilaterals and the four oriented pentagons.

The *converse* of a digraph is obtained when the direction of every arc is reversed; a digraph is *self-converse* if it is isomorphic to its converse. The *outdegree*, od u, of a point u of a digraph, is the number of arcs from it; its *indegree*, id u, is the number of arcs to it.

The complete bipartite graph is denoted K(m,n); the complete multipartite graph is $K(n_1,n_2,\cdots,n_r)$. For a connected graph G, nG is the graph with n components, each isomorphic to G. The *union* $G_1 \cup G_2$ of two graphs $G_1$ and $G_2$ is obtained by taking the union of their point sets and the union of their edge sets; their *join* $G_1 + G_2$ is the union of $G_1 \cup G_2$ with the complete bipartite graph joining their point sets. Obviously for two disjoint complete graphs,

$$K_m + K_n = K_{m+n} = K_m \cup K_n \cup K(m,n),$$

with the point set of K(m,n) chosen appropriately.

For two graphs G and H, if there exists a decomposition of E(H) into isomorphic copies of G, we write $G|H$ following [4] and we call this an *isomorphic factorisation*; if not, we write $G\nmid H$.

2. PRELIMINARIES

We require a few previous results concerning decompositions into cycles and semicycles. Rosa and Huang [8] have discussed the decomposition of $K_p$ into cycles, and in particular they prove the following fact.

THEOREM A. $C_4|K_p$ if and only if $p \equiv 1 \pmod 8$.

For orientations of cycles, the following results are known; they are due to Hung and Mendelsohn [6,7] and Schönheim [9].

THEOREM B. If E is either orientation of $C_3$, then $E|DK_p$ if and only if $p \equiv 0$ or $1 \pmod 3$.

THEOREM C. If D is the directed 4-cycle, then $D|DK_p$ if and only if $p \equiv 0$ or $1 \pmod 4$ and $p \neq 4$.

There are exactly four different ways of orienting the quadrilateral $C_4$, as shown in Figure 1. We denote these by A,B,C,D; the later the letter, the longer the maximum path or cycle contained in it. We call A the *alternator*. More generally, if n is even, $A_n$ denotes the alternator of order n, the orientation of $C_n$ in which directions of arcs alternate.

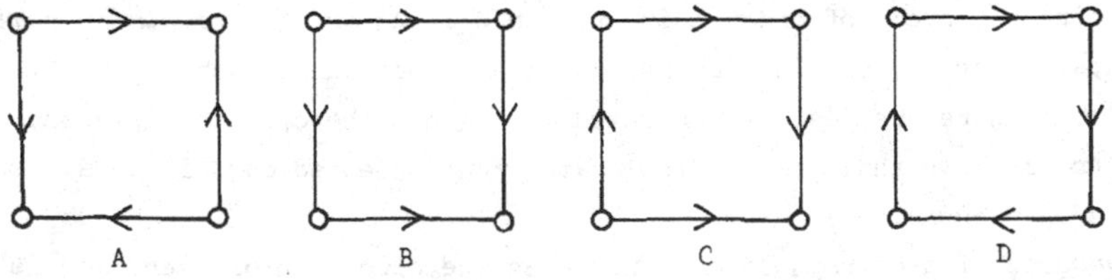


Figure 1. The four oriented quadrilaterals.

It is convenient to write $X|DG$ to mean that each of the semicycles A,B,C,D divides DG, and $X\nmid DG$ means that none of them does.

LEMMA 1. Every oriented quadrilateral is self-converse; hence $X|DC_4$ and so $C_4|G$ implies $X|DG$.

The next statement now follows from Theorem A.

THEOREM 1. If $p \equiv 1 \pmod 8$, then $X|DK_p$.

For complete bipartite graphs we find similarly that $C_4|K(2m,2n)$ and hence $X|DK(2m,2n)$; consequently $X|DK(2n_1,2n_2,\cdots,2n_r)$.

3. NONEXISTENCE

For any graphs $G_1$ and $G_2$ with $q_1$ and $q_2$ edges respectively, $G_1|G_2$ implies $q_1|q_2$. Applying the same reasoning to the directed case, we obtain the following condition.

THEOREM 2. If $p \equiv 2$ or $3 \pmod 4$ then $X\nmid DK_p$.

Consequently we need only consider the cases $p \equiv 0,1,4$ or $5 \pmod 8$. The case $p \equiv 1 \pmod 8$ was completely solved in Theorem 1.

By exhaustion it can be verified that the following isomorphic factorisations do and do not take place:

$$A\nmid DK_4 \qquad B\nmid DK_4 \qquad C|DK_4 \qquad D\nmid DK_4$$

$$A|DK_5 \qquad B\nmid DK_5 \qquad C\nmid DK_5 \qquad D|DK_5 .$$

The three factorisations in this list are exhibited explicitly in the next section.

A stronger statement than $A\nmid DK_4$ can be made.

THEOREM 3. For any even integers m and p, $A_m\nmid DK_p$.

PROOF. The alternator $A_m$ has half its points transmitters of od 2 and id 0; the other half are receivers with id 2 and od 0. Hence in any digraph whose arcs can be decomposed into copies of $A_m$, both the outdegree and the indegree of each point must be even.

In particular we have $A\nmid DK_p$ when $p \equiv 0$ or $4 \pmod 8$.

4. CONSTRUCTIONS

We begin by exhibiting the only decompositions of $DK_4$ and $DK_5$ into A,B,C or D, as promised. We note that the decomposition of ${}_2K_5$ into $C_4$ is unique.

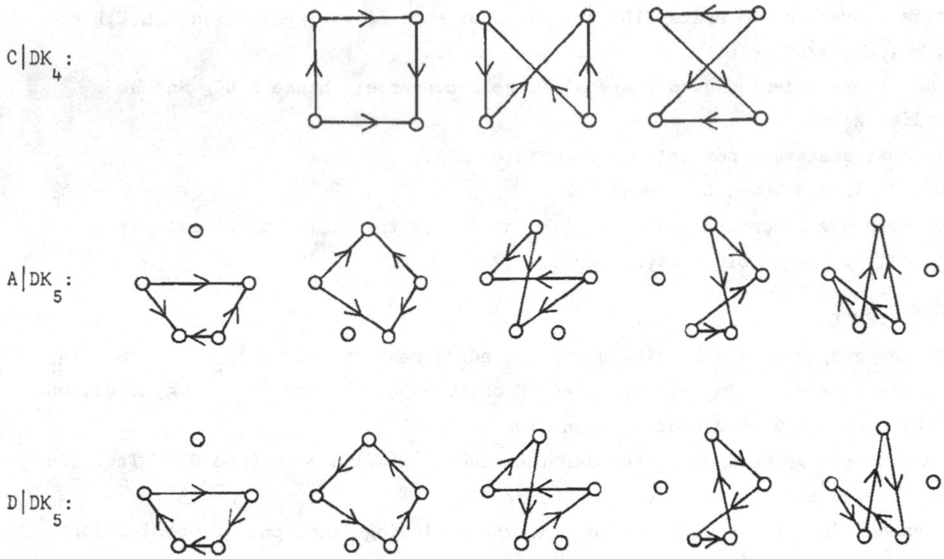


Figure 2. The factorisations $C|DK_4$, $A|DK_5$ and $D|DK_5$.

We now proceed toward the principal results.

LEMMA 4a. $B|DK_8$.

PROOF. We can decompose $K_8$ into four copies of $C_4$ and one copy of the cube $Q_3$, as shown in Figure 3. We see from Figure 4 that $B|DQ_3$; it follows that $B|DK_8$.

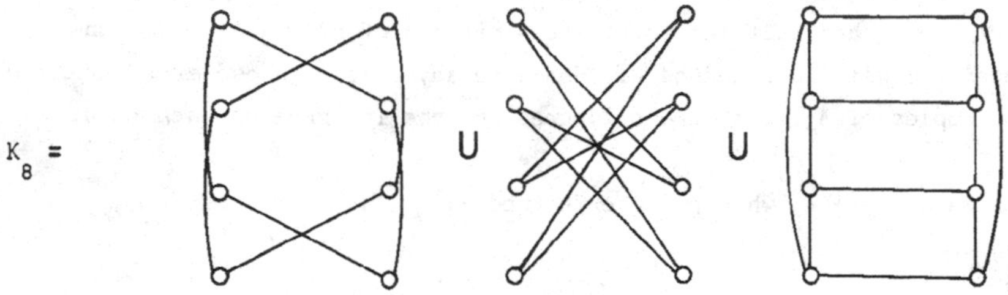


Figure 3. A decomposition of $K_8$.

$DQ_3$ =

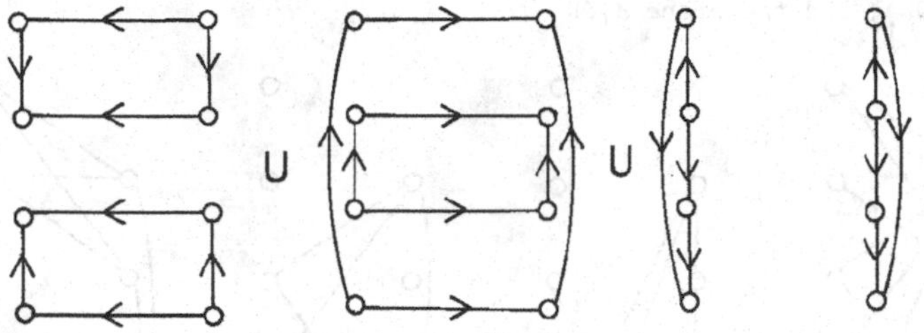

<u>Figure 4.</u> $B|DQ_3$.

We know that $C_4|K(4,8)$. We shall need the following fact, which is easily verified.

LEMMA 4b. It is possible to decompose $K(4,8)$ into copies of $C_4$ in such a way that three of the cycles form the graph $H_1$ shown in Figure 5; it can also be done so that three of the cycles form the graph $H_2$ of Figure 5.

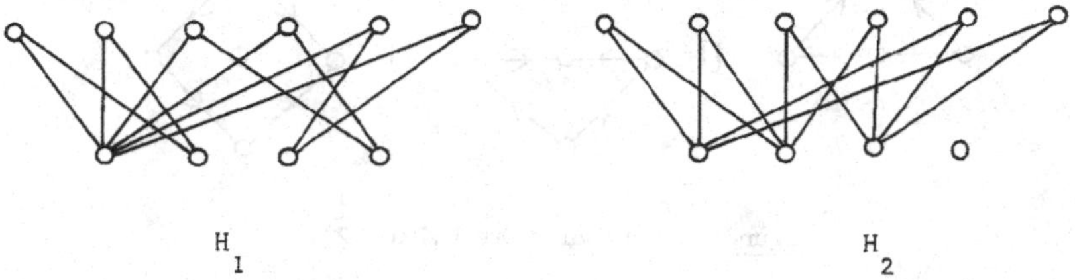


<u>Figure 5.</u> Two subgraphs of $K(4,8)$.

We write $R_1$ and $R_2$ for the complements of $H_1$ and $H_2$ in $K(4,8)$. Thus the lemma says that $C_4|R_1$ and $C_4|R_2$.

LEMMA 4c. $B|DK_{12}$.

PROOF. Consider the expansion

$$K_{12} = K_4 \cup K_8 \cup K(4,8) = K_4 \cup K_8 \cup H_1 \cup R_1 .$$

From Lemma 4b, $C_4|R_1$, so $B|DR_1$, and from Lemma 4a, $B|DK_8$. Thus it is sufficient to prove that $B|L_1$, where $L_1$ is $K_4 \cup H_1$ with the point sets of $K_4$ and $H_1$ chosen appropriately. Now $K(3,2)|L_1$ as shown in Figure 6, and $B|DK(3,2)$ as we see in

Figure 7. So $B|DL_1$, and therefore $B|DK_{12}$.

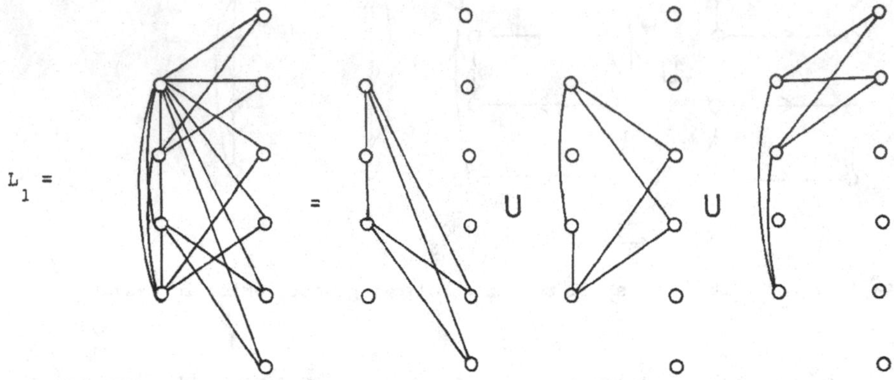

Figure 6. Proof that $K(3,2)|L_1$.

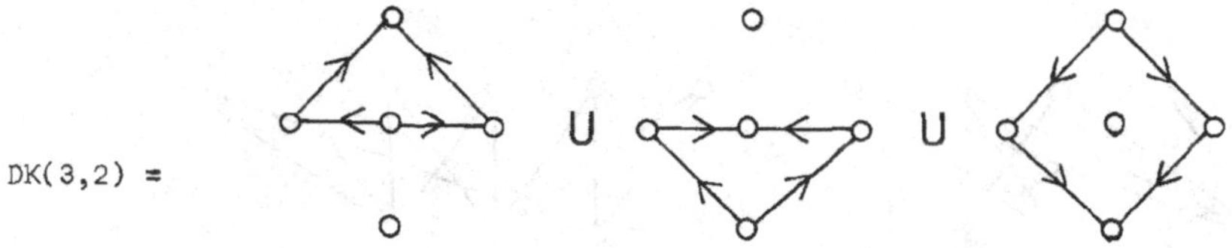

Figure 7. Proof that $B|DK(3,2)$.

THEOREM 4. $B|DK_{4n}$, when $n > 1$.

PROOF. If n is even, say $n = 2s$, then $K_{4n} = sK_8 \cup K(8,8,\cdots,8)$. If n is odd, say $n = 2s + 1$, then $K_{4n} = K_{12} \cup (s-1)K_8 \cup K(12,8,8,\cdots,8)$. In each case we know that B divides the digraph of each of the subgraphs shown, so $B|K_{4n}$. To see that $C|DK_{4n}$ we merely note that $K_{4n} = nK_4 \cup K(4,4,\cdots,4)$, $C|DK_4$ and $C|DK(4,4,\cdots 4)$.

LEMMA 5a. $B|DK_{13}$.

PROOF. We use the decomposition

$$K_{13} = K_5 \cup K_9 \cup K(4,8) = K_5 \cup K_9 \cup H_2 \cup R_2,$$

where the point sets are chosen so that $K_5$ and $K_9$ have one common point and $K(4,8)$

is based on the sets of 4 and 8 points remaining when this common one is deleted. We define $L_2$ to be $K_5 \cup H_2$, with the points identified appropriately, and in Figure 8 we decompose $L_2$ into $K(2,3)$ and $C_4$; so $B|DL_2$. From Theorem 1, $B|DK_9$, so $B|DK_{13}$.

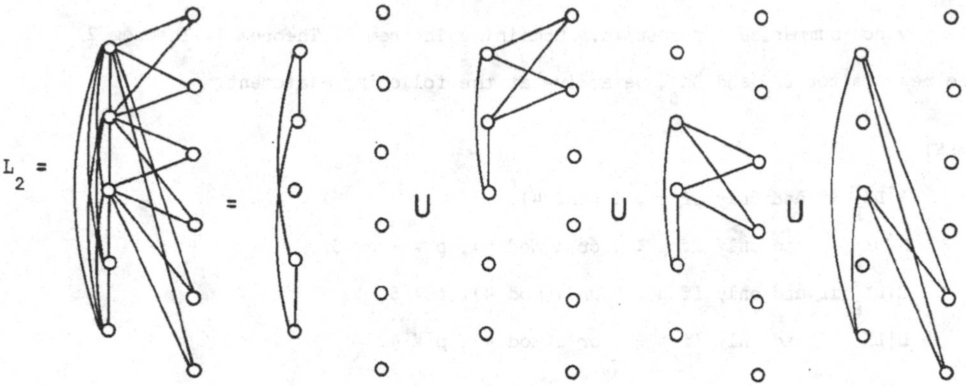

Figure 8. A decomposition of $L_2$.

LEMMA 5b. $C|DK_{13}$.

PROOF. We show in Figure 9 that $K(3,3,3,3)$ can be decomposed into copies of $C_4$ and $K_4$. Now, $K_{13} = K_1 + (4K_3 \cup K(3,3,3,3))$, so $C|DK_{13}$.

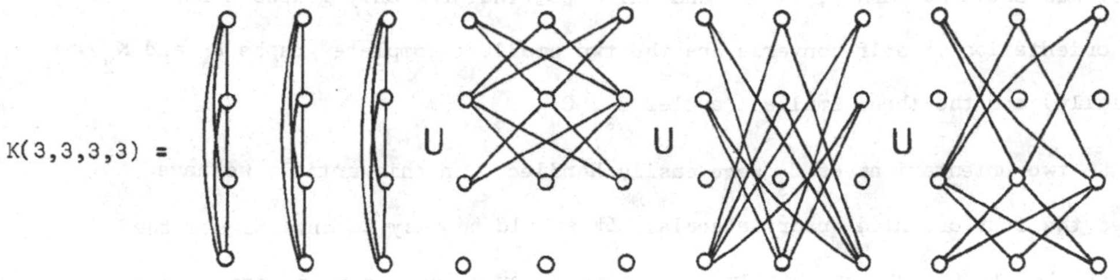

Figure 9. A decomposition of $K(3,3,3,3)$.

THEOREM 5. $A|DK_{8n+5}$, $B|DK_{8n+5}$, $C|DK_{8n+5}$, when $n \geq 1$.

PROOF. Since $K_{8n+5} = K_1 + (2n+1)K_4 \cup K(4,4,\cdots,4)$ and since $A|DK_5$ and $A|DK(4,4,\cdots,4)$ we have $A|DK_{8n+5}$. To prove that $B|DK_{8n+5}$ and $C|DK_{8n+5}$ we must think of $K_{8n+5}$ in the form $K_1 + (K_{12} \cup (n-1)K_8) \cup K(12,8,\cdots,8)$. We know that $B|DK_{13}$ and in section 2 we saw that $B|DK_9$ and $B|DK(12,8,\cdots,8)$. Hence $B|DK_{8n+5}$. Similar results hold for C, so $C|DK_{8n+5}$.

We may now summarize our results. Combining Theorem A, Theorem 1, Theorem 2 and the results for $DK_4$ and $DK_5$, we arrive at the following statement.

STATEMENT

(a) $A|DK_p$ if and only if $p \equiv 1 \pmod 4$.

(b) $B|DK_p$ if and only if $p \equiv 0$ or $1 \pmod 4$, $p \neq 4$ or 5.

(c) $C|DK_p$ if and only if $p \equiv 0$ or $1 \pmod 4$, $p \neq 5$.

(d) $D|DK_p$ if and only if $p \equiv 0$ or $1 \pmod 4$, $p \neq 4$.

## 5. PROBLEMS

For each of the four oriented quadrilaterals $X \in \{A,B,C,D\}$ we have seen that it is not difficult to specify those integers p such that $X|DK_p$. In the process of verifying these results, we made strong use of the fact that every oriented quadrilateral is self-converse.

It was shown by Harary, Palmer and Smith [3] that the only graphs G for which every orientation is self-converse are the two smallest complete graphs $K_1$ and $K_2$ (trivially) and the three smallest cycles $C_3$, $C_4$, $C_5$.

The two orientations of $C_3$ were easily handled. In this article we have studied the four oriented quadrilaterals. It should be easy to investigate the four oriented pentagons, especially as no alternator exists in that case. When $n \geq 6$, non-self-converse cases exist and the situation may well be more complex there. We would be interested to see further investigation of these problems. In particular, we pose the following problem.

*Is there any case where* $p \geq 2n$*, where* $n$ *divides* $p(p-1)$*, but where there exists some orientation* $Y$ *of the cycle* $C_n$ *for which* $Y \nmid DK_p$ *(other than the case where* $Y = A_n$ *and* $p$ *is even)?*

REFERENCES


[1] J.C. Bermond and D. Sotteau, Graph decompositions and G-designs. *Proc. 5th British Combinatorial Conf.*, U.M.P.I., Winnipeg (1976) 53-72.

[2] F. Harary, *Graph Theory*. Addison-Wesley, Reading (1969).

[3] F. Harary, E.M. Palmer and C.A.B. Smith, Which graphs have only self-converse orientations? *Canad. Math. Bull.* 10 (1967) 425-429.

[4] F. Harary, R.W. Robinson and N.C. Wormald, Isomorphic factorisations I: Complete graphs. *Trans. Amer. Math. Soc.* (to appear).

[5] F. Harary and W.D. Wallis, Isomorphic factorizations II: Combinatorial designs. *Proc. 8th S.E. Conf. Combinatorics, Graph Theory and Computing*, U.M.P.I., Winnipeg (to appear).

[6] S.H.Y. Hung and N.S. Mendelsohn, Directed triple systems. *J. Combinatorial Theory*, 14A (1973) 310-318.

[7] N.S. Mendelsohn, A natural generalization of Steiner triple systems. *Computers in Number Theory*, Academic Press, New York (1971) 323-338.

[8] A. Rosa and C. Huang, Another class of balanced graph designs: Balanced circuit designs. *Discrete Math.* 12 (1975) 269-293.

[9] J. Schönheim, Partition of the edges of the directed complete graph into 4-cycles. *Discrete Math.* 11 (1975) 67-70.

05B40, 52A45

# BRICK PACKING

D. A. Holton and J. A. Rickard
Department of Mathematics,
University of Melbourne,
Parkville, Victoria, 3052
Australia

ABSTRACT

Under certain circumstances it is possible to fit rectangles of size $m \times n$ into a larger rectangle of size $p \times q$ so that they fit exactly. When this is not the case the minimum wastage should be determined. A number of results are in the literature. We discuss the case where $m = 2$. The terms n, p, q are, of course, natural numbers.

## 1. INTRODUCTION

In a number of practical situations, it is desirable to cut a given rectangle into smaller rectangles, all of the same size, so that the cuts are made parallel to the edges of the larger, given rectangle. As this cannot always be done exactly, it is of some economic (as well as theoretic) interest to determine the minimum amount of the larger rectangle left after the smaller rectangles have been removed. These smaller rectangles are often referred to as *bricks* (in view of an obvious three-dimensional variant of the problem).

Suppose the rectangle has sides of length p, q and the bricks have sides of length m, n. Let the minimum amount of the $p \times q$ rectangle left after removing the bricks be called the *wastage*, $W_m$. Barnett and Kynch [1] have shown that

$$W_1 = \begin{cases} \alpha\beta & \text{if } \alpha + \beta \le n \\ (n-\alpha)(n-\beta) & \text{if } \alpha + \beta \ge n \end{cases},$$

where $p = kn + \alpha$, $q = \ell n + \beta$ and $k, \ell \ge 1$, $0 \le \alpha < n$, $0 \le \beta < n$. (Throughout we will assume that p, q, k, $\ell$, $\alpha$, $\beta$ are of this form.)

In this paper we discuss the case for $W_2$. Since Brualdi and Foregger [2] have already determined $W_m$ implicitly for harmonic bricks (where $m|n$), we only consider $W_2$ for n odd. We will show that for $p, q \ge 3n$, then

$$W_2 = \begin{cases} W_1 & \text{if } pq - W_1 \text{ is even} \\ W_1 + n & \text{if } pq - W_1 \text{ is odd} \end{cases}.$$

In other words, the wastage for $p, q \ge 3n$ is as small as can be expected given the result for $W_1$.

Now if p, q are both odd, then pq is odd. Hence $W_2$ must be odd. So $W_2$ here can never be smaller than $W_1$ if $W_1$ is odd and $W_1 + n$ if $W_1$ is even. Similarly, if one or both of p, q is even, then pq is even and $W_2$ is even. Then $W_2$ can never be smaller than $W_1$ if $W_1$ is even or $W_1 + n$ if $W_1$ is odd. Combining these results we see that

$$W_2 \geq \begin{cases} W_1 & \text{if } pq - W_1 \text{ is even} \\ W_1 + n & \text{if } pq - W_1 \text{ is odd} \end{cases}.$$

We show by the constructions of the following sections that the inequality reduces to equality for $p, q \geq 3n$.

## 2. p, q SMALL

In this section we establish the results of Table 1 via the constructions $T_1, T_2, \ldots, T_{12}$.

For instance, if $p = n + \alpha$, $q = n + \beta$ where $\alpha$ and $\beta$ are even and $\alpha + \beta < n$, then $W_1 = \alpha\beta$. We know that $W_2 \geq \alpha\beta + n$. The construction $T_1$ shows that a wastage of $\alpha\beta + n$ can in fact be achieved.

The constructions $T_1, T_2, \ldots, T_{12}$ are shown in Figure 2.1. We note that these constructions are, in general, not unique.

It should be pointed out at this stage that $W_2$ is not always equal to $W_1$ or $W_1 + n$. For example, if $p = q = 19$ and $n = 7$, then $W_2 = W_1 + 3n$. There are an infinite number of such examples with $p,q < 3n$. The proof of this and similar results will be given in a subsequent paper. For our present purposes though, we are able to avoid such cases.

## 3. p, q LARGE

If successive removal of bricks from a rectangle A yields a smaller rectangle A' we say we have performed a *reduction* of A. In this section we give four reductions, $R_1, R_2, R_3, R_4$ and show that these reductions enable us to reduce any rectangle with sides $p, q \geq 3n$ to a rectangle in Table 1. We will thus prove the main result of this paper.

$\underline{R_1}$: If p is even, then we can reduce the $p \times q$ rectangle to a $p \times (q - n)$ rectangle via the reduction of Figure 3.1.

$\underline{R_2}$: If $p, q > 2n$, then we can reduce the $p \times q$ rectangle to either a $(p - 2n) \times q$ rectangle or a $p \times (q - 2n)$ rectangle. This reduction is shown for q even and odd in Figure 3.2 (a) and (b), respectively, where the $p \times q$ rectangle is reduced to a $(p - 2n) \times q$ rectangle. A similar method achieves the $p \times (q - 2n)$ reduction.

$R_3$: If $p, q > 2n$, then we can reduce the $p \times q$ rectangle to a $(p-2n) \times (q-2n)$

| p | q | k | ℓ | α | β | α + β | $W_1$ | $W_2$ | Construction |
|---|---|---|---|---|---|---|---|---|---|
| Odd | Odd | 1 | 1 | Even | Even | < n | αβ | αβ + n | $T_1$ |
| " | " | " | " | " | " | > n | (n−α)(n−β) | (n−α)(n−β) | $T_2$ |
| " | " | 1 | 2 | Even | Odd | ≤ n | αβ | αβ + n | $T_3$ |
| " | " | " | " | " | " | ≥ n | (n−α)(n−β) | (n−α)(n−β) + n | $T_4$ |
| Even | Odd | 1 | 1 | Odd | Even | ≤ n | αβ | αβ | $T_5$ |
| " | " | 2 | 1 | Even | Even | < n | αβ | αβ | $T_6$ |
| " | " | " | " | " | " | > n | (n−α)(n−β) | (n−α)(n−β) + n | $T_7$ |
| Even | Even | 1 | 1 | Odd | Odd | < n | αβ | αβ + n | $T_8$ |
| " | " | 2 | 2 | Even | Even | < n | αβ | αβ | $T_9$ |
| " | " | " | " | " | " | > n | (n−α)(n−β) | (n−α)(n−β) + n | $T_{10}$ |
| Even | Odd | 3 | 3 | Odd | Even | ≥ n | (n−α)(n−β) | (n−α)(n−β) | $T_{11}$ |
| Even | Even | 3 | 3 | Odd | Odd | > n | (n−α)(n−β) | (n−α)(n−β) | $T_{12}$ |

Table 1

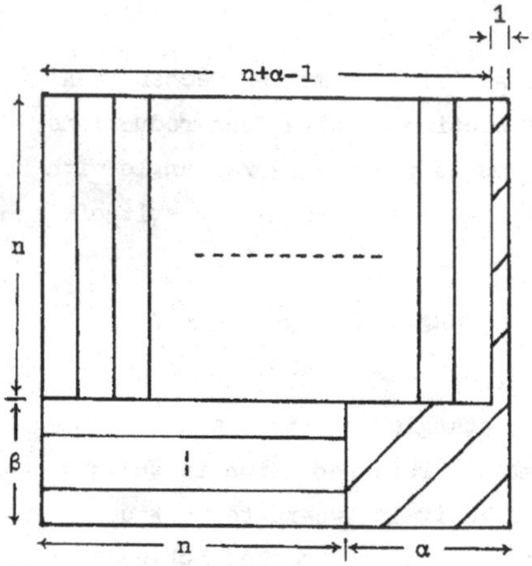


$T_1$

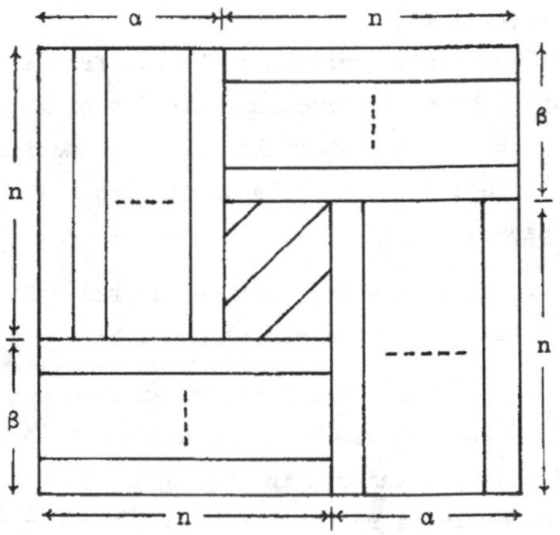


$T_2$

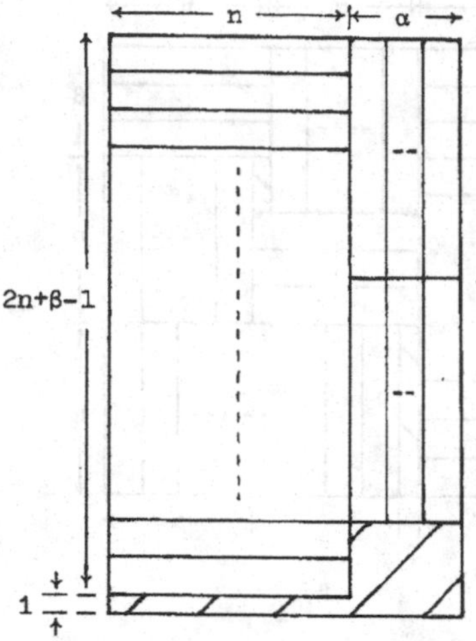


$T_3$

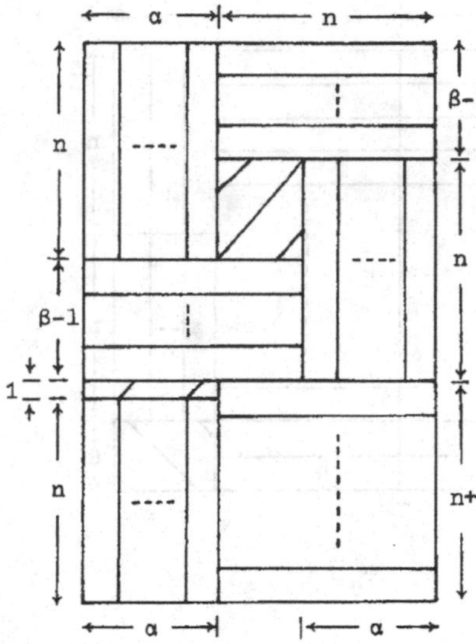


$T_4$

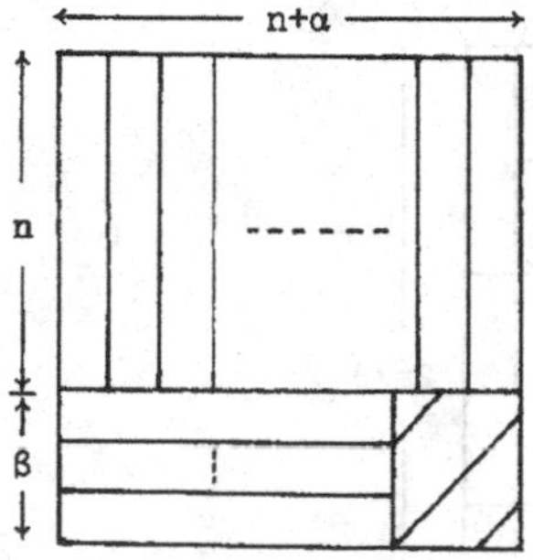


$T_5$

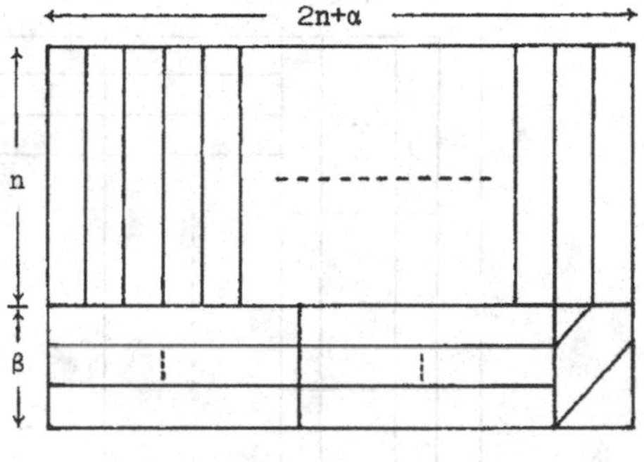


$T_6$

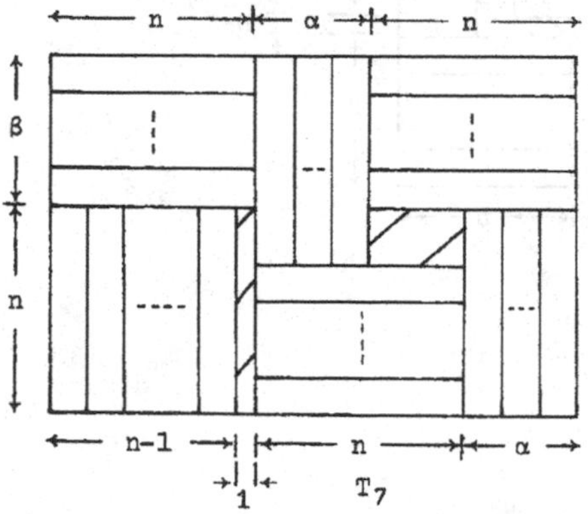


$T_7$

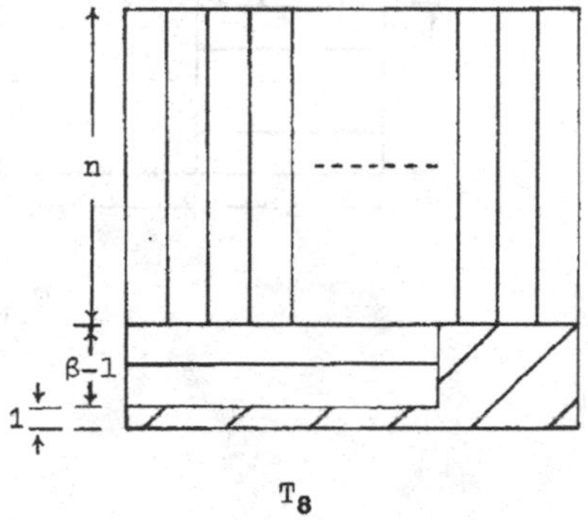


$T_8$

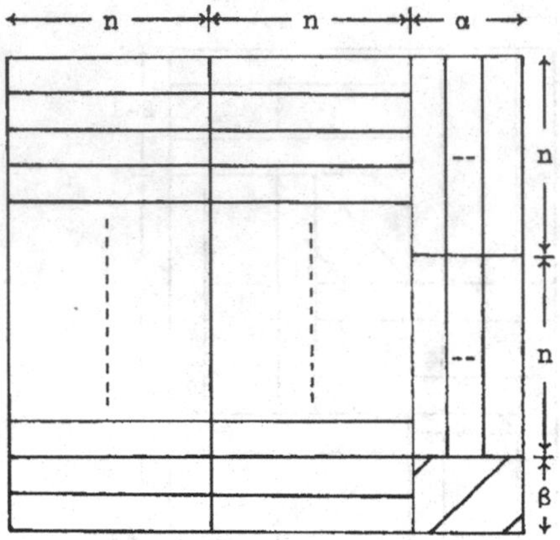


$T_9$

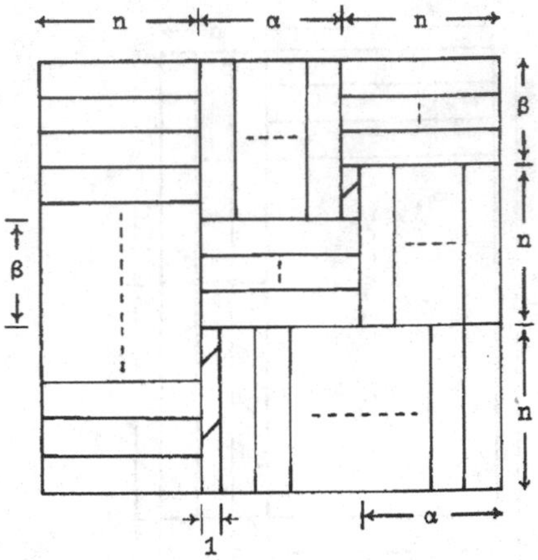


$T_{10}$

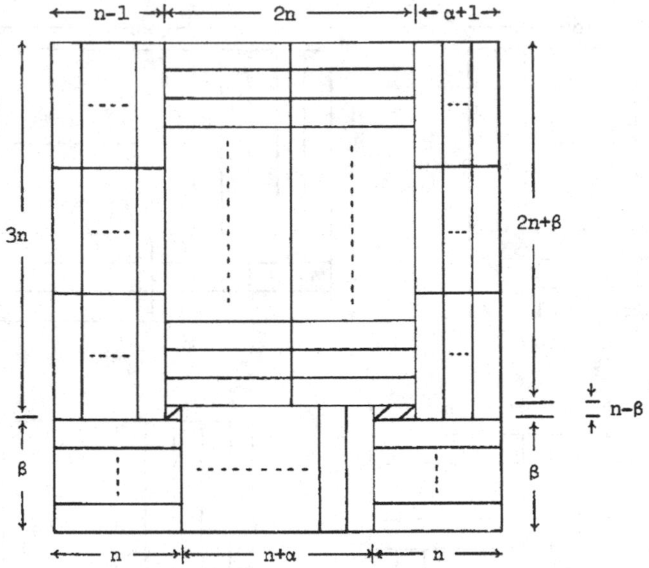


$T_{11}$

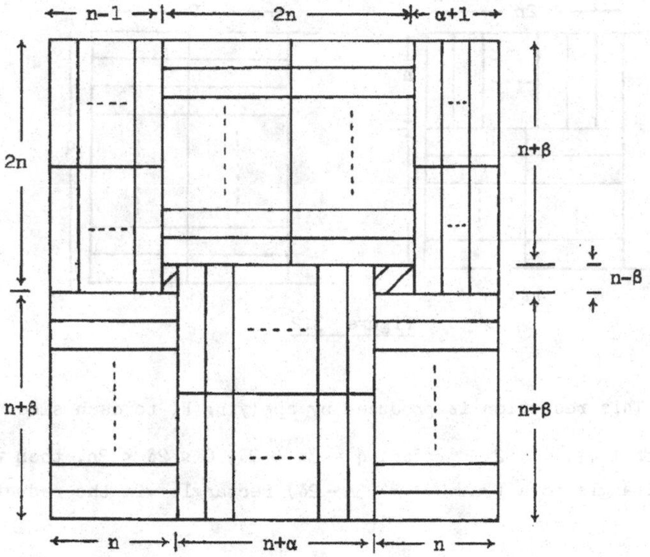


$T_{12}$

<u>Figure 2.1</u>

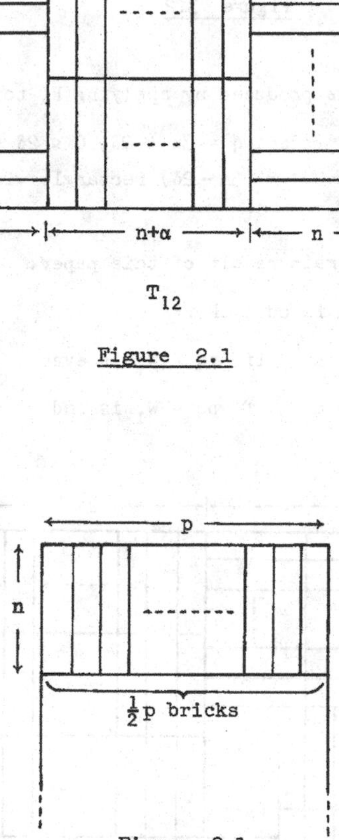


<u>Figure 3.1</u>

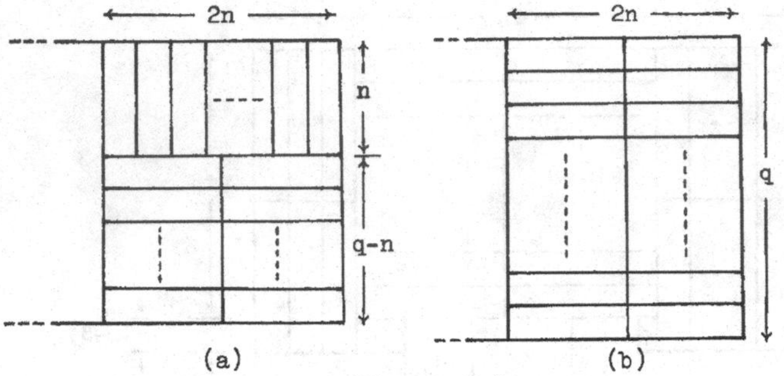


Figure 3.2

rectangle. This reduction is produced by applying $R_2$ to each side.

$R_4$: If $p = 3n + 2\gamma$, $0 < 2\gamma < 3n$ and $q = 3n + 2\delta$, $0 < 2\delta < 3n$, then we can reduce the $p \times q$ rectangle to a $(3n - 2\gamma) \times (3n - 2\delta)$ rectangle via the reduction of Figure 3.3.

We are now able to prove the main result of this paper.

THEOREM 3.1. If $p, q \geq 3n$ and $n$ is odd, then

$$W_2 = \begin{cases} W_1 & \text{if } pq - W_1 \text{ is even} \\ W_1 + n & \text{if } pq - W_1 \text{ is odd} \end{cases}.$$

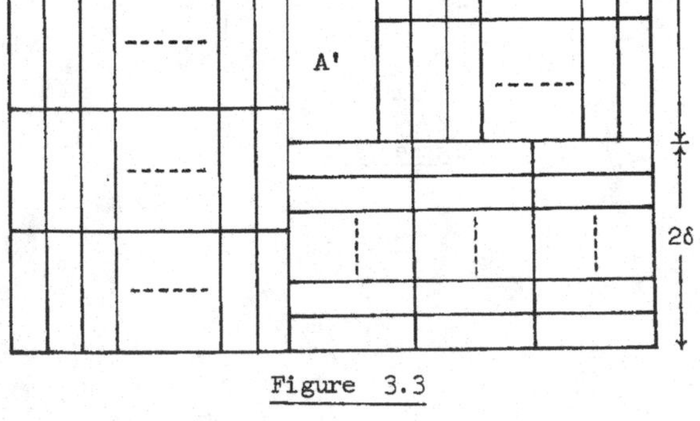

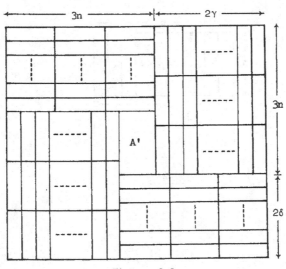

Figure 3.3

Proof: *Case 1*: p, q both odd.

*1.1*: If $3n \le p, q < 4n$, then $R_3$ reduces the $p \times q = (3n+\alpha) \times (3n+\beta)$ rectangle to an $(n+\alpha) \times (n+\beta)$ rectangle. The result now follows via $T_1$, if $\alpha + \beta < n$, or $T_2$, if $\alpha + \beta > n$.

*1.2*: If $p = 3n + \alpha$, $q = 4n + \beta$, then $R_3$ reduces the rectangle to $(n+\alpha) \times (2n+\beta)$ The result now follows via $T_3$ or $T_4$.

*1.3*: If $p = 4n + \alpha$, $q = 3n + \beta$, then proceed as in Case 1.2.

*1.4*: If $p = 4n + \alpha$, $q = 4n + \beta$, then $R_4$ reduces the rectangle to $(2n-\alpha) \times (2n-\beta) = \{n + (n-\alpha)\} \times \{n + (n-\beta)\}$. The result now follows by $T_1$ or $T_2$.

*1.5*: If $p \ge 5n$ and $3n \le q < 5n$, then repeated application of $R_2$ reduces the rectangle to one of the cases already considered above.

*1.6*: If $3n \le p < 5n$ and $q \ge 5n$, then proceed as in Case 1.5.

*1.7*: If $p, q \ge 5n$, then repeated application of $R_2$ and/or $R_3$ reduces the rectangle to one of the cases above.

*Case 2*: One of p, q odd, the other even. Without loss of generality, we assume p even, q odd.

*2.1*: If $p = 3n + \alpha$, $q = 3n + \beta$, then if $\alpha + \beta \ge n$ the result follows by $T_{11}$, while if $\alpha + \beta \le n$ the result follows by $R_3$ followed by $T_5$. Note that these alternatives give the same wastage if $\alpha + \beta = n$.

*2.2*: If $p = 4n + \alpha$, $q = 3n + \beta$, then the result follows via $R_3$ followed by $T_6$ or $T_7$.

*2.3*: If $p = 3n + \alpha$, $q = 4n + \beta$, then $R_1$ reduces the rectangle to $(3n+\alpha) \times (3n+\beta)$. If $\alpha + \beta > n$, then the result follows by $T_{12}$, while if $\alpha + \beta < n$, then $R_3$ followed by $T_8$ suffices.

*2.4*: If $p = 4n + \alpha$, $q = 4n + \beta$, then $R_1$ reduces the rectangle to $(4n+\alpha) \times (3n+\beta)$ with $\alpha$ even and $\beta$ odd. Applying $R_1$ again we get $(3n+\alpha) \times (3n+\beta)$ from which the result follows by Case 2.1.

*2.5*: If $p > 5n$, $3n \le q < 5n$, then repeated applications of $R_2$ reduce this case to one of the previous subcases of Case 2.

*2.6*: If $3n < p < 5n$, $q \ge 5n$, the result follows by the argument of Case 2.5.

*2.7*: If $p > 5n$, $q \ge 5n$, repeated application of $R_2$ and/or $R_3$ reduces the rectangle to one of the other subcases of Case 2.

*Case 3*: p, q both even.

*3.1*: If $p = 3n + \alpha$, $q = 3n + \beta$, then the result follows by the argument of Case 2.3.

*3.2*: If $p = 4n + \alpha$, $q = 3n + \beta$, then $R_1$ reduces the rectangle to $(3n+\alpha) \times (3n+\beta)$ with $3n + \alpha$ odd and $3n + \beta$ even. The result follows by Case 2.1.

*3.3*: If $p = 3n + \alpha$, $q = 4n + \beta$, then proceed as in Case 3.2.

*3.4*: If $p = 4n + \alpha$, $q = 4n + \beta$, then $R_3$ reduces the rectangle to $(2n+\alpha) \times (2n+\beta)$ and the result follows by $T_9$ or $T_{10}$.

*3.5*: If $p > 5n$, $3n < q < 5n$, then repeated application of $R_2$ reduces the rectangle to one of the subcases of Case 3 already considered.

*3.6*: If $3n < p < 5n$, $q > 5n$, then proceed as in Case 3.5.

*3.7*: If $p > 5n$, $q > 5n$, then repeated application of $R_2$ and/or $R_3$, reduces this case to one of the cases above.

We complete the $2 \times n$ case by deducing $W_2$ for harmonic bricks from Theorem 2.2 of [2].

THEOREM 3.2. For $p, q \geq 3n$, n even,

$$W_2 = \alpha\beta + (q - \beta)(\alpha - 2\left[\tfrac{\alpha}{2}\right]) + (p - \alpha)(\beta - 2\left[\tfrac{\beta}{2}\right])$$

$$- 2n \max\left(0, \left[\tfrac{\alpha}{2}\right] + \left[\tfrac{\beta}{2}\right] - \tfrac{n}{2}\right).$$

We note that for $\alpha$, $\beta$ even we get

$$W_2 = \begin{cases} \alpha\beta & \text{if } \alpha + \beta \leq n \\ (n - \alpha)(n - \beta) & \text{if } \alpha + \beta \geq n \end{cases}.$$

Note that this result was to be expected from Barnett and Kynch's [1] result, since the problem of fitting $2 \times 2n$ bricks into a $2p \times 2q$ box is clearly similar to that of packing $1 \times n$ bricks into a $p \times q$ box. In fact, the wastage in the former case is four times that in the latter. However for one or both of $\alpha$, $\beta$ odd the excess of $W_2$ above $W_1$ is, in general, much greater than n. This is due to the fact that with n even, no odd number can be expressed in the form $2s + nt$.

REFERENCES

[1] S. Barnett and G. J. Kynch, "Solution of a simple cutting problem", *Operations Research* 15 (1967), 1051-1056.

[2] Richard A. Brualdi and Thomas H. Foregger, "Packing boxes with harmonic bricks", *J. Comb. Th.* (B) 17 (1974), 81-114.

## COLOUR SYMMETRY IN CRYSTALLOGRAPHIC SPACE GROUPS

R. Hubbard

Department of Mathematics and Computer Science
Queensland Institute of Technology
P.O. Box 246, North Quay, Queensland 4000
Australia

20B25, 20H15, 50B30, 05B45

ABSTRACT. The feasibility of applying the restricted permutation representation method of Macdonald and Street to colouring the fundamental regions of 3-dimensional crystallographic groups is discussed. The tetragonal crystal class is used to illustrate a classification of colourings.

1. CONSTRUCTION OF SPACE GROUPS. Space groups have been constructed and classified in many ways using many different notations since the pioneering work of Bravais, 1850. Detailed classifications can be found in Buerger (2), Burckhardt (3), Hilton (4).

A space group is a 3-dimensional lattice into the cells of which symmetrical groups of objects such as molecules are introduced. The largest group of symmetries consistent with a given lattice and the objects in its cells is then a space group. There are 6 basic lattice types depending on the lengths and orientations of the three generators of minimum length of the lattice points. If the generating vectors are as shown in Figure 1, the 6 lattice types are given by the following conditions on $\underline{a}$, $\underline{b}$, $\underline{c}$, $\alpha$, $\beta$, $\gamma$.

| | | |
|---|---|---|
| Triclinic | $\lvert\underline{a}\rvert \neq \lvert\underline{b}\rvert \neq \lvert\underline{c}\rvert$ | $\alpha \neq \beta \neq \gamma$ |
| Monoclinic | $\lvert\underline{a}\rvert \neq \lvert\underline{b}\rvert \neq \lvert\underline{c}\rvert$ | $\alpha = \beta = \frac{\pi}{2}$ |
| Orthorhombic | $\lvert\underline{a}\rvert \neq \lvert\underline{b}\rvert \neq \lvert\underline{c}\rvert$ | $\alpha = \beta = \gamma = \frac{\pi}{2}$ |
| Tetragonal | $\lvert\underline{a}\rvert = \lvert\underline{b}\rvert$ | $\alpha = \beta = \gamma = \frac{\pi}{2}$ |

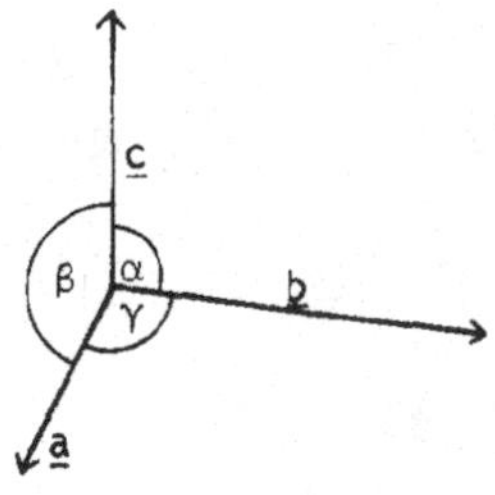

Hexagonal $|\underline{a}| = |\underline{b}|$ $\alpha = \beta = \frac{\pi}{2},\ \gamma = \frac{\pi}{3}$

Cubic $|\underline{a}| = |\underline{b}| = |\underline{c}|$ $\alpha = \beta = \gamma = \frac{\pi}{2}$

If a lattice cell generated as above contains no interior points it is called *primitive*, P, and the lattice points can be denoted by the set

$T = \{\underline{a}^{\lambda}\ \underline{b}^{\mu}\ \underline{c}^{\nu} | \lambda,\mu,\nu,\ \text{integers}\}$.

In the 32 crystallographic groups of point symmetries the generating rotations are restricted to certain combinations of the angles $2\pi,\ \pi,\ \frac{2\pi}{3},\ \frac{\pi}{2},\ \frac{\pi}{3}$. These 32 point groups can be combined with the appropriate lattices to form 230 three-dimensional space groups.

2. THE TETRAGONAL CRYSTAL CLASS. There are just 7 point groups which induce automorphisoms of the tetragonal lattice, but not of lattices of lower symmetry.

$4,\ \bar{4},\ \frac{4}{m},\ \bar{4}m2,\ 4mm,\ 422$ and $\frac{4}{m}\frac{2}{m}\frac{2}{m}$.

In each case the symbols denote the generators of the group, 4 denotes a 4-fold rotation, 2 a 2 - fold rotation, (with axis perpendicular to the 4-fold one) m in the numerator denotes a reflection in a plane through the axis of the 4-fold rotation, m in the denominator a reflection in a plane perpendicular to the axis of the 4-fold rotation. $\bar{4}$ is generated by a roto-reflection, i.e., a 4-fold rotation combined with a reflection in a plane perpendicular to its axis. Each of these point groups is consistent with both the primitive and the body - centred tetragonal lattice, giving the space groups $P4,\ \bar{P}4, \ldots, P\frac{4}{m}\frac{2}{m}\frac{2}{m},\ I4,\ I\bar{4}, \ldots$ $I\frac{4}{m}\frac{2}{m}\frac{2}{m}$. However this does not exhaust all the possibilities. There exist two further composite symmetry operations, screws and glides. A screw is a combination of a rotation and a translation parallel to the axis of rotation. If C is a 4-fold rotation with axis parallel to the lattice generator $\underline{c}$ and A is a 2-fold rotation with axis parallel to lattice generator $\underline{a}$ then the screws consistent with the tetragonal class are $\underline{c}^{1/4}C,\ \underline{c}^{1/2}C,\ \underline{c}^{3/4}C,\ \underline{a}^{1/2}\underline{b}^{1/2}A$. These appear in space group names as $4_1,\ 4_2,\ 4_3,\ 2_1$ respectively. A glide is a combination of a reflection and a translation parallel to the reflection plane and is indicated in the name of a space group by replacing the letter m by a, b, c or n depending on the direction of the translation part of the glide with respect to the lattice generators.

3. COLOURING SPACE GROUPS. Two-coloured space groups have appeared in Russian journals under the title *"Shubnikov Groups"* (1). Greater than two-colourings have not appeared. The only colourings considered in the sequel are those in which the number of colours is restricted by the requirement that rotations with axes which can be obtained from each other by a translation of the group have the same effect on the colours.

Let $G$ be a space group, $T$ the set of translations in $G$, and $H$ the set of rotations in $G$. As in the case of friezes and plane groups (7) the above restriction means that the subgroup $K$ of $G$ whose cosets are to provide the permutation representations must contain the commutator subgroup $[H,T]$. In those cases where $\frac{G}{[H,T]}$ is a finite group we will obtain all restricted colourings. All space groups for which $\frac{G}{[H,T]}$ is still an infinite group can be embedded as subgroups of the restricted colour groups. Since a translation parallel to the axis of a rotation commutes with that rotation, we will obtain a finite factor group $\bar{G} = \frac{G}{[H,T]}$ just when $H$ has two generators which are rotations. The group $T$ of translations and the commutator group $[H,T]$ are shown in Table I for those crystal classes which have two rotations as generators. The groups $[H,T]$ follow from those for the corresponding plane groups given by Macdonald and Street in (7).

| Crystal Class | [H,T] | T |
|---|---|---|
| $222,\ \frac{2}{m}\frac{2}{m}\frac{2}{m}$ | $\langle 2\underline{a}, 2\underline{b}\ 2\underline{c} \rangle$ | $\langle \underline{a},\underline{b},\underline{c} \rangle$ |
| $422,\ \bar{4}2m,\ \frac{4}{m}\frac{2}{m}\frac{2}{m}$ | $\langle 2\underline{a}, 2\underline{c} \rangle$ | $\langle \underline{a},\underline{c} \rangle$ |
| $32,\ \bar{3}\frac{2}{m}$ | $\langle 3\underline{a}, 2\underline{c} \rangle$ | $\langle \underline{a},\underline{c} \rangle$ |
| $622,\ \bar{6}m2,\ \frac{6}{m}\frac{2}{m}\frac{2}{m}$ | $\langle \underline{a}, 2\underline{c} \rangle$ | $\langle \underline{a},\underline{c} \rangle$ |
| $23,\ 432,\ \frac{2}{m}\bar{3}$; $\bar{4}3m,\ \frac{4}{m}\bar{3}\frac{2}{m}$ | $\langle \underline{a} \rangle$ | $\langle \underline{a} \rangle$ |

Table I.

4. COLOURING TETRAGONAL GROUPS. This class is considered in more detail. All primitive tetragonal space groups are subgroups of groups in the class $\frac{4}{m}\frac{2}{m}\frac{2}{m}$

except $P4_122$, $P4_322$, $P4_1$ and $P4_3$. These last are equivalent in pairs, only the directions of the screws being reversed. Generators and relations for all primitive space groups in the class $\frac{4}{m}\,\frac{2}{m}\,\frac{2}{m}$ and for $P4_122$ and $P4_12_12$ appear in Appendix 1. Many of these groups are equivalent as abstract groups but since the generators represent different symmetries they could produce some different patterns. It is clearly impossible to classify all these colourings here so we give an example in which we can demonstrate the embedding of several tetragonal groups in a group containing the maximum number of symmetries.

Using the factor group $\bar{G}$ in Appendix 1 for $P\frac{4}{m}\frac{2}{m}\frac{2}{m}$ we can select the normal subgroup $\bar{K} = \langle \gamma a, \alpha, \mu \rangle$ of index 4 in $\bar{G}$ which will give a unique pattern. The cosets of $\bar{K}$ are $\bar{K}$, $\bar{K}a$, $\bar{K}c$, $\bar{K}ac$ and the generators of $\bar{G}$ will have the permutations representations $\alpha \to (1)$, $\mu \to (1)$, $\gamma \to (12)(34)$, $a \to (12)(34)$, $c \to (13)(24)$ on these cosets. Let $S$ be any tetragonal subgroup of $P\frac{4}{m}\frac{2}{m}\frac{2}{m}$ and $\bar{S} = \dfrac{S}{\langle 2\underline{a}, 2\underline{c} \rangle}$. The generators of the $S$ and $\bar{S}$ are given in Table 2 together with suitable subgroups $\bar{R} < \bar{S}$ such that the cosets of $\bar{R}$ in $\bar{S}$ correspond to the cosets of $\bar{K}$ in $\bar{G}$. Note also that in each case $\bar{K}\bar{S} = \bar{G}$, so that the permutation representation will be transitive on the fundamental regions.

| S | $\bar{S}$ | $\bar{R}$ |
|---|---|---|
| $P422 = \langle C, A, \underline{a}, c \rangle$ | $\langle \gamma, \alpha, a, c \rangle$ | $\langle \gamma a, \alpha \rangle$ |
| $P\bar{4}m2 = \langle CJ, AJ, \underline{a}, \underline{c} \rangle$ | $\langle \gamma\mu, \alpha\mu, a, c \rangle$ | $\langle \gamma\mu a, \alpha\mu \rangle$ |
| $P\bar{4}2m = \langle CJ, A, \underline{a}, \underline{c} \rangle$ | $\langle \gamma\mu, \alpha, a, c \rangle$ | $\langle \gamma\mu a, \alpha \rangle$ |
| $P4mm = \langle C, AJ, \underline{a}, \underline{c} \rangle$ | $\langle \gamma, \alpha\mu, a, c \rangle$ | $\langle \gamma a, \alpha\mu \rangle$ |
| $P\frac{4}{m} = \langle C, C^2J, \underline{a}, \underline{c} \rangle$ | $\langle \gamma, \gamma^2\mu, a, c \rangle$ | $\langle \gamma a, \gamma^2\mu \rangle$ |
| $P\bar{4} = \langle CJ, \underline{a}, \underline{c} \rangle$ | $\langle \gamma\mu, a, c \rangle$ | $\langle \gamma\mu a \rangle$ |
| $P4 = \langle C, \underline{a}, \underline{c} \rangle$ | $\langle \gamma, a, c \rangle$ | $\langle \gamma a \rangle$ |

Table 2.

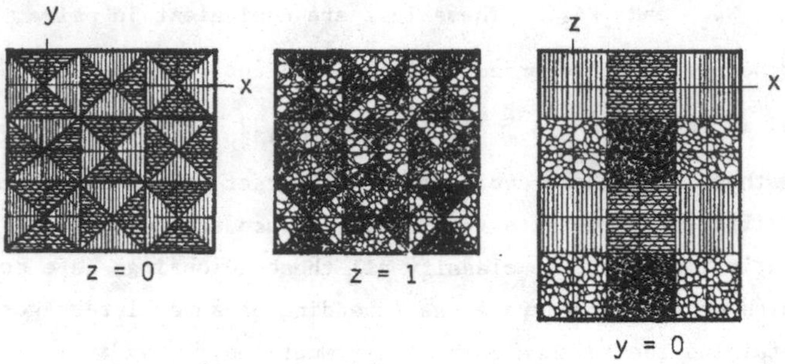


FIGURE 2.

The colouring in Figure 2 satisfies all the groups in Table 2 and $P\frac{4}{m}\frac{2}{m}\frac{2}{m}$ for the fundamental regions in Table 3.

| | |
|---|---|
| $P\frac{4}{m}\frac{2}{m}\frac{2}{m}$ | $m = \frac{a}{2}$, $n = \frac{c}{2}$ |
| P422 | $m = \frac{a}{2}$, $n = c$ or $r = a$, $n = \frac{c}{2}$ |
| P$\bar{4}$m2 | $m = \frac{a}{2}$, $n = c$ or $r = a$, $n = \frac{c}{2}$ |
| P$\bar{4}$2m | $m = \frac{a}{2}$, $n = c$ or $r = a$, $n = \frac{c}{2}$ |
| P4mm | $m = \frac{a}{2}$, $n = c$ or $r = a$, $n = \frac{c}{2}$ |
| $P\frac{4}{m}$ | $r = a$, $n = \frac{c}{2}$ |
| P$\bar{4}$ | $r = a$, $n = c$ |
| P4 | $r = a$, $n = c$ |

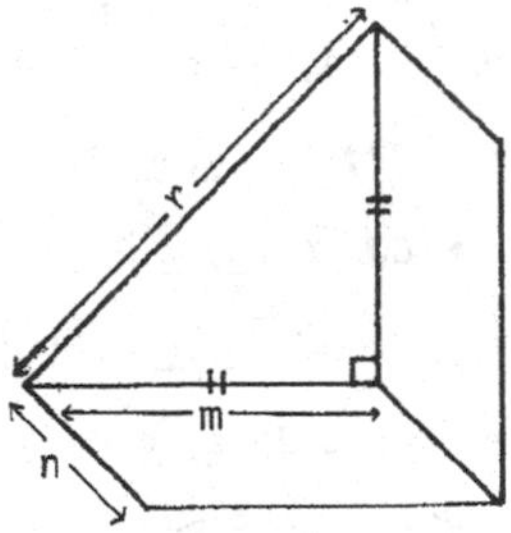


TABLE 3.

5. COLOURING CUBIC SPACE GROUPS. From Table 1. we see that for all cubic space groups $[H,T] = \langle \underline{a} \rangle$ i.e., when the space group contains no screws or glides the factor group $\bar{G}$ is isomorphic to the point group of G. Thus the cubic group with the maximum number of symmetries with no screws or glides $P\frac{4}{m}\bar{3}\frac{2}{m}$ has the factor group $\bar{G} \simeq S_4 \times C_2$. The only permissible screws and glides for the class $\frac{4}{m}\bar{3}\frac{2}{m}$ are formed by combining the point group operations with combinations of $\underline{a}^{½}, \underline{b}^{½}, \underline{c}^{½}$. But in the factor group the images of the translation parts of these operations will be self-inverse and hence the factor groups for $P\frac{4}{n}\bar{3}\frac{2}{n}$, $P\frac{4_2}{m}\bar{3}\frac{2}{n}$ and $P\frac{4_2}{n}\bar{3}\frac{2}{m}$ will also be isomorphic to $S_4 \times C_2$.

The only primitive cubic space groups which are not subgroups of these four are $P4_132$ and $P4_332$, with screws containing quarter translations. However from $G = P4_132 = \langle \underline{a}^{¼}\,\underline{b}^{¼}\,\underline{c}^{-¼}\ C, D \rangle$ we obtain the factor group $\bar{G} = \langle \gamma,\delta \mid \gamma^4 = \delta^3 = (\gamma\delta)^2 = 1 \rangle$ which is isomorphic to $S_4$. Hence all primitive cubic space groups have factor groups isomorphic to cubic point groups.

## APPENDIX 1.

In the listing below, J denotes the central inversion symmetry so that $JC^2 = M$ a reflection in a plane perpendicular to the axis of the 4-fold rotation C.

$JA = N$ a reflection in a plane parallel to the axis of C.

In each mapping $G \to \bar{G}$ we have

$$\underline{c}^x C \to \gamma,\ \underline{a}^x\, b^x\, A \to \alpha,\ \underline{a}^x\, b^x\, c^y\, J \to \mu,$$

$$\underline{a} \to a,\ \underline{c} \to c.$$

1. $G = P\frac{4}{m}\frac{2}{m}\frac{2}{m} = \langle C, A, J, \underline{a}, \underline{c} \rangle$

$\bar{G} = \langle \gamma, \alpha, \mu, a, c \mid \gamma^4 = \alpha^2 = \mu^2 = a^2 = c^2 = (\gamma\alpha)^2 = [\gamma,\mu] = [\mu,a] = [\mu,c] =$

$= [\alpha,\mu] = [\alpha,a] = [\alpha,c] = [a,c] = [\mu,a] = [\mu,c] = 1 \rangle$

2. $G = P\frac{4}{m}\frac{2}{c}\frac{2}{c} = \langle C, A, \underline{c}^{1/2} J, \underline{a} \rangle$

$\bar{G} = \langle \gamma, \alpha, \mu, a \mid \gamma^4 = \alpha^2 = \mu^2 = a^2 = (\gamma\alpha)^2 = (\alpha\mu)^4 = [\gamma,\mu] = [\gamma,a] = [\alpha,a] =$

$= [\mu,a] = 1 \rangle$

3. $G = \frac{4}{n}\frac{2}{b}\frac{2}{m} = \langle C, A, \underline{a}^{1/2}\, \underline{b}^{1/2} J, \underline{c} \rangle$

$\bar{G} = \langle \gamma, \alpha, \mu, c \mid \gamma^4 = \alpha^2 = \mu^2 = c^2 = (\gamma\alpha)^2 = (\alpha\mu)^4 = [\gamma^2,\mu] = [(\alpha\mu)^2,\alpha] =$

$= [(\alpha\mu)^2,\gamma] = [(\alpha\mu)^2,\mu] = [\gamma,c] = [\alpha,c] = [\mu,c] = 1,\ (\gamma\mu)^2 = (\alpha\mu)^2\gamma^2 \rangle$

4. $G = P\frac{4}{n}\frac{2}{n}\frac{2}{c} = \langle C, A, \underline{a}^{1/2}\, \underline{b}^{1/2}\, \underline{c}^{1/2} J \rangle$

$\bar{G} = \langle \gamma,\alpha,\mu \mid \gamma^4 = \alpha^2 = \mu^2 = (\gamma\alpha)^2 = (\alpha\mu)^4 = (\gamma\mu)^4 = [(\alpha\mu)^2,\gamma] = [(\alpha\mu)^2,\alpha] =$

$= [(\alpha\mu)^2,\mu] = [(\gamma\mu)^2,\gamma] = [(\gamma\mu)^2,\alpha] = [(\gamma\mu)^2,\mu] = [\gamma^2,\mu] = 1 \rangle$

5. $G = P\frac{4}{m}\frac{2_1}{b}\frac{2}{m} = \langle C, \underline{a}^{1/2}\, \underline{b}^{1/2} A, J, \underline{c} \rangle$

$\bar{G} = \langle \gamma, \alpha, \mu, c \mid \gamma^4 = \alpha^4 = \mu^2 = c^2 = (\gamma\alpha)^2 = [\gamma,\mu] =$

$= [\alpha^2,\mu] = [(\alpha\mu)^2,\mu] = [(\alpha\mu)^2,\gamma] = [\gamma,c] = [\alpha,c] = [\mu,c] =$

$= [\alpha^2,\gamma] = 1,\ \alpha^2(\alpha\mu)^2 = \gamma^2 \rangle$

6. $G = P\frac{4}{m}\frac{2_1}{n}\frac{2}{c} = \langle C, \underline{a}^{1/2}\, \underline{b}^{1/2} A, \underline{c}^{1/2} J \rangle$

$\bar{G} = \langle \gamma, \alpha, \mu \mid \gamma^4 = \alpha^4 = \mu^2 = (\gamma\alpha)^2 = (\alpha\mu)^4 = [\gamma,\mu] = [\alpha^2,\mu] =$

$= [(\alpha\mu)^2,\gamma] = [(\alpha\mu)^2,\alpha] = [\alpha^2,\gamma] = 1,\ (\gamma\mu)^2 = \alpha^2(\alpha\mu)^2 \rangle$

7. $G = P\,\frac{4}{n}\,\frac{21}{m}\,\frac{2}{m} = \langle C,\ \underline{a}^{1/2}\,\underline{b}^{1/2}\,A,\ \underline{a}^{1/2}\,\underline{b}^{1/2}\,J,\ \underline{c} \rangle$

$\bar{G} = \langle \gamma, \alpha, \mu, c \mid \gamma^4 = \alpha^4 = \mu^2 = c^2 = (\gamma\alpha)^2 = (\alpha\mu)^2 = [\alpha^2,\gamma] = [\alpha^2,\mu] =$

$= [\gamma^2,\mu] = [\gamma,c] = [\alpha,c] = [\mu,c] = 1,\ \gamma^2\alpha^2 = (\gamma\mu)^2 \rangle$

8. $G = P\,\frac{4_2}{m}\,\frac{2}{m}\,\frac{2}{c} = \langle \underline{c}^{1/2}\,C,\ A,\ J,\ \underline{a} \rangle$

$\bar{G} = \langle \gamma, \alpha, \mu, a \mid \gamma^4 = \alpha^2 = \mu^2 = a^2 = (\gamma\alpha)^2 = (\gamma\mu)^4 = [\alpha,\mu]$

$= [(\gamma\mu)^2,\gamma] = [(\gamma\mu)^2,\alpha] = [(\gamma\mu)^2,\mu] = [\gamma,a] = [\alpha,a] = [\mu,a] = 1 \rangle$

9. $G = P\,\frac{4_2}{m}\,\frac{2}{c}\,\frac{2}{m} = \langle \underline{c}^{1/2}\,C,\ A,\ \underline{c}^{1/2}\,J,\ \underline{a} \rangle$

$\bar{G} = \langle \gamma, \alpha, \mu, a \mid \gamma^4 = \alpha^2 = \mu^2 = a^2 = (\gamma\alpha)^2 = (\alpha\mu)^4$

$= [(\alpha\mu)^2,\gamma] = [(\alpha\mu)^2,\mu] = [(\alpha\mu)^2,\alpha] = [\gamma^2,\mu] = [\gamma,a]$

$= [\alpha,a] = [\mu,a] = 1,\ \gamma^2(\alpha\mu)^2 = (\gamma\mu)^2 \rangle$

10. $G = P\,\frac{4}{n}\,\frac{2_1}{c}\,\frac{2}{c} = \langle C,\ \underline{a}^{1/2}\,\underline{b}^{1/2}\,A,\ \underline{a}^{1/2}\,\underline{b}^{1/2}\,\underline{c}^{1/2}\,J \rangle$

$\bar{G} = \langle \gamma, \alpha, \mu \mid \gamma^4 = \alpha^4 = \mu^2 = (\gamma\alpha)^2 = (\alpha\mu)^4 = (\gamma\mu)^4 = [(\alpha\mu)^2,\gamma] = [(\alpha\mu)^2,\alpha] =$

$= [(\alpha\mu)^2,\mu] = [\alpha^2,\gamma] = [\gamma^2,\mu] = 1 \rangle$

11. $G = P\,\frac{4_2}{n}\,\frac{2}{b}\,\frac{2}{c} = \langle \underline{c}^{1/2}\,C,\ A,\ \underline{a}^{1/2}\,\underline{b}^{1/2}\,J \rangle$

$\bar{G} = \langle \gamma, \alpha, \mu \mid \gamma^4 = \alpha^2 = \mu^2 = (\gamma\alpha)^2 = (\alpha\mu)^4 = (\gamma\mu)^4$

$= [(\alpha\mu)^2,\gamma] = [(\alpha\mu)^2,\alpha] = [(\alpha\mu)^2,\mu] = [(\gamma\mu)^2,\gamma] = [\gamma^2,\mu] = 1 \rangle$

12. $G = P\,\frac{4_2}{n}\,\frac{2}{n}\,\frac{2}{m} = \langle \underline{c}^{1/2}\,C,\ A,\ \underline{a}^{1/2}\,\underline{b}^{1/2}\,\underline{c}^{1/2}\,J \rangle$

$\bar{G} = \langle \gamma, \alpha, \mu \mid \gamma^4 = \alpha^2 = \mu^2 = (\gamma\alpha)^2 = (\gamma\mu)^4 = (\alpha\mu)^4 =$

$= [(\gamma\mu)^2,\gamma] = [(\gamma\mu)^2,\alpha] = [(\gamma\mu)^2,\mu] = [(\alpha\mu)^2,\gamma] = [(\alpha\mu)^2,\alpha] = 1 \rangle$

13. $G = P\,\frac{4_2}{m}\,\frac{2_1}{b}\,\frac{2}{c} = \langle \underline{c}^{1/2}\,C,\ \underline{a}^{1/2}\,\underline{b}^{1/2}\,A,\ J \rangle$

$\bar{G} = \langle \gamma, \alpha, \mu \mid \gamma^4 = \alpha^4 = \mu^2 = (\gamma\alpha)^2 = (\gamma\mu)^4 = (\alpha\mu)^4 =$

$= [(\alpha\mu)^2,\gamma] = [(\alpha\mu)^2,\alpha] = [(\alpha\mu)^2,\mu] = [(\gamma\mu)^2,\alpha] = [\gamma^2,\mu] = [\alpha^2,\gamma] = 1 \rangle$

14. $G = P\,\frac{4_2}{m}\,\frac{2_1}{n}\,\frac{2}{m} = \langle \underline{c}^{1/2}\,C,\ \underline{a}^{1/2}\,\underline{b}^{1/2}\,A,\ \underline{c}^{1/2}\,J \rangle$

$\bar{G} = \langle \gamma, \alpha, \mu \mid \gamma^4 = \alpha^4 = \mu^2 = (\gamma\alpha)^2 = (\gamma\mu)^4 = (\alpha\mu)^4 =$

$= [(\alpha\mu)^2,\gamma] = [(\alpha\mu)^2,\alpha] = [(\alpha\mu)^2,\mu] = [\alpha^2,\gamma] = [(\gamma\mu)^2,\alpha] = [\gamma^2,\mu] = 1 \rangle$

15. $G = P\,\frac{4_2}{n}\,\frac{2_1}{c}\,\frac{2}{m} = \langle \underline{c}^{1/2}\, C,\ \underline{a}^{1/2}\, \underline{b}^{1/2}\, A,\ \underline{a}^{1/2}\, \underline{b}^{1/2}\, \underline{c}^{1/2}\, J \rangle$

$\bar{G} = \langle \gamma,\ \alpha,\ \mu \mid \gamma^4 = \alpha^4 = \mu^2 = (\gamma\alpha)^2 = (\gamma\mu)^4 = (\alpha\mu)^4 =$

$= [(\alpha\mu)^2,\gamma] = [(\alpha\mu)^2,\alpha] = [(\alpha\mu)^2,\mu] = [\alpha^2,\gamma] = [(\gamma\mu)^2,\alpha] = [\gamma^2,\mu] = 1 \rangle$

16. $G = P\,\frac{4_2}{n}\,\frac{2_1}{m}\,\frac{2}{c} = \langle \underline{c}^{1/2}\, C,\ \underline{a}^{1/2}\, \underline{b}^{1/2}\, A,\ \underline{a}^{1/2}\, \underline{b}^{1/2}\, J \rangle$

$\bar{G} = \langle \gamma,\ \alpha,\ \mu \mid \gamma^4 = \alpha^4 = \mu^2 = (\gamma\alpha)^2 = (\gamma\mu)^4 = (\alpha\mu)^2 = [(\gamma\alpha\mu)^2,\mu] = [\gamma^2,\alpha] = 1 \rangle$

17. $G = P4_122 = \langle \underline{c}^{1/4}\, C,\ A,\ \underline{a} \rangle$

$\bar{G} = \langle \gamma,\ \alpha,\ a \mid \gamma^8 = \alpha^2 = a^2 = (\gamma\alpha)^2 = [\gamma,a] = [\alpha,a] = 1 \rangle$

18. $G = P4_12_12 = \langle \underline{c}^{1/4}\, C,\ \underline{a}^{1/2}\, \underline{b}^{1/2}\, A \rangle$

$\bar{G} = \langle \gamma,\ \alpha \mid \gamma^8 = \alpha^4 = [\gamma^4,\alpha] = [\alpha^2,\gamma] = 1 \rangle$

## BIBLIOGRAPHY


1. N.V. Belov, N.N. Neronova and T.S. Smirnova, *"Shubnikov Groups,"* Akad. Nauk SSSR. Krystallografiya 2 (1957) 315 - 325.
2. M.J. Buerger, *Elementary Crystallography,* John Wiley and Sons, 1956.
3. J.J. Burckhardt, *Die Bewegungsgruppen der Kristallographie,* Birkhauser Verlag, 1966.
4. H. Hilton, *Mathematical Crystallography and the theory of Groups of Movements,* Dover, 1903.
5. M.A. Jaswon, *Mathematical Crystallography,* Longmans, 1965.
6. A.L. Loeb, *Color and Symmetry,* Wiley Interscience, 1971.
7. S. Oates Macdonald and A. Penfold Street, *"On Crystallographic Colour Groups,"* Combinatorial Mathematics IV, Proceedings of the Fourth Aust. Comb. Conf., Adelaide, Springer, 1975.

05B15

# GENERATION OF A FREQUENCY SQUARE ORTHOGONAL TO A 10×10 LATIN SQUARE

H. C. Kirton and Jennifer Seberry

N.S.W. Department of Agriculture, 150 Liverpool St., Sydney, 2000, and Department of Applied Mathematics, University of Sydney, N.S.W., 2006.

ABSTRACT

In general it is a difficult if not impossible task to find a latin square orthogonal to a given latin square. Because of a practical problem it was required to find a frequency square orthogonal to a given latin square. We describe a computer approach which was successful in finding a $(4,2^3)$ frequency square orthogonal to a given 10×10 latin square.

## 1. INTRODUCTION

In 1966 a Grapefruit Variety and Rootstock experiment was planted at the Horticultural Research Station, Dareton (in south western N.S.W.) to compare the effects of various rootstocks on grapefruit scions with particular regard to yield and quality under the fast growing conditions present at that locality.

The design used was a 10×10 latin square, with each of the five rootstock by two variety combinations being present as a single tree plot in each row and column. The source of the particular design used is unknown, and the person who proposed its use remained anonymous. In May 1977, the horticultural research officer currently responsible for this experiment indicated he would like to superimpose some new treatments that should improve the quality of the fruit by making them stay longer on the trees. These new treatments were to be various rates of application of a hormone spray, and it was felt that any superimposed design should leave about 40% of the trees untreated by this growth regulatory substance.

The problem was thus to find a frequency square $(4,2^3)$ that was orthogonal to rows, columns and existing treatments in the latin square that had already been used for 11 years:-

| | | | | | | | | | |
|---|---|---|---|---|---|---|---|---|---|
| D | J | I | F | B | H | E | G | A | C |
| C | H | B | D | G | E | J | A | I | F |
| F | A | C | J | E | D | I | H | B | G |
| H | D | E | B | I | C | A | F | G | J |
| J | B | A | I | C | F | G | D | H | E |
| I | E | H | G | J | A | B | C | F | D |
| E | F | G | C | A | I | H | J | D | B |
| B | I | J | A | F | G | D | E | C | H |
| G | C | D | E | H | B | F | I | J | A |
| A | G | F | H | D | J | C | B | E | I |

## 2. DEFINITIONS

Hedayat [2] and Hedayat and Seiden [3] have defined an F-square as follows:

DEFINITION. Let $A = (a_{ij})$ be an $n \times n$ matrix and let $\Sigma = (c_1, \ldots, c_m)$ be an ordered set of m distinct elements or symbols of A. In addition, suppose that for each $k = 1, 2, \ldots, m$, $c_k$ appears $\lambda_k$ times ($\lambda_k \geq 1$) in each row and in each column of A. Then A will be called a *frequency square* or, more concisely, an *F-square* on $\Sigma$ of order n and frequency vector $(\lambda_1, \lambda_2, \ldots, \lambda_m)$ and will be denoted by $F(n; \lambda_1, \ldots, \lambda_m)$. Note that $\lambda_1 + \lambda_2 + \cdots + \lambda_m = n$ and that where $\lambda_k = 1$ for every k and $m = n$, a latin square results.

As with latin squares, one may consider orthogonality of a pair or a set of F-squares of the same order. The above cited authors give the following definition covering these cases:

DEFINITION. Given an F-square $F_1(n; \lambda_1, \lambda_2, \ldots, \lambda_k)$ on a set $\Sigma = (a_1, a_2, \ldots, a_k)$ and an F-square $F_2(n; \mu_1, \mu_2, \ldots, \mu_t)$ on a set $\Omega = (b_1, b_2, \ldots, b_t)$ we say $F_2$ is an *orthogonal* mate for $F_1$ (and write $F_2 \perp F_1$) if on superposition of $F_2$ on $F_1$, $a_i$ appears $\lambda_i \mu_j$ times with $b_j$.

Federer [1] has recently written a most interesting paper indicating how Hadamard matrices can be used to obtain $(4t-1)^2$ mutually orthogonal $F(4t; 2t, 2t)$-squares (a complete set).

## 3. THE METHOD AND RESULTS

A computer program was written to provide a sufficient search of possible frequency squares involving t treatments that are orthogonal to a given latin square. Each cell of the latin square has associated with it the following parameters:-

(a) The original treatment from the latin square design.

(b) A vector whose $i^{th}$ element denotes the number of free choices of frequency square treatments remaining for this cell at level i of the design generation process. The program caters for values of i up to 44, with the problem being declared "too big" if i exceeds 44.

(c) A matrix whose (i,j) element takes values as follows (k is a positive integer $\le t$):

(i) $(i,1) = k$ means the cell has new treatment k,

and (ii) $(i,j) = -k$ where $1 < j \le t$ means the cell cannot have new treatment k, at level i of the design generation.

The process for design generation consists of choosing the cell with the fewest free choices, and placing the smallest available k as the frequency square treatment for that cell at the current level of choice if no freedom exists for choice of k, and at the next level of choice otherwise. The implications of this choice are then checked for other cells in the same column, row and latin treatment. When the $(i,1)^{th}$ element for any cell would be negative, the level of choice is decreased by one step, and the last chosen k is eliminated from the set of available values for the cell where the choice was made. The generation process terminates when (i) (i,1) element for each cell is positive (successful completion), (ii) i becomes less than one (all possibilities rejected), or (iii) the problem is "too big".

The frequency square orthogonal to the "de-randomized" 10×10 latin square was generated in two runs of the program. In the first run, each cell of the latin square that received a treatment coded F to J was assumed to have a new treatment number 99 (a dummy). The program then searched for a $(2,1^3)$ frequency square to be superimposed on the cells containing latin square treatments coded A to E.

The program thus searched for a $(2,1^3)$ frequency square orthogonal to the equivalent latin square

| | | | | | | | | | |
|---|---|---|---|---|---|---|---|---|---|
| A | B | C | D | E | F | G | H | I | J |
| B | E | G | F | I | J | H | D | C | A |
| C | F | H | B | D | A | E | G | J | I |
| D | A | B | E | H | C | J | I | G | F |
| E | D | I | A | C | H | B | J | F | G |
| F | J | D | I | B | G | C | A | H | E |
| G | I | J | H | A | B | F | C | E | D |
| H | C | E | J | G | I | D | F | A | B |
| I | G | F | C | J | D | A | E | B | H |
| J | H | A | G | F | E | I | B | D | C |

starting with (k = 1,2,3, or 4)

| | | | | | | | | | |
|---|---|---|---|---|---|---|---|---|---|
| k | k | k | k | k | 99 | 99 | 99 | 99 | 99 |
| k | k | 99 | 99 | 99 | 99 | 99 | k | k | k |
| k | 99 | 99 | k | k | k | k | 99 | 99 | 99 |
| k | k | k | k | 99 | k | 99 | 99 | 99 | 99 |
| k | k | 99 | k | k | 99 | k | 99 | 99 | 99 |
| 99 | 99 | k | 99 | k | 99 | k | k | 99 | k |
| 99 | 99 | 99 | 99 | k | k | 99 | k | k | k |
| 99 | k | k | 99 | 99 | 99 | k | 99 | k | k |
| 99 | 99 | 99 | k | 99 | k | k | k | k | 99 |
| 99 | 99 | k | 99 | 99 | k | 99 | k | k | k |

After 14 seconds (Univac 1108), it found a solution

| | | | | | | | | | |
|---|---|---|---|---|---|---|---|---|---|
| 1 | 1 | 2 | 3 | 4 | 99 | 99 | 99 | 99 | 99 |
| 4 | 1 | 99 | 99 | 99 | 99 | 99 | 1 | 3 | 2 |
| 3 | 99 | 99 | 1 | 1 | 4 | 2 | 99 | 99 | 99 |
| 2 | 3 | 4 | 1 | 99 | 1 | 99 | 99 | 99 | 99 |
| 1 | 4 | 99 | 2 | 1 | 99 | 3 | 99 | 99 | 99 |
| 99 | 99 | 1 | 99 | 2 | 99 | 1 | 4 | 99 | 3 |
| 99 | 99 | 99 | 99 | 3 | 2 | 99 | 1 | 4 | 1 |
| 99 | 2 | 3 | 99 | 99 | 99 | 4 | 99 | 1 | 1 |
| 99 | 99 | 99 | 4 | 99 | 3 | 1 | 2 | 1 | 99 |
| 99 | 99 | 1 | 99 | 99 | 1 | 99 | 3 | 2 | 4 |

This solution is such that each new treatment (1,2,3,4) occurs with appropriate frequencies in each row, column and twice as often with each of the latin square treatments A to E.

The second half of the design was generated in the second run by assigning a dummy new treatment (99) to each cell of the latin square that received a treatment A to E. After 4 seconds, this gave

| | | | | | | | | | |
|---|---|---|---|---|---|---|---|---|---|
| 99 | 99 | 99 | 99 | 99 | 1 | 1 | 2 | 3 | 4 |
| 99 | 99 | 3 | 1 | 4 | 2 | 1 | 99 | 99 | 99 |
| 99 | 3 | 4 | 99 | 99 | 99 | 99 | 1 | 1 | 2 |
| 99 | 99 | 99 | 99 | 1 | 99 | 2 | 1 | 4 | 3 |
| 99 | 99 | 1 | 99 | 99 | 4 | 99 | 3 | 2 | 1 |
| 1 | 4 | 99 | 2 | 99 | 3 | 99 | 99 | 1 | 99 |
| 2 | 1 | 1 | 3 | 99 | 99 | 4 | 99 | 99 | 99 |
| 3 | 99 | 99 | 1 | 2 | 1 | 99 | 4 | 99 | 99 |
| 4 | 1 | 2 | 99 | 3 | 99 | 99 | 99 | 99 | 1 |
| 1 | 2 | 99 | 4 | 1 | 99 | 3 | 99 | 99 | 99 |

The two runs of the program were thus able to generate the $(4,2^3)$ frequency square

| | | | | | | | | | |
|---|---|---|---|---|---|---|---|---|---|
| 1 | 1 | 2 | 3 | 4 | 1 | 1 | 2 | 3 | 4 |
| 4 | 1 | 3 | 1 | 4 | 2 | 1 | 1 | 3 | 2 |
| 3 | 3 | 4 | 1 | 1 | 4 | 2 | 1 | 1 | 2 |
| 2 | 3 | 4 | 1 | 1 | 1 | 2 | 1 | 4 | 3 |
| 1 | 4 | 1 | 2 | 1 | 4 | 3 | 3 | 2 | 1 |
| 1 | 4 | 1 | 2 | 2 | 3 | 1 | 4 | 1 | 3 |
| 2 | 1 | 1 | 3 | 3 | 2 | 4 | 1 | 4 | 1 |
| 3 | 2 | 3 | 1 | 2 | 1 | 4 | 4 | 1 | 1 |
| 4 | 1 | 2 | 4 | 3 | 3 | 1 | 2 | 1 | 1 |
| 1 | 2 | 1 | 4 | 1 | 1 | 3 | 3 | 2 | 4 |

By applying the same randomization to this frequency square as had been applied to the original latin square, a feasible design was generated.

REFERENCES

(1) Walter T. Federer, "On the existence and construction of a complete set of orthogonal F(4t;2t,2t)-squares", Paper No. BU-564-M in the Biometrics Unit Mimeo Series, Department of Plant Breeding and Biometry, Cornell University, Ithaca, New York, 1975.

(2) A. Hedayat, *On the Theory of the Existence, Non-existence and the Construction of Mutually Orthogonal F-squares and Latin Squares*, Ph.D. Dissertation, Cornell University, 1969.

(3) A. Hedayat and E. Seiden, "F-square and orthogonal F-squares design: generation of latin square and orthogonal latin squares design", *Ann. Math. Statistics* 41 (1970), 2035-2044.

# FACTORIZATION IN THE MONOID OF LANGUAGES

J.L Lassez[+] and H.J. Shyr[++]
[+]Department of Computer Science,
University of Melbourne.
[++]Department of Mathematics,
University of Western Ontario.

ABSTRACT.

It is shown that the equation $Y_1Y_2 = Y_3Y_4$ over $\hat{P} \cup Q$ where $P$ is the set of irreducible prefix codes and $Q$ is the set of primitive words admits non-trivial solutions only when $Y_1 = Y_3 = Q$.

## 1. INTRODUCTION AND DEFINITIONS.

Let $X$ be a finite alphabet and let $X^*$ be the free monoid generated by $X$. Every element of $X^*$ will be called a *word*. We let $X^+ = X^* - \{1\}$, where $1$ is the empty word. Any subset of $X^+$ or $\{1\}$ will be called a *language*. The family of all languages over $X$ will be denoted by $M$. For any $A, B \in M$, the *concatenation* of $A$ and $B$ is the set $AB = \{xy | x \in A, y \in B\}$. With the concatenation operation, $M$ is a monoid, called *the monoid of languages over* $X$. For any $A \in M$, let $A^* = \bigcup_{i=0}^{\infty} A^i$, $A^+ = \bigcup_{i=1}^{\infty} A^i$. We can extend the notion of concatenation operation to subsets of $M$, namely, for any $\alpha, \beta \subseteq M$, $\alpha\beta = \{AB | A \in \alpha, B \in \beta\}$. Also $\alpha^* = \bigcup_{i=0}^{\infty} \alpha^i$ and $\alpha^+ = \bigcup_{i=1}^{\infty} \alpha^i$.

A non-empty subset $\alpha$ of $M$ is called a *code* if and only if $A_1A_2\ldots A_m = B_1B_2\ldots B_n$, $A_i$, $B_j \in \alpha$ implies $m = n$ and $A_i = B_i$, $i = 1,2,\ldots,n$. $\alpha$ is called a *prefix set* if $A$, $AB \in \alpha$ implies $B = \{1\}$. A *prefix code* is a prefix set which is also a code. (see first part of [3] for a survey of the basic properties.) For any $A \in M$, i.e., $A \subseteq X^+$ or $A = \{1\}$, the set $\tilde{A} = \{\{x\} | x \in A\} \subseteq M$. Hence every language $A$ over $X$ can be identified as a subset of $M$. Thus the notions of code, prefix code on $M$ generalise the notions of code and prefix code over $X$. In particular $X^*$ is a free submonoid of $M$.

A word $f \in X^+$ is called *primitive* if $f = g^n$, $g \in X^+$ implies $n = 1$ [4].

The set of all primitive words over $X$ will be denoted by $Q$. Throughout this paper we assume that $X$ consists of at least two letters. It is known [6] that the *family of prefix codes* $P(X)$ over $X$ is a free monoid and hence the *generating* set $\hat{P}(X) = (P(X) - \{1\}) - (P(X) - \{1\})^2$ is a code.

For any $x \in X^+$, let $lg(x)$ be the length of $x$. In particular $lg(1) = 0$. For any $A \subseteq X^+$, let $lg(A) = \min\{lg(x) \mid x \in A\}$. Then it is easy to see that $lg(AB) = lg(A) + lg(B)$ for all $A, B \in M$. Hence $M$ is a *monoid with length* [1], however as $M$ is not a free monoid it is not possible to generalise all results on codes from $X^*$ to $M$. In particular a prefix set in $M$ is not necessarily a code, as we will see as a consequence of the following results which link the notion of code to the existence of non trivial solutions of equations on languages.

## 2. PRELIMINARY RESULTS.

<u>Proposition 1</u>: [5] Let $X$ be an alphabet. Let $f, g \in Q$, $f \neq g$. Then for any $m, n \geq 2$, $f^n g^m \in Q$.

We call a language $A$, $A \neq \emptyset$, *irreducible* if $A = BC$, $B, C \in M$ implies $B = \{1\}$ or $C = \{1\}$. Let $IRR(M) = \{A \in M \mid A \neq \{1\}$ and $A$ is irreducible$\}$.

<u>Proposition 2</u>: [7] Let $S$ be a submonoid of $M$. Then there exists a unique irreducible generating set $S_I$, contained in every other generating set and containing all the irreducible elements of $S$. i.e., $S_I = (S - \{1\}) - (S - \{1\})^2$ and $S = S_I^*$. In particular $(IRR(M - \{1\}))^+ = M - \{1\}$.

For any non-empty language $A \subseteq X^+$, let $\underline{A} = \{x \in A \mid lg(x) \leq lg(y)$ for all $y \in A\}$.

<u>Proposition 3</u>: [10] Let $A, B, C,$ and $D$ be non-empty languages over $X$. If $AB = CD$, then the following are true:

(1) $lg(A) + lg(B) = lg(C) + lg(D) = lg(AB)$;

(2) $\underline{A}\,\underline{B} = \underline{C}\,\underline{D} = \underline{AB}$;

(3) If $lg(A) = lg(C)$, then $\underline{A} = \underline{C}$ and $\underline{B} = \underline{D}$.

<u>Proposition 4</u>: Let $QY_2 = Y_3Q$ , where $Y_2, Y_3 \in P(X)$ . Then $Y_2 \cap H = \emptyset$ and $Y_3 \cap H = \emptyset$ , where $H = \{a^n | n \geq 1, a \in X\}$ .

<u>PROOF</u>: We establish the following two results first. (I) for every $a \in X$ and for any $n \geq 1$, $a^n \in Y_2$ if and only if $a^n \in Y_3$ ; (II) for every $a \neq b$, $a, b \in X$ and for any $n \geq 1$, $a^n \in Y_2$ if and only if $ba^{n-1} \in Y_3$ .

(I) Suppose $a^n \in Y_2$ , $n \geq 1$ . Then $a^{n+1} \in QY_2 = Y_3Q$ . It follows that $a^n \in Y_3$ . Similarly we can show that $a^n \in Y_3$ implies that $a^n \in Y_2$ .

(II) Suppose $a \neq b$, $a, b \in X$ and $a^n \in Y_2$, $n \geq 1$. Then $ba^n \in QY_2 = Y_3Q$ and so $ba^{n-1} \in Y_3$ . Conversely, let $ba^{n-1} \in Y_3$ , $n \geq 1$ . Then $ba^n \in Y_3Q = QY_2$ . This implies that $a^m \in Y_2$ , $m \leq n$ . If $n = 1$, then $a \in Y_2$ and the result holds. Now if $n \neq 1$ and $a^m \in Y_2$ , then by the first part of the proof, we have $ba^{m-1} \in Y_3$ . So if $m \neq n$ , then $ba^{n-1} \in Y_3$ and $ba^{m-1} \in Y_3$ , which contradicts the fact that $Y_3$ is a prefix code. Thus $m = n$ must hold and hence $a^n \in Y_2$ .

Now suppose $a^n \in Y_2$ where $a \in X$ and $n \geq 1$ . Then by (II) $ba^{n-1} \in Y_3, b \in X$ . Since $ba^2a^n \in QY_2 = Y_3Q$ , we have $ba^{n+1} \in Y_3$ , which is a contradiction. Hence for all $a \in X$ , $a^n \notin Y_2$, $n \geq 1$ . Similarly we can show that $a^n \notin Y_3$ for all $a \in X$ and $n \geq 1$ . Thus $Y_2 \cap H = \emptyset$ and $Y_3 \cap H = \emptyset$.#

<u>Proposition 5</u>: $\{A,Q\}$ is a code for every $A \in \hat{P}(X)$ .

<u>Proof</u>: (i) $A \subseteq X$ . Suppose $A_1A_2\ldots A_m = B_1B_2\ldots B_n$ for some m and n where $A_i, B_j \in \{A,Q\}$ . Then by Proposition 3,

$$\underline{A}_1\underline{A}_2\ldots\underline{A}_m = \underline{B}_1\underline{B}_2\ldots\underline{B}_n \, .$$

Since $\underline{A}_i = \underline{B}_j$ for every i and j, we have $m = n$ . Now if $A_k = Q$ for some k, $1 \leq k \leq n$ , then

$$a^{k-1}(ab)b^{n-k} \in A_1A_2\ldots A_m = B_1B_2\ldots B_n$$

where $ab \in A_k$. Since $B_i \in \{A,Q\}$ for all $i$, we have $ab \in B_k$. Thus $B_k = Q$. Similarly, we can show that $B_k = Q$ implies $A_k = Q$.

(ii) $A \not\subseteq X$, $A \in \hat{P}(X)$. Let $A_1A_2\ldots A_m = B_1B_2\ldots B_n$ for some $m$ and $n$, where $A_i, B_j \in \{A,Q\}$. Again by Proposition 3,

$$\underline{A}_1\underline{A}_2\ldots\underline{A}_m = \underline{B}_1\underline{B}_2\ldots\underline{B}_n .$$

Since $A \in \hat{P}(X)$ and $A \not\subseteq X$, we have $\underline{A} \subsetneq X^p$ for some $p \geq 1$. Hence $\underline{A}_i = X$ if and only if $A_i = Q$ ($\underline{B}_i = X$ if and only if $B_i = Q$). If $\underline{A} \in \hat{P}(X)$ we have

$$m = n \text{ and } \underline{A}_i = \underline{B}_i \text{ for } i = 1,2,\ldots,n .$$

This in turn implies that $A_i = B_i$ for $i = 1,2,\ldots,n$.

Now if on the other hand $\underline{A} \notin \hat{P}(X)$, then $\underline{A} = C_1C_2\ldots C_q$, $C_i \in \hat{P}(X)$, $i = 1,2,\ldots,q$ and there exists at least one $u$, $1 \leq u \leq q$ such that $C_u \neq X$. In this case we have

$$\overbrace{A_{11}\ldots A_{1r_1}}^{\underline{A}_1}\ \overbrace{A_{21}\ldots A_{2r_2}}^{\underline{A}_2}\ \ldots\ \overbrace{A_{m1}\ldots A_{mr_m}}^{\underline{A}_m} = \overbrace{B_{11}\ldots B_{1s_1}}^{\underline{B}_1}\ \overbrace{B_{21}\ldots B_{2s_2}}^{\underline{B}_2}\ \ldots\ \overbrace{B_{n1}\ldots B_{ns_n}}^{\underline{B}_n}$$

where each $A_{ir_i}$ and $B_{js_j}$ is in $\hat{P}(X)$. Therefore the corresponding $A_{ir_i}$ is equal to the corresponding $B_{js_j}$.

Let $\underline{A}_1 = \underline{B}_1$ ..., $\underline{A}_{i-1} = \underline{B}_{i-1}$ and $\underline{A}_i \neq \underline{B}_i$. We may assume $A_i = A$ and $B_i = Q$, i.e., $\underline{A}_i = A_{i1}\ldots A_{ir_i} = C_1\ldots C_q$ and $\underline{B}_i = X$. Then $A_{i1} = C_1 = X$ must hold. It follows that $B_{i+1} \neq A$, otherwise $C_u = X$, a contradiction. If $\underline{B}_{i+1} = X$, then $A_{i2} = X$. By arguing in this way, we always end up with $C_u = X$, a contradiction. Therefore $m = n$ and $A_i = B_i$ for $i = 1,2,\ldots n$. Hence $\{A,Q\}$ is a code for every $A \in \hat{P}(X)$.#

3. SOLUTIONS OF THE EQUATION: $Y_1Y_2 = Y_3Y_4$ over $\hat{P}(X) \cup \{Q\}$

In this section we will show that the family $F = \hat{P}(X) \cup \{Q\}$ is not a code by considering the solutions of the equation

$$Y_1Y_2 = Y_3Y_4 \qquad \ldots (1)$$

in $F$. A solution of the above equation is called *trivial* in $F$, if $Y_1 = Y_3$ and $Y_2 = Y_4$. We are only interested in the nontrival solutions in $F$. It is immediate that the equation (1) has no non-trivial solution in the subfamily $\hat{P}(X)$ of $F$, since $\hat{P}(X)$ is a code. Hence we consider the equation in which at least one of $Y_i$, $i = 1,2,3,4$, is $Q$. It suffices to consider the following cases:

| | | | |
|---|---|---|---|
| (i) | $QY_2 = Y_3Y_4$ ; | (v) | $Y_1Q = Y_3Q$ ; |
| (ii) | $Y_1Q = Y_3Y_4$ ; | (vi) | $QY_2 = Y_3Q$ ; |
| (iii) | $QQ = Y_3Y_4$ ; | (vii) | $QY_2 = QQ$ ; |
| (iv) | $QY_2 = QY_4$ ; | (viii) | $Y_1Q = QQ$ ; |

where $Y_i \in \hat{P}(X)$, $i = 1,2,3,4$.

That (i), (ii) and (iii) have no non-trivial solution in $\hat{P}(X)$ is immediate, since $QY_2$, $Y_1Q$, $QQ$, are not prefix codes for any $Y_1$ in $\hat{P}(X)$, while $Y_3Y_4$ is a prefix code for every $Y_3$ and $Y_4$ in $P(X)$.

The equation of type (iv) $QY_2 = QY_4$ has a non-trivial solution: Indeed, if we let $B = \{b^na | n \geq 1\} \cup \{ab^na | n \geq 1\}$ and let $C = B - \{ab^4a\}$, then both $B$ and $C$ are in $\hat{P}(X)$. We now show that $QB = QC$. It sufficies to show that for any $f \in Q$, then $fab^4a \in QC$. This is true, for if $fab \in Q$, then $fab^4a = (fab)(b^3a) \in QC$ and on the other hand if $fab \notin Q$, then by Proposition 1, $(fab)(b^2) \in Q$. Hence $fab^4a = (fab^3)(ba) \in Q$.

We now show that (v) $Y_1Q = Y_3Q$ has no non-trivial solution. Suppose there exist two prefix codes $Y_1$ and $Y_3$ such that $Y_1Q = Y_3Q$. Then by Proposition 3, $\underline{Y_1Q} = \underline{Y_3Q}$ and $Y_1 = Y_3$ hold:

Let

$$A = Y_1 - Y_3 \quad , \quad B = Y_3 - Y_1$$

and let

$$C = A \cup B \ .$$

If $C = \emptyset$ , then $Y_1 = Y_3$ and we are done. Suppose $C \neq \emptyset$ . Let $x \in \underline{C}$ . We note that in this case if $y \in Y_1 \cup Y_3$ and $lg(y) < lg(x)$ , then $y \in Y_1 \cap Y_3$ . If $x \in Y_1$ (the proof for the case $x \in Y_3$ is similar), then $xa \in X$ . Hence $xa = zq$ for some $z \in Y_3$ , $q \in Q$ . We have $q \neq a$ , because if $q = a$ , then $x = z \in Y_1 \cap Y_3$ , which contradicts $x \in \underline{C}$ . Thus $lg(q) > 1$, and $lg(z) < lg(x)$ . But this implies that $z \in Y_1 \cap Y_3$ and $x = zz'$ , $z' \in X^+$ . This is also not possible, because $Y_1$ is a prefix code.

Case (vii). Suppose $QY_2 = QQ$ for some $Y_2 \in P(X)$ . Then for any $a \in X$ , $aa \in QQ = QY_2$ . Hence $a \in Y_2$ and $X \subseteq Y_2$ hold. Since $Y_2$ is a prefix code, we have $Y_2 = X$ . Thus $QX = QQ$ . But this is not possible, for $(aba)(ba) \in QQ$ while $ababa \notin QX$ .

Similarly we can show that case (viii) has no non-trivial solution.

Finally we consider case (vi). Suppose $QY_2 = Y_3Q$ for some $Y_2, Y_3 \in P(X)$ . Then by Proposition 3, $lg(Y_2) = lg(Y_3)$ . By Proposition 4 we see that $lg(Y_2) \neq 1$ . Now suppose $lg(Y_2) = lg(Y_3) = 2$ . If $ab \in \underline{Y}_2$ , where $\underline{Y}_2 = \{x \in Y_2 \mid lg(x) \leq lg(y) \text{ for all } y \in Y_2\}$ , then $aab \in Q\underline{Y}_2 \subseteq Y_3Q$ . This implies that $a^2 \in Y_3$ , which is a contradiction. Similarly $ba \notin \underline{Y}_2$ . Therefore we may assume $lg(Y_2) = lg(Y_3) = m \geq 3$ .

Let us assume now $x = r_1r_2 \ldots r_m \in \underline{Y}_3$ , $r_i \in X, m \geq 3$ . Then for any $a \in X$ , $xa = r_1r_2 \ldots r_m a \in \underline{Y}_3Q \subseteq QY_2$ , and $r_2r_3 \ldots r_m a \in \underline{Y}_2$ . Let $b \in X, b \neq a$ . Then $xab = r_1r_2 \ldots r_m ab \in \underline{Y}_3Q \subseteq QY_2$ . Since $Y_2 \in P(X)$ , $r_2r_3 \ldots r_m ab \notin Y_2$ and hence $r_3r_4\ldots r_m ab \in Y_2$ . Again from $xabb = r_1r_2\ldots r_m abb \in \underline{Y}_3Q \subseteq QY_2$ and $Y_2 \in \hat{P}(X)$ , we have $r_4\ldots r_m abb \in Y_2$(if $m = 3$ , then $abb \in \underline{Y}_2$) . Continuing this process we have eventually $abb\ldots b \in \underline{Y}_2$ . It is also easy to see that $aab\ldots b \in \underline{Y}_3$ .

From the above result we have $ab\ldots b \in \underline{Y}_2$ and $aab\ldots b \in \underline{Y}_3$ . Since $aab\ldots b\ abb\ldots b \in Q$ , we have $aab\ldots b\ abb\ldots b\ ab\ldots b \in QY_2 \subseteq Y_3Q$ . This implies that $aab\ldots b$, $aab\ldots bay \in Y_3$ , $y \in X^*$ , which is a contradiction, for $Y_3$ is a prefix code. Therefore the equation $QY_2 = Y_3Q$ has no non-trivial solution.#

REFERENCES.


[1] Clifford, A.H. and Preston, G.B., "The Algebraic Theory of Semigroups", Vol. I, II, Amer. Math. Soc., Providence, RI. (1961).

[2] Lassez, J.L., A Correspondence on strongly prefix codes, IEEE Transactions on Information Theory, May (1975), 344-345.

[3] Lassez, J.L., Prefix codes, trees and automata, Information Sciences 8 (1975), 155-171.

[4] Lentin, A. and Schutzenberger, M.P., A Combinatorial Problem in the Theory of Free Monoid, in "Combinatorial Mathematics and its Applications" (R.C. Bose and T.A. Dowling, Eds.), North Carolina Press, Chapell Hill, NC. (1967) 128-144.

[5] Lyndon, R.C. and Schutzenberger, M.P., The Equation $a^M = b^N c^P$ in a Free Group, Michigan Math. J. 9 (1962), 289-298.

[6] Perrin, D., Codes Conjugues, Information and Control, 20 (1972), 222-231.

[7] Sevrin, L.N., On Subsemigroups of Free Semigroups, Soviet Math. Dokl. (1960) 892-894.

[8] Shyr, H.J., Codes and Factor Theorems for Subsets of a Free Monoid, Utilitas Mathematica, Vol3, (1973), 153-159.

[9] Shyr, H.J., Left Cancellative Subsemigroup of a Semigroup, Soochow J. of Math. & Natural Sci., Vol.2, (1976), 25-33.

[10] Wood, D., A Factor Theorem for Subsets of a Free Monoid, Information and Control, 21, (1972), 21-26.

# ON GRAPHS AS UNIONS OF EULERIAN GRAPHS

05C99

Charles H.C. Little

DEPARTMENT OF MATHEMATICS AND COMPUTER SCIENCE,
ROYAL MELBOURNE INSTITUTE OF TECHNOLOGY,
MELBOURNE, VIC. 3000, AUSTRALIA

ABSTRACT

We present a criterion for a graph to be the union of a specified number of Eulerian graphs.

TEXT

We denote the vertex and edge sets of a graph G by VG and EG respectively. Furthermore, the *union* of graphs $H_1, H_2, \ldots\ldots, H_r$ is defined to be the graph with vertex set

$$\bigcup_{i=1}^{r} VH_i \quad \text{and edge set} \quad \bigcup_{i=1}^{r} EH_i .$$

In a recent paper, Matthews (2) defines the Eulericity $\mathcal{E}(G)$ of a graph G to be the smallest number of Eulerian subgraphs whose union is G. (An Eulerian graph is defined to be one in which every vertex has even valency.) It is known that the four colour theorem is equivalent to the statement that $\mathcal{E}(G) \leq 2$ for every planar graph G without an isthmus. In this paper, we present a criterion for $\mathcal{E}(G) \leq r$ where r is a positive integer.

If T is a spanning tree of a connected graph G, then the edges of EG - ET will be referred to as the *chords* of T.

For all positive integers $n$, we denote by $V_n$ the vector space of dimension $n$ over the field of residue classes modulo 2.

We say that subgraphs $H_1, H_2, \ldots\ldots, H_r$ of a graph $G$ are *independent* if there is no subset $H$ of $\{H_1, H_2, \ldots\ldots, H_r\}$ such that every edge of $G$ belongs to an even number of subgraphs in $H$. It is clear that if $G$ is the union of Eulerian subgraphs $H_1, H_2, \ldots\ldots, H_{E(G)}$, then these subgraphs are independent. Indeed, suppose that $H$ is a subset $\{H_1, H_2, \ldots\ldots, H_r\}$ of $\{H_1, H_2, \ldots\ldots, H_{E(G)}\}$ and that each edge of $G$ belongs to an even number of subgraphs in $H$. Then

$$EH_1 \subseteq \bigcup_{i=2}^{r} EH_i ,$$

so that $G$ is the union of $H_2, H_3, \ldots\ldots, H_{E(G)}$, in contradiction to the minimality of $E(G)$. It follows that $G$ is the union of $s$ Eulerian subgraphs if and only if it is the union of $r$ independent Eulerian subgraphs for some integer $r \leq s$. Therefore we need only a criterion for $G$ to be the union of $r$ independent Eulerian subgraphs.

We are now ready for our theorem.

*Theorem* : Let $T$ be a spanning tree of a connected graph $G$, and let $c_1, c_2, \ldots\ldots, c_\beta$ be the chords of $T$. For each $k$ we denote by $C_k$ the unique circuit for which $EC_k \cap (EG - ET) = \{c_k\}$. Let $EG = \{e_1, e_2, \ldots\ldots, e_m\}$, and for each $j \in \{1, 2, \ldots\ldots, m\}$ let $X_j = (x_{j1}, x_{j2}, \ldots\ldots, x_{j\beta})$, where $x_{jk} = 1$ if $e_j \in EC_k$ and $x_{jk} = 0$ otherwise.

Let $X = \{X_1, X_2, \ldots\ldots, X_m\}$, and let $r$ be a positive integer. If $\beta \geq r$, then $G$ fails to be the union of $r$ independent Eulerian graphs if and only if every subspace of $V_\beta$ of dimension $\beta - r$ contains a vector of $X$.

Proof : Let $H_1, H_2, \ldots\ldots, H_r$ be independent Eulerian subgraphs of G. For all $i \in \{1, 2, \ldots\ldots, r\}$, let $A_i$ be the vector $(a_{i1}, a_{i2}, \ldots\ldots, a_{im})$ such that, for all j, $a_{ij} = 1$ if $e_j \in EH_i$ and $a_{ij} = 0$ otherwise. Then the vectors $A_1, A_2, \ldots\ldots, A_r$ are distinct linear combinations of the vectors $Y_1, Y_2, \ldots\ldots, Y_\beta$, where, for all k, $Y_k = (x_{1k}, x_{2k}, \ldots\ldots, x_{mk})$. (See (1, pp.37 - 40).)

Let $Y = \{Y_1, Y_2, \ldots\ldots, Y_\beta\}$, and let $S_1, S_2, \ldots\ldots, S_r$ be the subsets of Y such that, for all i, the modulo 2 sum of the vectors in $S_i$ is $A_i$. For all i, let $S_i = \{Y_{i_1}, Y_{i_2}, \ldots\ldots, Y_{i_{\sigma(i)}}\}$ where $\sigma(i) = |S_i|$. Now an edge $e_j$ belongs to $EG - EH_i$ if and only if $a_{ij} = 0$, that is, if and only if

$$\sum_{k=1}^{\sigma(i)} x_{ji_k} = 0 .$$

Since $H_1, H_2, \ldots\ldots, H_r$ are independent, we see that $A_1, A_2, \ldots\ldots, A_r$ are linearly independent vectors, and hence the solution space of this system of r equations (obtained by letting i run through 1, 2, ....., r), is a subspace Z(H) of $V_\beta$ of dimension $\beta - r$, where H denotes the set $\{H_1, H_2, \ldots\ldots, H_r\}$. Therefore an edge $e_j$ belongs to $EG - \bigcup_{i=1}^{r} EH_i$ if and only if $X_j \in Z(H)$. Hence G is the union of r independent Eulerian subgraphs if and only if $X \cap Z = \phi$ for some subspace Z of the form Z(H). On the other hand, if Z is a subspace of $V_\beta$ of dimension $\beta - r$, then it is clear by reversing the above argument that there exist independent Eulerian subgraphs $H_1, H_2, \ldots\ldots, H_r$ for which $Z = Z(H)$, where $H = \{H_1, H_2, \ldots., H_r$ Hence G is the union of r independent Eulerian subgraphs if and only if $X \cap Z = \phi$ for some subspace Z of $V_\beta$ of dimension $\beta - r$.

REFERENCES:

(1) F. Harary, Graph Theory, Addison - Wesley, London, 1969

(2) K.R. Matthews, "On The Eulericity Of A Graph", submitted for publication.

05 B 45 20 B 25 20 H 15 50 B 30

# THE ANALYSIS OF COLOUR SYMMETRY

Sheila Oates Macdonald and Anne Penfold Street

Department of Mathematics,
University of Queensland,
St. Lucia, Q 4067, Australia

ABSTRACT

The plane coloured crystals are classified.

## 1. INTRODUCTION

In this paper, we continue the work begun in (4) and (5) on coloured crystals, that is, regular repeated coloured patterns which fill a space of given dimension. In (5) we dealt with colourings of friezes and in (4) with the reasons for coincidences of coloured patterns. Here we show how to find all plane coloured crystals.

As usual, we consider two crystals to be equivalent if they have the same symmetry group and furthermore, we consider two colourings of the same crystal to be equivalent if one can be obtained from the other by a symmetry of the underlying crystal or by a permutation of the colours or both.

If a crystal has the symmetry group G, then each colouring of that crystal corresponds to a permutation representation P of the group G on the cosets of one of its subgroups. If two subgroups are conjugate in G, they correspond to equivalent permutation representations of G (but not necessarily to equivalent colourings, as assumed in (9)). In fact, each colouring depends on three things: the representation P; the assignment of colours to the orbits of the subgroup on the fundamental regions; the embedding of G as a proper normal subgroup of other symmetry groups (see (4) and (5)).

This paper fully classifies what we call *restricted colourings* of the plane crystals, that is, colourings which arise subject to the requirement that rotations and translations commute in their effect on the colours, provided of course that the crystal has non-trivial rotations in its symmetry group. If the crystal has no rotational symmetry, then we classify those of its colourings which can be derived by regarding it as a subgroup of a group which does have rotations. All of this generalises the work of Loeb (3) who assumed that translations and rotations together generate a cyclic group, and relates to Senechal's discussion of colour symmetry in (8). We also indicate briefly how to find the *unrestricted colourings* of the crystals, where we place no requirement on the permutation representation P, except that it be of given degree.

## 2. THE PLANE SYMMETRY GROUPS

We have found it convenient to use the classification of the plane symmetry groups due to Schwarzenberger (7) and we work in terms of a standard presentation for each group which corresponds to his derivation. In each case, translations are denoted by x and y (or some combination of them) and Table 1 shows the relative magnitudes ($|y|/|x|$) and orientations ($\measuredangle xy$) of the translations.

| G | $\|y\|/\|x\|$ | $\measuredangle xy$ | G | $\|y\|/\|x\|$ | $\measuredangle xy$ |
|---|---|---|---|---|---|
| p1,p2 | arbitrary | arbitrary | p4 | 1 | $\pi/2$ |
| pm,pg,cm* | " | $\pi/2$ | p6 | 1 | $\pi/3$ |
| p2mm,p2mg | " | $\pi/2$ | p3m1*,p31m | 1 | $\pi/3$ |
| p2gg,c2mm* | " | $\pi/2$ | p4mm,p4mg | $1/\sqrt{2}$ | $\pi/4$ |
| p3 | 1 | $2\pi/3$ | p6mm | $1/\sqrt{3}$ | $\pi/6$ |

*In these cases, x and y are not themselves basis vectors.

Table 1: Relative lengths and angles for groups.

In Table 2 we list the presentations: first p1 with point group H = {1}; next p2, p3, p4 and p6, with $H = H_o$, the group of rotations; then pm, pg and cm with $H_o = \{1\}$; finally the nine remaining groups with $\{1\} < H_o < H$. We denote glides by p and q (along the x and y directions respectively), reflections by r and s (again in the x and y axes respectively) and rotations by u. We list also the generators of the group $[H_o,T]$, where T is the group of translations, and a presentation of the quotient group $\overline{G} = G/[H_o,T]$, since it is the subgroups of $\overline{G}$ which give rise to the restricted colourings. (Where $\overline{G}$ has a simple form, this is also listed.) In $\overline{G}$, X represents the image of x, Y of y and so on.

In Tables 3 and 4 we list embeddings of plane groups as subgroups of each other; these were first derived by Moser (6). The group G is always assumed to be in its standard presentation as in Table 2; K denotes a proper subgroup of given structure, and we always choose K to have minimal index of any subgroup of that structure. Table 3 deals with case (i), where G and K both have trivial rotation groups or both have non-trivial rotation groups; in these cases, we list simply the generators of K and the index of K in G. (Note that the index is shown in italics if K is not

| G | Generators | Relations | $[H_o,T]$ | $\bar{G}=G/[H_o,T]$ |
|---|---|---|---|---|
| p1 | $x,y$ | $[x,y]=1$ | $\{1\}$ | $G$ |
| p2 | $x,y,u$ | $[x,y]=xx^u=yy^u=u^2=1$ | $\langle x^2,y^2\rangle$ | $\langle X,Y,U \mid X^2=Y^2=U^2=[X,Y]=[X,U]=[Y,U]=1\rangle\simeq Z_2\times Z_2\times Z_2$ |
| p3 | $x,y,u$ | $[x,y]=y^{-1}x^u=xyy^u=u^3=1$ | $\langle x^{-1}y,xy^2\rangle$ | $\langle X,U \mid X^3=U^3=[X,U]=1\rangle\simeq Z_3\times Z_3$ |
| p4 | $x,y,u$ | $[x,y]=y^{-1}x^u=xy^u=u^4=1$ | $\langle x^{-1}y,xy\rangle$ | $\langle X,U \mid X^2=U^4=[X,U]=1\rangle\simeq Z_2\times Z_4$ |
| p6 | $x,y,u$ | $[x,y]=y^{-1}x^u=xy^{-1}y^u=u^6=1$ | $\langle x,y\rangle$ | $\langle U \mid U^6=1\rangle\simeq Z_6$ |
| pm | $x,y,s$ | $[x,y]=[y,s]=xx^s=s^2=1$ | $\{1\}$ | $G$ |
| pg | $x,y,q$ | $[x,y]=xx^q=y^{-1}q^2=1$ | $\{1\}$ | $G$ |
| cm | $xy,y^2,s$ | $[xy,y^2]=[y^2,s]=xyy^2(xy)^s=s^2=1$ | $\{1\}$ | $G$ |
| p2mm | $x,y,r,s$ | $[x,y]=[x,r]=[y,s]=xx^s=yy^r=r^2=s^2=(rs)^2=1$ | $\langle x^2,y^2\rangle$ | $\langle X,Y,R,S \mid X^2=Y^2=R^2=S^2=[X,Y]=[X,R]=[X,S]=[Y,R]=[Y,S]=[R,S]=1\rangle\simeq Z_2\times Z_2\times Z_2\times Z_2$ |
| p2mg | $x,y,q,r$ | $[x,y]=[x,r]=xx^q=yy^r=y^{-1}q^2=r^2=(qr)^2=1$ | $\langle x^2,y^2\rangle$ | $\langle X,Q,R \mid X^2=[X,R]=[X,Q]=Q^4=R^2=(QR)^2=1\rangle\simeq Z_2\times D_4$ |
| p2gg | $x,y,p,q$ | $[x,y]=x^{-1}p^2=xx^q=y^{-1}q^2=yy^p=(pq)^2=1$ | $\langle x^2,y^2\rangle$ | $\langle P,Q \mid P^4=Q^4=[P^2,Q]=[P,Q^2]=(PQ)^2=1\rangle$ |
| c2mm | $xy,y^2,r,s$ | $[xy,y^2]=[y^2,s]=(xy)^{-1}y^2(xy)^r=y^2(y^2)^r=xyy^{-2}(xy)^s=r^2=s^2=(rs)^2=1$ | $\langle (xy)^2,y^4\rangle$ | $\langle XY,Y^2,R,S \mid (XY)^2=Y^4=R^2=S^2=(RS)^2=[XY,Y^2]=[Y^2,R]=[Y^2,S]=1,(XY)^R=(XY)^S=XYY^2\rangle\simeq Z_2\times D_4$ |
| p3m1 | $xy,y^3,r,s$ | $[xy,y^3]=[y^3,s]=(xy)^{-2}y^3(xy)^r=r^2=s^2=(rs)^3=(xy)^{-3}y^6(y^3)^r=xyy^{-3}(xy)^s=1$ | $\langle (xy)^3,y^3\rangle$ | $\langle XY,R,S \mid (XY)^3=R^2=S^2=(RS)^3=1,(XY)^R=(XY)^S=(XY)^2\rangle$ |
| p31m | $x,y,r,s$ | $[x,y]=[x,r]=[y,s]=xy^{-1}x^s=x^{-1}yy^r=r^2=s^2=(rs)^3=1$ | $\langle xy,x^2y^{-1}\rangle$ | $\langle X,R,S \mid X^3=R^2=S^2=(RS)^3=[X,R]=[X,S]=1\rangle\simeq Z_3\times D_3$ |
| p4mm | $x,y,r,s$ | $[x,y]=[x,r]=[y,s]=xy^{-2}x^s=x^{-1}yy^r=r^2=s^2=(rs)^4=1$ | $\langle x,x^{-1}y^2\rangle$ | $\langle Y,R,S \mid Y^2=R^2=S^2=(SR)^4=[Y,R]=[Y,S]=1\rangle\simeq Z_2\times D_4$ |
| p4mg | $x,y,q,r$ | $[x,y]=[x,r]=y^{-1}q^2=xy^{-2}x^q=x^{-1}yy^r=r^2=(qr)^4=1$ | $\langle x,x^{-1}y^2\rangle$ | $\langle Q,R \mid R^2=Q^4=(QR)^4=[R,Q^2]=1\rangle$ |
| p6mm | $x,y,r,s$ | $[x,y]=[x,r]=[y,s]=xy^{-3}x^s=x^{-1}yy^r=r^2=s^2=(rs)^6=1$ | $\langle x^{-1}y,y\rangle$ | $\langle R,S \mid R^2=S^2=(RS)^6=1\rangle\simeq D_6$ |

Table 2: The plane symmetry groups

| G | K | Generators | Index |
|---|---|---|---|
| p1 | p1 | $\langle x^2,y\rangle$ | 2 |
| p2 | p2 | $\langle x^2,y,u\rangle$ | 2 |
| p3 | p3 | $\langle xy^{-1},xy^2,u\rangle$ | 3 |
| p4 | p2 | $\langle x,y,u^2\rangle$ | 2 |
| | p4 | $\langle xy,x^{-1}y,u\rangle$ | 2 |
| p6 | p2 | $\langle x,y,u^3\rangle$ | 3 |
| | p3 | $\langle x,y,u^2\rangle$ | 2 |
| | p6 | $\langle x^2,y^2,u\rangle$ | 4 |
| pm | p1 | $\langle x,y\rangle$ | 2 |
| | pm | $\langle x^2,y,s\rangle$ | 2 |
| | pg | $\langle x,y^2,sy\rangle$ | 2 |
| | cm | $\langle xy,y^2,s\rangle$ | 2 |
| pg | p1 | $\langle x,y\rangle$ | 2 |
| | pg | $\langle x^2,y,q\rangle$ | 2 |
| cm | p1 | $\langle xy,y^2\rangle$ | 2 |
| | pm | $\langle x^2,y^2,s\rangle$ | 2 |
| | pg | $\langle x^2,y^2,sxy\rangle$ | 2 |
| | cm | $\langle xy.y^2,(y^2)^3,s\rangle$ | 3 |
| p2mm | p2 | $\langle x,y,rs\rangle$ | 2 |
| | p2mm | $\langle x,y^2,r,s\rangle$ | 2 |
| | p2mg | $\langle x,y^2,r,sy\rangle$ | 2 |
| | p2gg | $\langle x^2,y^2,rx,sy\rangle$ | 4 |
| | c2mm | $\langle xy,y^2,r,s\rangle$ | 2 |
| p2mg | p2 | $\langle x,y,qr\rangle$ | 2 |
| | p2mg | $\langle x,y^3,q^3,r\rangle$ | 3 |
| | p2gg | $\langle x^2,y,rx,q\rangle$ | 2 |
| p2gg | p2 | $\langle x,y,pq\rangle$ | 2 |
| | p2gg | $\langle x,y^3,p,q^3\rangle$ | 3 |
| c2mm | p2 | $\langle xy,y^2,rs\rangle$ | 2 |
| | p2mm | $\langle x^2,y^2,r,s\rangle$ | 2 |
| | p2mg | $\langle x^2,y^2,sxy,r\rangle$ | 2 |
| | p2gg | $\langle x^2,y^2,rxy,sxy\rangle$ | 2 |
| | c2mm | $\langle (xy)^2,(y^2)^2,r,s\rangle$ | 4 |

| G | K | Generators | Index |
|---|---|---|---|
| p3m1 | p3 | $\langle xy,(xy)^{-2}y^3,rs\rangle$ | 2 |
| | p3m1 | $\langle (xy)^2,(y^3)^2,r,s\rangle$ | 4 |
| | p31m | $\langle (xy)^3y^{-3},y^3,r,s\rangle$ | 3 |
| p31m | p3 | $\langle x,x^{-1}y,rs\rangle$ | 2 |
| | p3m1 | $\langle xy,y^3,r,s\rangle$ | 3 |
| | p31m | $\langle x^2,y^2,r,s\rangle$ | 4 |
| p4mm | p2 | $\langle x,y,(rs)^2\rangle$ | 4 |
| | p4 | $\langle y,x^{-1}y,rs\rangle$ | 2 |
| | p2mm | $\langle xy^{-1},y,rsr,s\rangle$ | 2 |
| | p2mg | $\langle xy^{-1},y^2,sy,rsr\rangle$ | 4 |
| | p2mg | $\langle x^{-1}y^2,x^2,rx,srs\rangle$ | 4 |
| | p2gg | $\langle x,x^{-1}y^2,ry,srsy\rangle$ | 4 |
| | c2mm | $\langle y,x,srs,r\rangle$ | 2 |
| | p4mm | $\langle y^2,x,s,r\rangle$ | 2 |
| | p4mg | $\langle y^2,x,ry,sx^{-1}y\rangle$ | 2 |
| p4mg | p2 | $\langle x,y,(qr)^2\rangle$ | 4 |
| | p4 | $\langle y,xy^{-1},qr\rangle$ | 2 |
| | p2mm | $\langle x,x^{-1}y^2,r,r^q\rangle$ | 4 |
| | p2mg | $\langle x^{-1}y^2,x,ry,r^q\rangle$ | 4 |
| | p2gg | $\langle y,xy^{-1},q,q^r\rangle$ | 2 |
| | c2mm | $\langle y,x,r^q,r\rangle$ | 2 |
| | p4mg | $\langle x^3,y^3,qy,r\rangle$ | 9 |
| p6mm | p2 | $\langle x,y,(rs)^3\rangle$ | 6 |
| | p3 | $\langle y,x^{-1}y,(rs)^2\rangle$ | 4 |
| | p6 | $\langle y,x^{-1}y^2,rs\rangle$ | 2 |
| | p2mm | $\langle x^2y^{-3},y,r^{sr},s\rangle$ | 6 |
| | p2mg | $\langle x^2y^{-3},y,sxy^{-1},r^{sr}\rangle$ | 6 |
| | p2gg | $\langle x^2y^{-3},y,r^{sr}xy^{-1},sxy^{-1}\rangle$ | 6 |
| | c2mm | $\langle xy^{-1},y,r^{sr},s\rangle$ | 3 |
| | p3m1 | $\langle y,x,r^s,r\rangle$ | 2 |
| | p31m | $\langle xy^{-1},y,s^r,s\rangle$ | 2 |
| | p6mm | $\langle y^3,x,s,r\rangle$ | 3 |

Table 3: Subgroups of minimal index: case (i)

normal in G.) Table 4 deals with case (ii), where G has a non-trivial rotation group, but K has no proper rotations; here we need more detailed information since we use these embeddings to give the restricted colourings of the four groups with no rotations, so we give not only the generators of K, but also $K[H_o,T]/[H_o,T]$, using the notation of Table 2. An asterisk alongside the image indicates that $[H_o,T] \nleq K$, so that the image of K is larger than might be expected. Noteworthy perhaps is the fact that the two different embeddings of pg in p4mm have different size images.

In Table 5 we list the outer automorphisms of G induced by embeddings of G as a normal subgroup of another symmetry group for the groups with $H_o \neq \{1\}$, since these automorphisms, as shown in (4), may induce extra coincidences of patterns. The generators of the subgroup are shown in standard notation and the effects of the automorphisms on these generators is given. For instance, considering p2mm as a subgroup of itself we have $x = \tilde{x}$, $y = \tilde{y}^2$, $r = \tilde{r}$, $s = \tilde{s}$ where $\tilde{x}$, $\tilde{y}$, $\tilde{r}$, $\tilde{s}$ are the generators of the larger group. Conjugation by y changes these to $\tilde{x}$, $\tilde{y}$, $\tilde{r}\tilde{y}^2$, $\tilde{s}$, or in terms of the generators of the subgroup, x, y, ry, s yielding the automorphism $\zeta$. As far as possible we have used the same symbol to denote (i) the same automorphism of G induced by different embeddings and (ii) the effect of the same inner automorphism of a given symmetry group on various normal subgroups. However, we have treated all automorphisms of the groups without rotations which merely change the generators of the translation group as essentially trivial, as there is no preferred direction for these; consequently we denote all these by $\iota$.

3. RESTRICTED COLOURINGS

We first find all restricted colourings of the thirteen plane crystallographic groups which contain non-trivial rotations. As can be seen from Table 2, in all these cases $\overline{G} = G/[H_o,T]$ is finite and so has only finitely many subgroups. The permutation representations of $\overline{G}$ on the cosets of these yield corresponding representations of G.

For the four groups with no rotations $\overline{G} = G$ is infinite, so we consider instead the colourings obtained by embedding these in the thirteen groups already considered In order that the representation of $K \leq G$ induced by the permutation representation of G on the cosets of L be transitive it is necessary that $LK = G$, that is, $\overline{L}\overline{K} = \overline{G}$, where $\overline{L}$ and $\overline{K}$ are the images of L and K in the homomorphism $G \to \overline{G}$. The representation of $\overline{K}$ (and thus of K) obtained is the same as the representation of $\overline{K}$ on the cosets of $\overline{L} \cap \overline{K}$ and so the same representation of $\overline{K}$ will be obtained for any $\overline{L}_1$, $\overline{L}_2$

| G \ K | p1 | pm | pg | cm |
|---|---|---|---|---|
| p2 | $\langle x,y\rangle \twoheadrightarrow \langle X,Y\rangle \simeq Z_2\times Z_2$ | | | |
| p3 | $\langle x,y\rangle \twoheadrightarrow \langle X,X\rangle \simeq Z_3$ | | | |
| p4 | $\langle x,y\rangle \twoheadrightarrow \langle X,X\rangle \simeq Z_2$ | | | |
| p6 | $\langle x,y\rangle \twoheadrightarrow \langle 1,1\rangle \simeq \{1\}$ | | | |
| p2mm | $\langle x,y\rangle \twoheadrightarrow \langle X,Y\rangle \simeq Z_2\times Z_2$ | $\langle x,y,s\rangle \twoheadrightarrow \langle X,Y,S\rangle \simeq Z_2\times Z_2\times Z_2$ | $\langle x,y^2,sy\rangle \twoheadrightarrow \langle X,1,SY\rangle \simeq Z_2\times Z_2$ | $\langle xy,y^2,s\rangle \twoheadrightarrow \langle XY,1,S\rangle \simeq Z_2\times Z_2$ |
| p2mg | $\langle x,y\rangle \twoheadrightarrow \langle X,Y\rangle \simeq Z_2\times Z_2$ | $\langle x,y,r\rangle \twoheadrightarrow \langle X,Y,R\rangle \simeq Z_2\times Z_2\times Z_2$ | $\langle x,y,q\rangle \twoheadrightarrow \langle X,Y,Q\rangle \simeq Z_2\times Z_4$ | $\langle xy,x^2,r\rangle \twoheadrightarrow \langle XY,1,R\rangle \simeq Z_2\times Z_2$ |
| p2gg | $\langle x,y\rangle \twoheadrightarrow \langle X,Y\rangle \simeq Z_2\times Z_2$ | | $\langle x,y,p\rangle \twoheadrightarrow \langle X,Y,P\rangle \simeq Z_2\times Z_4$ | |
| c2mm | $\langle xy,y^2\rangle \twoheadrightarrow \langle XY,Y^2\rangle \simeq Z_2\times Z_2$ | $\langle x^2,y^2,s\rangle \twoheadrightarrow \langle Y^{-2},Y^2,S\rangle \simeq Z_2\times Z_2$ | $\langle x^2,y^2,sxy\rangle \twoheadrightarrow \langle Y^{-2},Y^2,SXY\rangle \simeq Z_4$ | $\langle xy,y^2,s\rangle \twoheadrightarrow \langle XY,Y^2,S\rangle \simeq D_4$ |
| p3m1 | $\langle xy,y^3\rangle \twoheadrightarrow \langle XY,1\rangle \simeq Z_3$ | $\langle (xy)^2y^{-3},y^3,r\rangle \twoheadrightarrow \langle (XY)^2,1,R\rangle \simeq D_3^*$ | $\langle (xy)^2y^{-3},y^3,rxy\rangle \twoheadrightarrow \langle (XY)^2,1,RXY\rangle \simeq D_3^*$ | $\langle xy,y^3,r\rangle \twoheadrightarrow \langle XY,1,R\rangle \simeq D_3$ |
| p31m | $\langle x,y\rangle \twoheadrightarrow \langle X,X^{-1}\rangle \simeq Z_3$ | $\langle x,xy^{-2},r\rangle \twoheadrightarrow \langle X,1,R\rangle \simeq Z_6^*$ | $\langle x^2,xy^{-2},rx\rangle \twoheadrightarrow \langle X^2,1,RX\rangle \simeq Z_6^*$ | $\langle x,y,r\rangle \twoheadrightarrow \langle X,Y,R\rangle \simeq Z_6$ |
| p4mm | $\langle x,y\rangle \twoheadrightarrow \langle 1,Y\rangle \simeq Z_2$ | $\langle xy^{-1},y,s\rangle \twoheadrightarrow \langle Y^{-1},Y,S\rangle \simeq Z_2\times Z_2$ | $\langle xy^{-1},y^2,sy\rangle \twoheadrightarrow \langle Y^{-1},1,SY\rangle \simeq Z_2\times Z_2^*$<br>$\langle x,xy^{-2},ry\rangle \twoheadrightarrow \langle 1,1,RY\rangle \simeq Z_2$ | $\langle x,y,r\rangle \twoheadrightarrow \langle 1,Y,R\rangle \simeq Z_2\times Z_2$ |
| p4mg | $\langle x,y\rangle \twoheadrightarrow \langle 1,Y\rangle \simeq Z_2$ | $\langle x,x^{-1}y^2,r\rangle \twoheadrightarrow \langle 1,1,R\rangle \simeq Z_2$ | $\langle xy^{-1},y,q\rangle \twoheadrightarrow \langle Y^{-1},Y,Q\rangle \simeq Z_4$ | $\langle x,y,r\rangle \twoheadrightarrow \langle 1,Y,R\rangle \simeq Z_2\times Z_2$ |
| p6mm | $\langle x,y\rangle \twoheadrightarrow \langle 1,1\rangle \simeq \{1\}$ | $\langle y,x^{-2}y^3,s\rangle \twoheadrightarrow \langle 1,1,S\rangle \simeq Z_2^*$ | $\langle y,x^{-2}y^3,sxy^{-1}\rangle \twoheadrightarrow \langle 1,1,S\rangle \simeq Z_2^*$ | $\langle y,xy^{-1},s\rangle \twoheadrightarrow \langle 1,1,S\rangle \simeq Z_2$ |

* In these cases $[H_o,T] \not\leq K$

Table 4: Subgroups of minimal index: case (ii) and their images in $G/[H_o,T]$

| | | p2 | p4 | p6 | p2mm | p2mg | p2gg | c2mm | p4mm | p4mg | | | p6mm | |
|---|---|---|---|---|---|---|---|---|---|---|---|---|---|---|
| | | $\alpha$ | $\iota$ | $\iota$ | $\iota$ | $\alpha$ | $\iota$ | $\iota$ | $\iota$ | $\iota$ | $\alpha$ | $\alpha$ | $\iota$ | $\iota$ |
| p2 | x | x | y | y | $x^{-1}$ | $x^{-1}$ | x | $xy^{-1}$ | x | $x^{-1}y^2$ | $x^{-1}y^2$ | x | x | $x^{-1}y^3$ |
| | y | y | $x^{-1}$ | $x^{-1}y$ | y | y | $y^{-1}$ | $y^{-1}$ | $xy^{-1}$ | y | y | $xy^{-1}$ | $xy^{-1}$ | y |
| | u | ux | u | u | u | uy | $uxy^{-1}$ | u | u | u | $ux^{-1}y^2$ | $ux^{-1}y^2$ | u | u |

| | | p3 | p6 | p3m1 | p31m | p6mm | |
|---|---|---|---|---|---|---|---|
| | | $\beta$ | $\iota$ | $\gamma$ | $\gamma$ | $\gamma$ | $\gamma$ |
| p3 | x | x | y | $y^{-1}$ | x | $y^{-1}$ | x |
| | y | y | $x^{-1}y$ | $x^{-1}$ | $y^{-1}$ | $x^{-1}$ | $x^{-1}y^{-1}$ |
| | u | ux | u | $u^{-1}$ | $u^{-1}$ | $u^{-1}$ | $u^{-1}$ |

| | | p4 | p4mm | p4mg |
|---|---|---|---|---|
| | | $\delta$ | $\varepsilon$ | $\delta$ |
| p4 | x | x | $y^{-1}$ | x |
| | y | y | $x^{-1}$ | $y^{-1}$ |
| | u | $uxy^{-1}$ | $u^{-1}$ | $ux^{-1}y^2$ |

| | | p6mm | |
|---|---|---|---|
| | | $\gamma$ | $\gamma$ |
| p6 | x | $xy^{-1}$ | y |
| | y | $y^{-1}$ | $xy^{-1}$ |
| | u | $u^{-1}$ | $u^{-1}$ |

| | | p2mm | | c2mm | p4mm | p4mg |
|---|---|---|---|---|---|---|
| | | $\zeta$ | $\alpha$ | $\zeta\alpha$ | $\varepsilon$ | $\zeta\varepsilon$ |
| p2mm | x | x | x | x | y | y |
| | y | y | y | y | x | x |
| | r | ry | r | ry | s | s |
| | s | s | sx | sx | r | ry |

| | | p2mm | c2mm |
|---|---|---|---|
| | | $\zeta$ | $\zeta\alpha$ |
| p2mg | x | x | x |
| | y | y | y |
| | q | q | qx |
| | r | ry | ry |

| | | p2mm | | p2mg | c2mm | p4mm | | p4mg |
|---|---|---|---|---|---|---|---|---|
| | | $\zeta$ | $\alpha$ | $\alpha$ | $\zeta\alpha$ | $\zeta\alpha$ | $\varepsilon$ | $\zeta\varepsilon$ |
| p2gg | x | x | x | x | x | x | y | y |
| | y | y | y | y | y | y | x | x |
| | p | py | p | p | py | py | q | s |
| | q | q | qx | qx | qx | qx | p | ry |

| | | p2mm | p4mm | p4mg |
|---|---|---|---|---|
| | | $\zeta$ | $\varepsilon$ | $\varepsilon\zeta$ |
| c2mm | xy | xy | xy | xy |
| | $y^2$ | $y^2$ | $(xy)^2y^{-2}$ | $(xy)^2y^{-2}$ |
| | r | $ry^2$ | s | $s(xy)^2y^{-2}$ |
| | s | s | r | r |

| | | p31m | p6mm |
|---|---|---|---|
| | | $\eta$ | $\gamma$ |
| p3m1 | xy | xy | xy |
| | $y^3$ | $y^3$ | $(xy)^3y^{-3}$ |
| | r | $r(xy)^{-1}$ | s |
| | s | s | r |

| | | p6mm |
|---|---|---|
| | | $\gamma$ |
| p31m | x | y |
| | y | x |
| | r | s |
| | s | r |

| | | p4mm |
|---|---|---|
| | | $\theta$ |
| p4mm | x | x |
| | y | y |
| | r | r |
| | s | $sxy^{-1}$ |

| | | p4mm |
|---|---|---|
| | | $\theta$ |
| p4mg | x | x |
| | y | y |
| | q | $qx^{-1}y^2$ |
| | r | r |

Table 5: Automorphisms induced by embeddings

such that $\bar{L}_1 \cap \bar{K} = \bar{L}_2 \cap \bar{K}$. Of course, for any particular subgroup $\bar{M}$ of $\bar{K}$ it will not in general be possible to find a supplement $\bar{L}$ of $\bar{K}$ in $\bar{G}$ such that $\bar{L} \cap \bar{K} = \bar{M}$. However, in all these cases we find that each possible representation of $\bar{K}$ occurs from some embedding, though not in general in every embedding. For instance, in the embedding of pm in p2mg y is contained in every supplement of pm, and so has trivial image in each induced representation, but the embedding of pm in p2mm gives a representation in which the image of y is non-trivial.

We illustrate some of the methods outlined above by constructing the six restricted colourings of p3 corresponding to the six subgroups of $\bar{G} = Z_3 \times Z_3$. These are shown in Figure 1. From Table 5 we see that p3 has an outer automorphism $\beta$ (as subgroup of p3) which maps u to ux. Hence the three subgroups $\langle U \rangle$, $\langle UX \rangle$, $\langle UX^2 \rangle$ of $\bar{G}$ which are normal in $\bar{G}$ are equivalent under this automorphism. Inspection of the corresponding diagrams in Figure 1 shows that indeed these three patterns resemble one another in that each has a centre of three fold rotation which fixes all colours.

## 4. UNRESTRICTED COLOURINGS

To find all possible n-colourings of a given crystal, we need to be able to find all the subgroups of index n in its symmetry group.

(a) For $G \cong$ p1, the problem is essentially solved (2, Ch.I). We state the relevant theorem in a convenient form.

Theorem. Let K be a subgroup of index n in $G \cong$ p1. Then $K \cong$ p1 and has generators $x^\alpha y^\beta$, $y^\gamma$ where $\alpha\gamma = n$, $\alpha > 0$ and $0 \le \beta \le \gamma-1$. Each such set of generators corresponds to a unique subgroup, so that the number of distinct subgroups of index n in p1 is given by $\sum_{d|n} d$, where d is a positive integer which divides n.

For each of the remaining symmetry groups G, we must consider the greatest common divisor of the index n and the order $|H|$ of the point group of G. If $(n,|H|) = 1$, then the subgroup K must have its point-group $H_K$ isomorphic to H; if $|H|\,|\,n$, then K may be isomorphic to any of the possible subgroups of G; if $1 < (n,|H|) = d < |H|$, then $|H_K|$ must be divisible by $|H|/d$. For certain values of n, and certain plane groups, no subgroup of index n exists. We consider some examples.

(b) If $G \cong$ p2gg, then $|H| = 4$. Hence if $[G:K] = n$ and n is odd, then $|H_K| = 4$. Thus $K \cong$ p2gg also, and has generators $p^\alpha$, $q^\beta$ where $\alpha\beta = n$. Thus the number of

(i) $\bar{K} = \bar{G}$, $X \to (1)$, $U \to (1)$

(ii) $\bar{K} = \langle U \rangle$, $X \to (123)$, $U \to (1)$

(iii) $\bar{K} = \langle X \rangle$, $X \to (1)$, $U \to (123)$

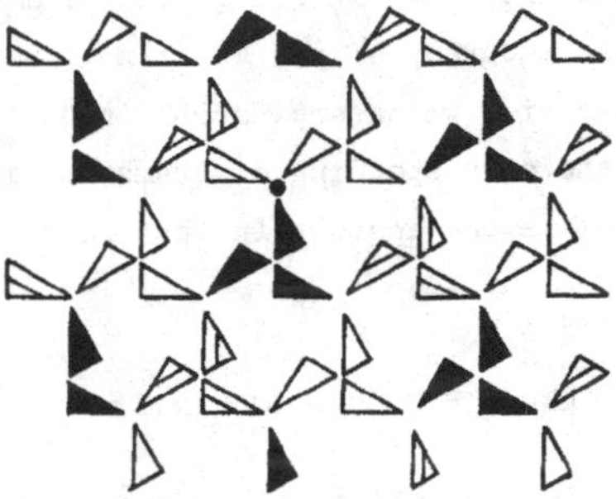

(iv) $\bar{K} = \langle UX \rangle$, $X \to (132)$, $U \to (123)$

(v) $\bar{K} = \langle UX^2 \rangle$, $X \to (123)$, $U \to (123)$

(vi) $\bar{K} = \{\bar{1}\}$, $X \to (147)(258)(369)$, $U \to (123)(456)(789)$

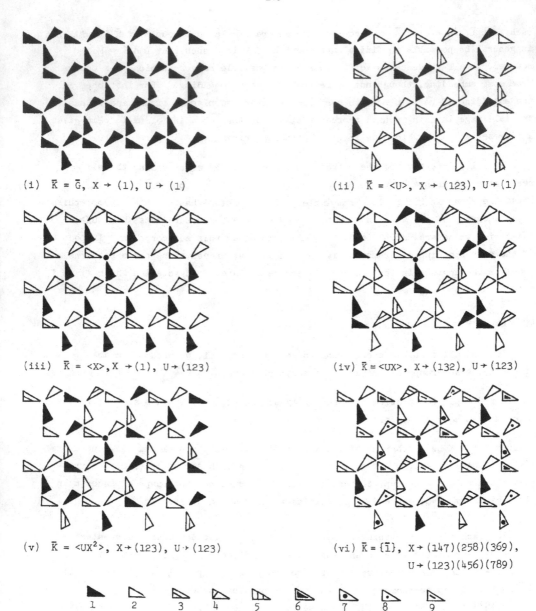

In each case, the triangle immediately below the origin is assigned colour 1.

Figure 1: Restricted colourings of p3.

such subgroups equals the number of positive divisors of n. If $n = 2m$, where m is odd, then $|H_K| = 2$ or 4, and $K \approx p2$, pg or p2gg. If $K \approx p2gg$, then it is generated as before by $p^\alpha, q^\beta$ with $\alpha\beta = n$. If $K \approx pg$, it is either $\langle x^\alpha, y^\beta, p^\alpha \rangle$ or $\langle x^\alpha, y^\beta, q^\beta \rangle$ with $\alpha\beta = m$. If $K \approx p2$, it must be $\langle x^\alpha y^\beta, y^\gamma, pq \rangle$ where $\alpha\gamma = m$, $0 < \alpha$ and $0 \le \beta \le \gamma-1$ Otherwise $n = 4m$, for some m; then $K \approx p2$, pg or p2gg as above, or else $K \approx p1$ and is generated by $x^\alpha y^\beta$, $y^\gamma$ where $\alpha\gamma = m$, $0 < \alpha$ and $0 \le \beta \le \gamma-1$.

(c) Finally let $G \approx p3$, and let K be a subgroup of index n in G. If $n = 3m$, then either $K \approx p1$, in which case $K = \langle x^\alpha y^\beta, y^\gamma \rangle$, where $0 < \alpha$, $\alpha\gamma = m$ and $0 \le \beta \le \gamma-1$ as usual, or else $K \approx p3$; if $n \not\equiv 0 \pmod 3$, then certainly $K \approx p3$. If $K \approx p3$, what generators can we choose for K? We could certainly take $x^\alpha y^\beta$, $x^\gamma y^\delta$, $ux^\varepsilon y^\zeta$ as such a set, since $(ux^\varepsilon y^\zeta)^3 = 1$. Now $u^x = uxy^{-1}$ and $u^y = uxy^2$, so

$$(ux^\varepsilon y^\zeta)^{x^\mu y^\nu} = u.x^{\mu+\nu+\varepsilon}y^{-\mu+2\nu+\zeta}.$$

If $\varepsilon + \zeta \equiv 0 \pmod 3$, then we can choose integers $\mu$ and $\nu$ such that $\mu + \nu + \varepsilon = -\mu + 2\nu + \zeta = 0$, and we may assume that K is conjugate to a subgroup generated by $x^\alpha y^\beta$, $x^\gamma y^\delta$, u. Similarly, if $\varepsilon + \zeta \equiv 1 \pmod 3$, then K is conjugate to $\langle x^\alpha y^\beta, x^\gamma y^\delta, ux \rangle$, and if $\varepsilon + \zeta \equiv 2 \pmod 3$, then K is conjugate to $\langle x^\alpha y^\beta, x^\gamma y^\delta, ux^2 \rangle$.

In any case, $(x^\alpha y^\beta)^u = (x^\alpha y^\beta)^{ux} = (x^\alpha y^\beta)^{ux^2} = x^{-\beta}y^{\alpha-\beta}$, so that $x^\gamma y^\delta = x^{-\beta}y^{\alpha-\beta}$ and again $(x^\gamma y^\delta)^u = x^{\beta-\alpha}y^{-\alpha}$. Hence the generators of K (or its conjugate) are of the form $x^\alpha y^\beta$, $x^{-\beta}y^{\alpha-\beta}$, $ux^A$, where A = 0, 1 or 2, and we see that the value of the determinant

$$\begin{vmatrix} \alpha & -\beta \\ \beta & \alpha-\beta \end{vmatrix} = \alpha^2 - \alpha\beta + \beta^2 = n. \qquad (*)$$

To see when (*) has a solution in integers, we use a classical result in number theory; see for instance (1, §§37.1-37.3).

Theorem. The Diophantine equation

$$\alpha^2 - \alpha\beta + \beta^2 = n$$

has a solution if and only if

$$n = 2^{2b} 3^a p_1^{a_1} \cdots p_k^{a_k} q_1^{2b_1} \cdots q_\ell^{2b_\ell},$$

where $p_1, \ldots, p_k$ are primes congruent to 1(mod 6) and $q_1, \ldots, q_\ell$ are primes congruent to 5(mod 6). Moreover, when n is of this form, there are $[\tfrac{1}{2}\{(a_1+1)\ldots(a_k+1)+1\}]$ inequivalent solutions where the solutions equivalent to $\alpha = c$, $\beta = d$ are (c,d), (-d,c-d), (d-c,-c), (c-d,-d), (d,c), (-c,d-c) and their negatives.

Notice that since $(x^\alpha y^\beta)^u = x^{-\beta}y^{\alpha-\beta}$ and $(x^{-\beta}y^{\alpha-\beta})^u = x^{\beta-\alpha}y^{-\alpha}$, the solutions (c,d), (-d,c-d), (d-c,-c) and their negatives all generate the same subgroup K.

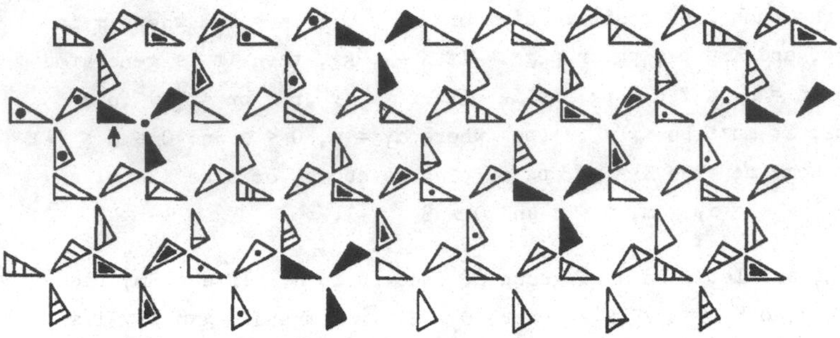
[1]

[2]

[3]

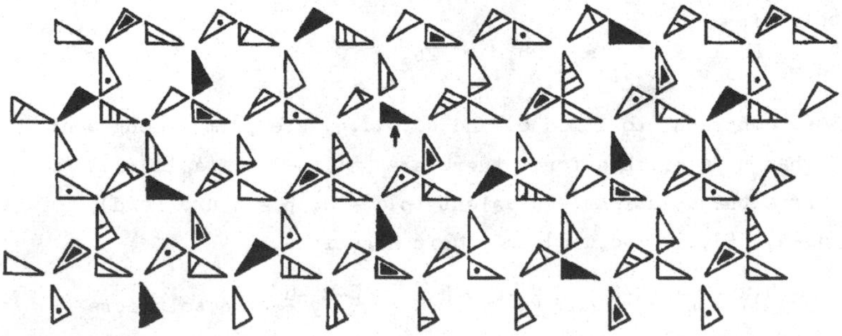
[4]

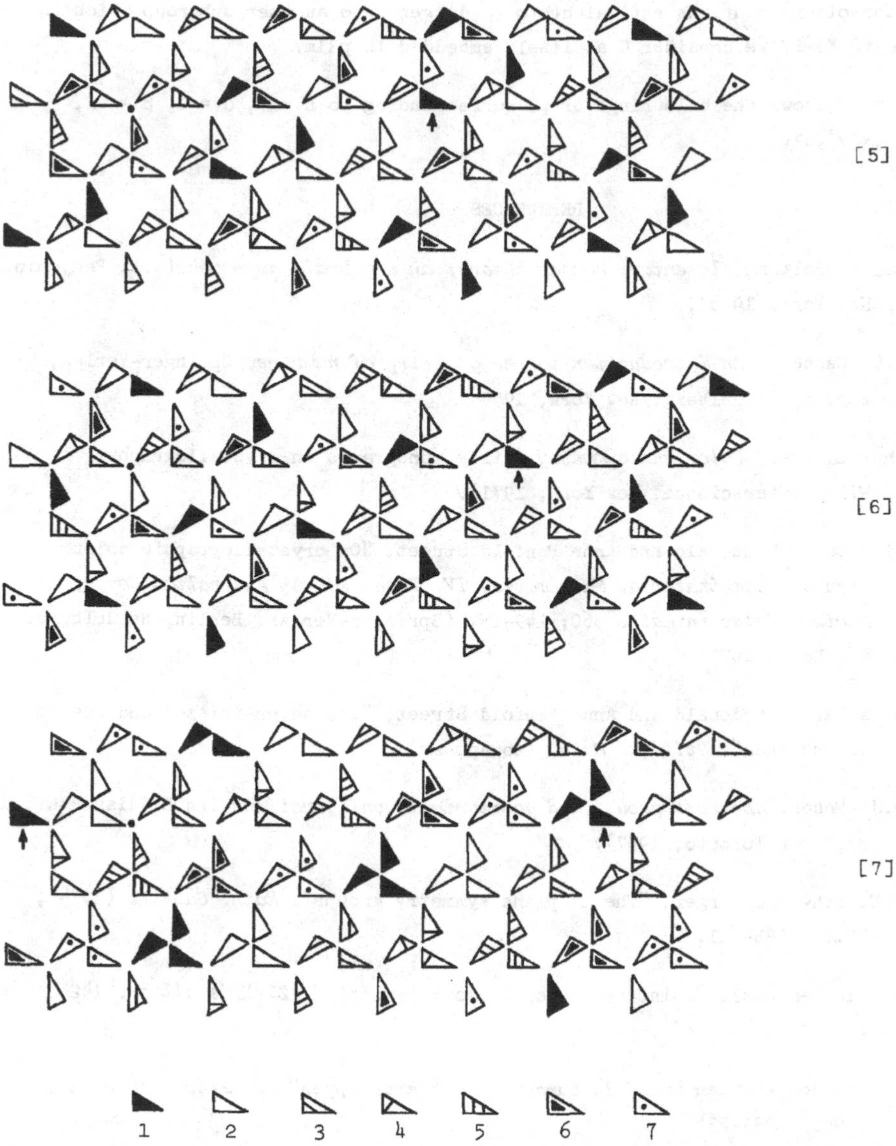


In each case, the triangle indicated by the arrow is assigned colour 1. Note that the position of this triangle in colouring [k] is obtained from that in colouring [1] by the application of $x^{k-1}$.

Figure 2: Unrestricted colourings of p3, with $K = \langle xy^{-2}, x^2y^3, u\rangle$, $x \to (1234567)$, $u \to (253)(467)$.

However the other solutions equivalent to (c,d) generate another subgroup which is conjugate to K, if we consider G as itself embedded in p31m.

Figure 2 shows the colourings of p3 corresponding to $n = 7$, $\alpha = 1$, $\beta = -2$, $K = \langle xy^{-2}, x^2y^3, u \rangle$.

REFERENCES

(1) Ethan D. Bolker, *Elementary number theory; an algebraic approach* (W.A. Benjamin, New York, 1970).

(2) J.W.S. Cassels, *An introduction to the geometry of numbers* (Springer-Verlag, Berlin, Heidelberg, New York, 1959).

(3) Arthur L. Loeb, *Color and symmetry* (Wiley monographs in crystallography, Wiley-Interscience, New York, 1971).

(4) Sheila Oates Macdonald and Anne Penfold Street, "On crystallographic colour groups", *Combinatorial Mathematics IV, Proc. Fourth Australian Conf., Lecture Notes in Math.* 560, 149-157 (Springer-Verlag, Berlin, Heidelberg, New York, 1977).

(5) Sheila Oates Macdonald and Anne Penfold Street, "The seven friezes and how to colour them", *Utilitas Math.* (to appear).

(6) W.O.J. Moser, *Abstract groups and geometrical configurations* (Ph.D. dissertation, Univ. of Toronto, 1957).

(7) R.L.E. Schwarzenberger, "The 17 plane symmetry groups", *Math. Gazette* (1974), LVIII, 123-131.

(8) Marjorie Senechal, "Point groups and colour symmetry", *Z. Kristallogr.* 142 (1975), 1-23.

(9) B.L. van der Waerden and J.J. Burckhardt, "Farbgruppen", *Z. Kristallogr.* 115 (1961), 231-234.

# COMPUTING AUTOMORPHISMS AND CANONICAL LABELLINGS OF GRAPHS

Brendan D. McKay
Department of Mathematics,
University of Melbourne,
Parkville, Victoria, 3052
Australia

ABSTRACT

A new algorithm is presented for the related problems of canonically labelling a graph or digraph and of finding its automorphism group. The automorphism group is found in the form of a set of less than n generators, where n is the number of vertices. An implementation is reported which is sufficiently conserving of time and space for it to be useful for graphs with over a thousand vertices.

## 1. INTRODUCTION

Let V be the finite set $\{1, 2, \ldots, n\}$. Define $\underset{\sim}{G}(V)$ to be the set of all (labelled) graphs with vertex set V. Let $S_n$ be the symmetric group acting on V. For $G \in \underset{\sim}{G}(V)$ and $g \in S_n$, define $G^g \in \underset{\sim}{G}(V)$ to be the graph in which vertices $v^g$ and $w^g$ are adjacent exactly when v and w are adjacent in G. The *automorphism group* of G, Aut(G), is the group $\{g \in S_n \mid G^g = G\}$.

The *canonical label problem* is to find a map *canon*: $\underset{\sim}{G}(V) \to \underset{\sim}{G}(V)$ such that for $G \in \underset{\sim}{G}(V)$ and $g \in S_n$,

(1) *canon* (G) is isomorphic to G, and (2) *canon* $(G^g)$ = *canon* (G).

Note that there may be many functions *canon* satisfying (1) and(2).

If $G, H \in \underset{\sim}{G}(V)$, we see that G and H are isomorphic if and only if *canon* (G) = *canon* (H).

In this paper we present a new algorithm for computing *canon* (G) which will also find a set of fewer than n automorphisms which generate Aut(G). With only minor modifications which we will indicate, the algorithm is equally applicable to digraphs. Undefined graph theoretic or group theoretic concepts can be found in [1] or [7] respectively.

## 2. EQUITABLE PARTITIONS

Let V = $\{1, 2, \ldots, n\}$. A *partition* of V is a collection $\pi$ of disjoint non-empty subsets of V whose union is V. The elements of $\pi$ are called its *cells*. An *ordered partition* of V is a *sequence* $(C_1, C_2, \ldots, C_k)$ for which $\{C_1, C_2, \ldots, C_k\}$ is a partition. The sets of all partitions of V, and of all ordered partitions of V will be denoted by $\Pi(V)$ and $\underset{\sim}{\Pi}(V)$ respectively.

Define $\Pi^*(V) = \Pi(V) \cup \underline{\Pi}(V)$. Let $\pi_1, \pi_2 \in \Pi^*(V)$. We write $\pi_1 \leq \pi_2$ ($\pi_1$ is *finer* than $\pi_2$, $\pi_2$ is *coarser* than $\pi_1$) if every cell of $\pi_1$ is contained in some cell of $\pi_2$. If both $\pi_1 \leq \pi_2$ and $\pi_2 \leq \pi_1$, we write $\pi_1 \cong \pi_2$. If $\pi \in \Pi^*(V)$, the number of cells of $\pi$ is denoted by $|\pi|$. $\pi$ is called *discrete* if $|\pi| = n$.

Let $\pi \in \Pi^*(V)$ and $g \in S_n$. Then $\pi^g \in \Pi^*(V)$ is formed by replacing each cell $C \in \pi$ by $C^g$. If $\pi = \pi^g$, g is said to *fix* $\pi$. Denote by $\pi \vee g$ the finest partition of V which is coarser than $\pi$ but fixed by g. The existence of $\pi \vee g$ follows from the fact that $(\Pi(V), \leq)$ is a lattice [3].

Choose a fixed $G \in \underset{\sim}{G}(V)$. If $W \subseteq V$ and $v \in V$, the number of vertices in W which are adjacent to v will be denoted by $d(v, W)$. Let $\pi \in \Pi^*(V)$. $\pi$ is said to be *equitable* (for G) if for any $C_1, C_2 \in \pi$ and $v_1, v_2 \in C_1$ we have $d(v_1, C_2) = d(v_2, C_2)$. For an arbitrary $\pi$, the coarsest equitable partition which is finer than $\pi$ will be denoted by $\xi(\pi)$. Similarly, $\theta(\pi)$ denotes the partition whose cells are the orbits of the subgroup of Aut(G) which fixes $\pi$. The proof of the following lemma can be found in [3].

LEMMA 1. Let $\pi \in \Pi(V)$. Then

(i) $\theta(\pi) \leq \xi(\pi)$,

(ii) $\theta(\pi)$ is equitable, and

(iii) if $\pi$ is equitable, and $n - |\pi| \leq 5$, the smallest cells of $\pi$ of size $\geq 2$ are cells of $\theta(\pi)$. (Not true if G is a digraph.) □

Corneil proved in [2] that for any $\pi$, $\theta(\pi) = \xi(\pi)$ if G is a tree. This can be generalised to uni-cyclic graphs and many others. See [3] for further details.

Algorithms for computing $\xi(\pi)$ have been used many times in graph isomorphism programs ([2], [5], [6]). For our own purposes, however, the following system appears to be more efficient. Let $\pi \in \Pi(V)$ and let $\alpha$ be a subset of $\pi$.

ALGORITHM 1: Compute $\tilde{\pi} = \mathcal{R}(G, \pi, \alpha)$

(1) $\tilde{\pi} \leftarrow \pi$

(2) If $\alpha = \emptyset$ or $\tilde{\pi}$ is discrete, *stop*.
Choose any non-null subset $\beta$ of $\alpha$.
$\alpha \leftarrow \alpha \backslash \beta$, $i \leftarrow 1$
(Suppose $\tilde{\pi} = \{C_1, C_2, \ldots, C_k\}$ and $\beta = \{W_1, W_2, \ldots, W_r\}$.)

(3) Partition $C_i$ into subsets $D_1, D_2, \ldots, D_s$ according to the vectors
$(d(v, W_1), d(v, W_2), \ldots, d(v, W_r))$ for $v \in C_i$.
$\tilde{\pi} \leftarrow \tilde{\pi} \cup \{D_1, D_2, \ldots, D_s\} \backslash \{C_i\}$

$\alpha \leftarrow \alpha \cup \{D_2, \ldots, D_s\}$ (if $s \geq 2$)

$i \leftarrow i+1$

If $i \leq k$ go to (3); otherwise go to (2).

THEOREM 1. For any $\pi \in \Pi(V)$, $\mathcal{R}(G, \pi, \pi) \simeq \xi(\pi)$.

Proof: (a) In step (3) of the algorithm, $|\alpha|$ is not increased unless $\tilde{\pi}$ is made finer. Since $|\alpha|$ is reduced in step (2), the algorithm is sure to terminate.

(b) By definition, $\xi(\pi) \leq \pi$. Suppose that before some execution of step (3) we have $\xi(\pi) \leq \tilde{\pi}$. Since $\beta \subseteq \tilde{\pi}$, each element of $\beta$ is a union of cells of $\xi(\pi)$. Hence we will also have $\xi(\pi) \leq \tilde{\pi}$ after execution of step (3). Therefore $\xi(\pi) \leq \mathcal{R}(G, \pi, \pi) \leq \pi$.

(c) Suppose that $\mathcal{R}(G, \pi, \pi)$ is not equitable. Then for some $C_1, C_2 \in \mathcal{R}(G, \pi, \pi)$ there are $v, w \in C_1$ such that $d(v, C_2) \neq d(w, C_2)$. Since $\tilde{\pi}$ is made successively finer by the algorithm, $v$ and $w$ must always be in the same cell of $\tilde{\pi}$.

At step (1), $C_2$ is contained in some cell of $\alpha$. Hence $C_2$ must sometime be contained in an element $W$ of $\beta$.

(d) Since $v$ and $w$ are never separated, $d(v, W) = d(w, W)$. Hence there is a cell $C_3$ of $\mathcal{R}(G, \pi, \pi)$ other than $C_2$ for which $d(v, C_3) \neq d(w, C_3)$. Since $C_2$ and $C_3$ are different cells of $\mathcal{R}(G, \pi, \pi)$ they must be separated sometime in step (3). At least one of them, say $C_2$, will then be contained in a new element of $\alpha$.

(e) Since the argument in (d) can be repeated indefinitely, the algorithm never terminates, contradicting (a). Hence $\mathcal{R}(G, \pi, \pi)$ is equitable, and so $\mathcal{R}(G, \pi, \pi) \simeq \xi(\pi)$. □

A considerable advantage which Algorithm 1 has over previous algorithms is that in some important situations, $\alpha$ can be a proper subset of $\pi$. Suppose $\pi_1$ is an equitable partition of $V$ coarser than $\pi$. Let $\alpha \subseteq \pi$ be such that for any $D \in \pi_1$, $C \subseteq D$ for at most one $C \in \pi \backslash \alpha$. Then it can be proved (see [3]) that $\mathcal{R}(G, \pi, \alpha) \simeq \xi(\pi)$.

When implemented on a computer, Algorithm 1 becomes an operation on *ordered* partitions, and produces an ordered result. We will presume an implementation such that for any $g \in S_n$, $\mathcal{R}(G^g, \pi^g, \alpha^g) = \mathcal{R}(G, \pi, \alpha)^g$.

## 3. THE BASIC STRUCTURE OF THE ALGORITHM

Our algorithm for computing *canon* (G) and generators for Aut(G) is based on a depth-first search through a tree whose nodes are equitable ordered partitions of V.

The first node, or *root*, is the equitable partition $\mathcal{R}(G, (V), (V))$, where $(V)$ is the ordered partition with one cell. Suppose that $\pi = (C_1, C_2, \ldots, C_r)$ is an arbitrary node of the search tree. If $\pi$ is discrete, it is an end-node of the tree and we will call it a *terminal partition*. If $\pi$ is not discrete, let $C_i$ be the first cell of $\pi$ with the smallest size $\geq 2$. For $v \in C_i$ define $\pi \circ v = (C_1, \ldots, C_i \setminus \{v\}, \ldots, C_r, \{v\})$. The successors of $\pi$ in the search tree are the partitions $\mathcal{R}(G, \pi \circ v, \{v\})$, for $v \in C_i$, which are equitable by the remark following Theorem 1.

Let X be the set of terminal partitions in the search tree. Given any $\tau \in X$, we can form the graph $G^\tau$ by labelling the vertices of G in the order they appear in $\tau$. Define the equivalence relation $\sim$ on X by $\tau_1 \sim \tau_2$ if and only if $G^{\tau_1} = G^{\tau_2}$. It is easy to show ([3] or [4]) that for any fixed $\varepsilon \in X$, $\mathrm{Aut}(G) = \{g \mid \tau = \varepsilon^g, \tau \in X, \tau \sim \varepsilon\}$. Furthermore, by comparing the adjacency matrices lexicographically we can define an order on $\underset{\sim}{G}(V)$ and define *canon* $(G) = \max\{G^\tau \mid \tau \in X\}$.

Since $|X|$ is a multiple of $|\mathrm{Aut}(G)|$ it is usually not feasible to generate the entire search tree. This problem is largely overcome by the methods described in [3] or [4] which use discovered automorphisms of G to eliminate sections of the search tree from consideration. However, while these methods will reduce the size of each equivalence class of terminal partitions to a manageable size (usually n or less), they will not reduce the number of classes. One way of doing this is to define a function $\Lambda(G, \pi)$ for each $G \in \underset{\sim}{G}(V)$ and $\pi \in \underset{\sim}{\Pi}(V)$ such that $\Lambda(G^g, \pi^g) = \Lambda(G, \pi)$ for any $g \in S_n$. An example of such a function would be the number of edges of G whose end-points are both in the same cell of $\pi$. Now suppose that $\tau \in X$ and that $\pi_1 \geq \pi_2 \geq \cdots \geq \pi_k = \tau$ is the sequence of equitable partitions from the root of the search tree to $\tau$. Then $\tau$ can be associated with the sequence $\underset{\sim}{\Lambda}(\tau) = (\Lambda(G, \pi_1), \Lambda(G, \pi_2), \ldots, \Lambda(G, \pi_k))$. Clearly $\tau_1$ and $\tau_2$ cannot be equivalent unless $\underset{\sim}{\Lambda}(\tau_1) = \underset{\sim}{\Lambda}(\tau_2)$. Further efficiency can be achieved by ordering the vectors $\underset{\sim}{\Lambda}(\tau)$ lexicographically and redefining *canon* (G) to be $\max\{G^\tau \mid \tau \in X, \underset{\sim}{\Lambda}(\tau) = \Lambda^*\}$, where $\Lambda^* = \max\{\underset{\sim}{\Lambda}(\tau) \mid \tau \in X\}$. By this means we can eliminate sections of the search tree which cannot contain either new automorphisms or *canon* (G).

In some cases, additional means of acceleration is provided by Lemma 1 (iii). If $\pi$ is an equitable partition in the search tree and $n - |\pi| \leq 5$, all the terminal partitions descended from $\pi$ are equivalent.

## 4. THE ALGORITHM

ALGORITHM 2: Compute generators for Aut(G), where G is a graph or digraph, and optionally compute *canon* $(G) = G^\beta$.

Notes: (i) Lower case Greek letters represent partitions. The variables $C_1, C_2, \ldots$ represent sets.

(ii) The minimum of an empty set is defined to be $\infty$, where $\infty$ is a number larger than any number it is compared to.

(iii) If x is a permutation (or partition), $\Omega(x)$ denotes the set consisting of the smallest element of each cycle (or cell) of x, and $\Phi(x)$ is the set of elements in trivial cycles (or cells of size 1) of x.

(iv) *lab* and *dig* are boolean variables. *lab* = *true* if *canon* (G) is required. *dig* = *true* if G is a digraph, or a graph with loops.

(1) $k \leftarrow 1$, $size \leftarrow 1$, $h \leftarrow 0$, $index \leftarrow 0$
$\theta \leftarrow$ discrete partition of V
$\pi_1 \leftarrow \mathcal{R}(G, (V), (V))$
If *dig* or $n - |\pi_1| \geq 6$, $q \leftarrow 2$, else $q \leftarrow 1$.
If $\pi_1$ is discrete go to (20).
$C_1 \leftarrow$ first cell of $\pi_1$ with smallest size $\geq 2$
$e_1 \leftarrow 0$, $v_1 \leftarrow \min C_1$, $w_1 \leftarrow v_1$

(2) $k \leftarrow k+1$
$\pi_k \leftarrow \mathcal{R}(G, \pi_{k-1} \circ v_{k-1}, \{v_{k-1}\})$
$z \leftarrow \Lambda(G, \pi_k)$
If $\pi_k$ is not discrete, $e_k \leftarrow 0$ and $C_k \leftarrow$ first cell of $\pi_k$ with smallest size $\geq 2$.
If $h = 0$, go to (6).
If $hx = k-1$ and $z = x_k$, $hx \leftarrow k$
If not *lab*, go to (4).
If $hy \neq k-1$, go to (3).
$qy \leftarrow z - y_k$
If $qy = 0$, $hy \leftarrow k$.

(3) If $qy > 0$, $y_k \leftarrow z$.

(4) If $hx = k$ or (*lab* and $qy \geq 0$), go to (5).
$k \leftarrow q-1$
Go to (9).

(5) If $\pi_k$ is discrete, go to (7).

$v_k = \min C_k$

If $h = 0$, $w_k \leftarrow v_k$.

If *dig* or $n - |\pi_k| \geq 6$, $q \leftarrow k+1$.

Go to (2).

(6) If *lab*, $y_k \leftarrow z$.

$x_k \leftarrow z$

Go to (5).

(7) If $h < q$, go to (15).

Compute the permutation g such that $\varepsilon^g = \pi_k$.

(8) ($g \in \mathrm{Aut}(G)$ : Write g if desired, and store $(\Phi(g), \Omega(g))$ if room is available.)

$\theta \leftarrow \theta \vee g$, $k \leftarrow h$

(9) If $k = 0$, *stop*.

If $k > h$, go to (13).

$h = \min \{k, h\}$

(10) If $v_k = w_k$ are in the same cell of $\theta$, index $\leftarrow$ index + 1.

$v_k \leftarrow \min \{v \in C_k \mid v > v_k\}$

If $v_k = \infty$, go to (12).

If $v_k \neq \Omega(\theta)$, go to (10).

(11) $q \leftarrow \min \{q, k+1\}$, $hx \leftarrow \min \{hx, k\}$

If not *lab*, go to (12).

$hb \leftarrow \min \{hb, k\}$

If $hy < k$, go to (2).

$hy \leftarrow k$, $qy \leftarrow 0$

Go to (2).

(12) size $\leftarrow$ size $\times$ index

index $\leftarrow 0$, $k \leftarrow k-1$

Go to (9).

(13) If $e_k = 0$, go to (14).

$e_k \leftarrow 1$

For any stored pairs $(\Phi(g), \Omega(g))$ such that $\{v_1, v_2, \ldots, v_{k-1}\} \subseteq \Phi(g)$, set $C_k \leftarrow C_k \cap \Omega(g)$.

(14) $v_k \leftarrow \min \{v \in C_k \mid v > v_k\}$
If $v_k = \infty$, set $k \leftarrow k-1$ and go to (9).
Go to (11).

(15) If $h = 0$, go to (20).
If $hx \neq k$, go to (16).
Compute the permutation $g$ such that $\varepsilon^g = \pi_k$.
If $g \in \mathrm{Aut}(G)$, go to (8).

(16) If $qy < 0$ or not *lab*, go to (18).
If $qy > 0$, go to (17).
If $G^\beta = G^{\pi_k}$, go to (19).
If $G^\beta > G^{\pi_k}$, go to (18).

(17) $\beta \leftarrow \pi_k$, $hy \leftarrow k$, $hb \leftarrow k$, $y_{k+1} \leftarrow \infty$, $qy \leftarrow 0$

(18) $k \leftarrow q-1$
Go to (9).

(19) $k \leftarrow hb$
If $k \neq h$, go to (9).
Compute the permutation $g$ such that $\beta^g = \pi_k$.
Go to (8).

(20) $h \leftarrow k$, $hx \leftarrow k$, $x_{k+1} \leftarrow \infty$, $\varepsilon \leftarrow \pi_k$
$k \leftarrow k+1$
If not *lab*, go to (9).
$\beta \leftarrow \pi_{k-1}$, $hy \leftarrow k+1$, $hb \leftarrow k+1$, $y_{k+2} \leftarrow \infty$, $qy \leftarrow 0$
Go to (9).

Let $G$ be a graph or a digraph and let $A = \mathrm{Aut}(G)$. If $W \subseteq V$, $A_W$ denotes the point-wise stabiliser of $W$ in $A$. Consider the instant when the first line of step (12) has been executed for a particular value of $k$. Define $\theta^{(k-1)}$, $\mathrm{index}^{(k-1)}$, $\mathrm{size}^{(k-1)}$ to be the current values of $\theta$, index and size, and let $\Sigma^{(k-1)}$ be the set of all elements of $\mathrm{Aut}(G)$ found by this stage.

THEOREM 2. (1) Let $K$ be the value of $k-1$ at the start of step (20). Define $A^{(0)} = A$ and $A^{(k)} = A_{\{w_1,\ldots,w_k\}}$ for $1 \le k \le K$. Then for $0 \le k \le K$,

(i) $\mathrm{size}^{(k)} = |A^{(k)}|$
(ii) $\mathrm{index}^{(k)} = |A^{(k)}|/|A^{(k+1)}| \quad (k < K)$
(iii) the cells of $\theta^{(k)}$ are the orbits of $A^{(k)}$
(iv) $\Sigma^{(k)}$ generates $A^{(k)}$
(v) $|\Sigma^{(k)}| \le n - \ell_k$, where $A^{(k)}$ has $\ell_k$ orbits.

(2) If *lab* is true, $G^{\beta}$ is a canonical labelling of G when the algorithm terminates.

Proof: Apart from minor complications, the theorem follows from the results in [3] and [4]. □

A simple method for generating Aut(G) from $\Sigma^{(0)}$ is given in [4], as are a few other facts about $\Sigma^{(0)}$, for example the following lemma.

LEMMA 2. Suppose that for some $W \subseteq V$, $A_W$ has exactly one non-trivial orbit. Then $\Sigma^{(0)}$ has a subset which generates a conjugate of $A_W$ in A. □

## 5. EXPERIMENTAL PERFORMANCE

The algorithm has been implemented in (partly non-standard) Fortran on a CDC Cyber 70 Model 73 computer. The graph G is represented by its adjacency matrix, stored one bit per entry. The storage of the partitions $\pi_i$ is facilitated by the easily verified fact that in any sequence of partitions $\pi_1 \geq \pi_2 \geq \cdots$, the total number of different cells is less than 2n. Suppose that n bits occupy m machine words, and that $\ell$ is the maximum value of k for which $\pi_k$ is ever computed (obviously, $\ell \leq n$). Then at most $n(2m + 8) + 2\ell m$ words of storage are required by the program, plus an extra $n(m + 2)$ words if *canon* (G) is required, and an optional $2m(n - 1)$ words to ensure that $(\Phi(g), \Omega(g))$ can always be stored at step (8). The function $\Lambda(G, \pi)$ used in the implementation has an integer value formed from the cell-sizes of $\pi$ and from various items remaining from the computation of $\pi$ by Algorithm 1.

The execution times for various common families of graphs are shown in Figure 1. For all cases except for the random graphs, the times are for computing *canon* (G) as well as Aut(G). We believe that both the execution times and their rate of increase with n are considerably superior to that of any previously published algorithm.

| | | |
|---|---|---|
| $\overline{K}_n$ | : | empty graph. |
| RD | : | randomly selected digraph with constant out-degree = $\frac{1}{2}n$. |
| RC | : | randomly selected circulant graph with degree = $\frac{1}{2}n$. |
| $Q_m$ | : | m-dimensional cube; $n = 2^m$. |
| RG | : | randomly selected graph with edge-density = $\frac{1}{2}$<br>(1): *canon* (G) found, (2): *canon* (G) not found. |
| SR25 | : | strongly regular graphs on 25 vertices (average time). |
| SR35 | : | strongly regular block intersection graphs of Steiner triple systems with 15 points and 35 blocks (average time). |

The dashed line marked P in Figure 1 gives the time required to perform a single permutation of an adjacency matrix with edge-density $\frac{1}{2}$. Since this is an essential

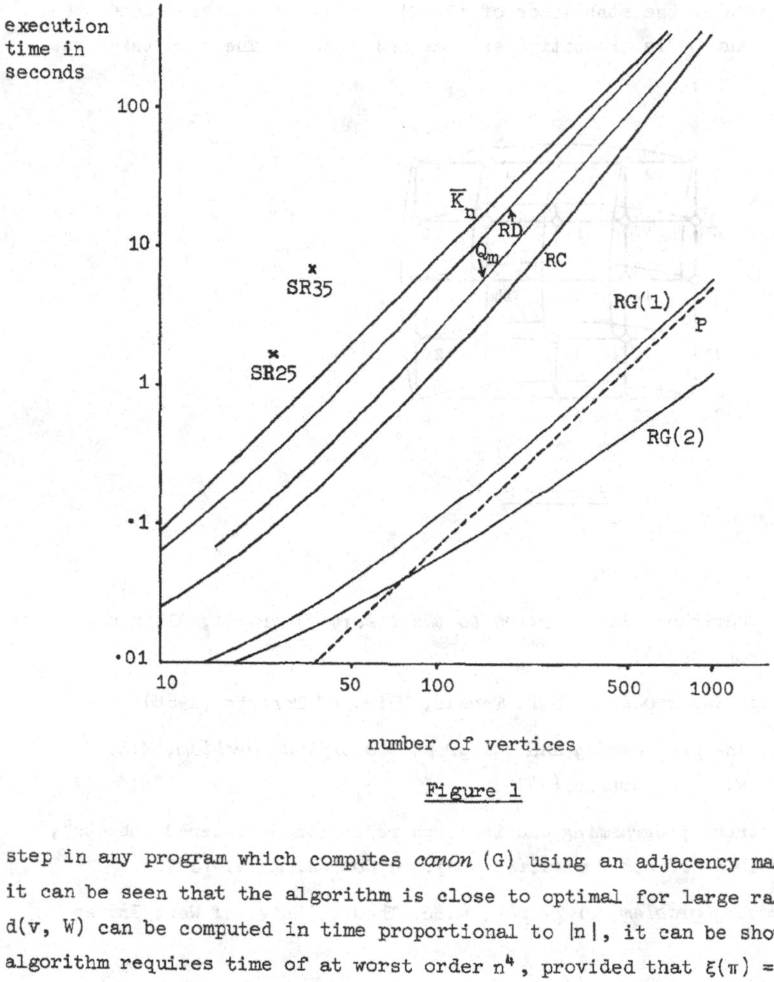


Figure 1

step in any program which computes *canon* (G) using an adjacency matrix representation, it can be seen that the algorithm is close to optimal for large random graphs. If d(v, W) can be computed in time proportional to $|n|$, it can be shown that the algorithm requires time of at worst order $n^4$, provided that $\xi(\pi) = \theta(\pi)$ for any partition $\pi$. However, no useful upper bound has been proved in general.

A listing of the program, plus suggestions for implementation, can be obtained from the author.

## 6. EXAMPLE

Let G be the graph $C_5 \times C_5$ labelled as shown in Figure 2.

The generators of Aut(G) found by the algorithm were

(6 21)(7 22)(8 23)(9 24)(10 25)(11 16)(12 17)(13 18)(14 19)(15 20),

(2 5)(3 4)(7 10)(8 9)(12 15)(13 14)(17 20)(18 19)(22 25)(23 24),

(2 6)(3 11)(4 16)(5 21)(8 12)(9 17)(10 22)(14 18)(15 23)(20 24) and

(1 2)(3 5)(6 7)(8 10)(11 12)(13 15)(16 17)(18 20)(21 22)(23 25),

of which the first generates the stabiliser of $\{1, 2\}$ and the first three generate the stabiliser of 1. Aut(G) is transitive and has order 200. The time taken was 0·16 seconds.

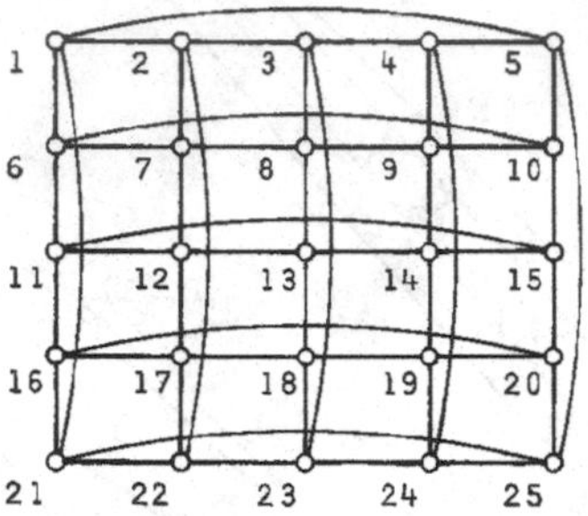


Figure 2

REFERENCES

[1] M. Behzad and G. Chartrand, *Introduction to the theory of graphs*, Allyn and Bacon, Boston (1971).

[2] D.G. Corneil, *Graph Isomorphism*, Ph.D. Thesis, Univ. of Toronto (1968).

[3] B.D. McKay, *Backtrack programming and the graph isomorphism problem*, M.Sc. Thesis, Univ. of Melbourne (1976).

[4] B.D. McKay, "Backtrack programming and isomorph rejection on ordered subsets", to appear in *Proc. 5th Australian Conf. on Combin. Math.* (1976).

[5] R. Parris, *The coding problem for graphs*, M.Sc. Thesis, Univ. of West Indies (1968).

[6] J.P. Steen, "Principle d'un algorithme de recherche d'un isomorphisme entre deux graphes", *RIRO*, R-3, 3 (1969), 51-69.

[7] H. Wielandt, *Finite permutation groups*, Academic Press, New York and London (1964).

05B05 05B25 62K10

# ON A RESULT OF BOSE AND SHRIKHANDE

Elizabeth J. Morgan

Department of Mathematics,
University of Queensland,
St. Lucia, Q. 4067, Australia.

ABSTRACT. Necessary and sufficient conditions are given for the extendability of a regular 2-component pairwise balanced design (PB2-design) to a balanced incomplete block design. This gives an alternative non graph-theoretic proof of a result of R.C. Bose and S.S. Shrikhande, showing extendability of a PB2-design with certain parameters to a projective plane of even order q, q > 6.

A *regular 2-component pairwise balanced design*, or PB2-design, with parameters

$$(V;\ b_1,b_2;\ r_1,r_2;\ k_1,k_2;\ \lambda) \tag{1}$$

is a collection of $b_1 + b_2$ subsets, called blocks, chosen from a V-set such that (i) every pair of elements belongs to $\lambda$ blocks, and (ii) for $i = 1,2$ there are $b_i$ blocks of size $k_i$, and these $b_i$ blocks together contain each element $r_i$ times. The $b_i$ blocks of size $k_i$ will be denoted by $B_i$ $(i = 1,2)$. We shall assume that $k_1 > k_2$. Straightforward counting arguments show that the parameters of a PB2-design satisfy

$$Vr_1 = b_1k_1, \qquad Vr_2 = b_2k_2, \tag{2}$$

$$\lambda(V - 1) = r_1(k_1 - 1) + r_2(k_2 - 1). \tag{3}$$

For definitions of partially balanced incomplete block design with m associate classes (PBIBD(m)), and balanced incomplete block design (BIBD, or $(v,b,r,k,\lambda)$-design) see for example [6,pp.63,121].

A PB2-design is said to be *extendable* if it is possible to adjoin a number of new elements to the blocks of $B_2$, increasing the block size from $k_2$ to $k_1$, in such a way that the resulting design is a BIBD. (The term *embedding* of a PB2-design into a BIBD is also used.) If we start from a PB2-design with parameters (1) and can add p new elements to obtain a BIBD, then the new design will have parameters

$$(V + p,\ b_1 + b_2,\ r_1 + r_2,\ k_1,\ \lambda). \tag{4}$$

A *maximal n-arc* in a BIBD is a set of elements with the property that each block of the design meets the set in n elements or no elements. It was shown in [5] that if there is a maximal n-arc in a $(v,b,r,k,\lambda)$-design, then it must contain $p = r(n - 1)/\lambda + 1$ elements, and deletion of these p elements leaves a PB2-design with parameters

$$\left(\frac{r(k - n)}{\lambda};\ \frac{rt(k - n)}{\lambda k},\ \frac{r(r - t)}{\lambda};\ t, r - t;\ k, k - n;\ \lambda\right), \tag{5}$$

where $t = (r - \lambda)/n$.

Conversely, given a PB2-design with parameters (1), when can it be extended to a BIBD with parameters (4), where $p = (r_1 + r_2)(k_1 - k_2 - 1)/\lambda + 1$? Clearly, whenever it can be extended, the p new elements form a maximal $(k_1 - k_2)$-arc in the BIBD. However we prove the following result.

THEOREM 1. *A PB2-design* $(V;\ b_1,b_2;\ r_1,r_2;\ k_1,k_2;\ \lambda)$ *with* $k_1 > k_2$ *is extendable to a* $(V + p,\ b_1 + b_2,\ r_1 + r_2,\ k_1,\ \lambda)$*-design (where* $p = (r_1 + r_2)(k_1 - k_2 - 1)/\lambda + 1$*) if and only if there exist* p *sets of blocks of* $B_2$*, say* $S_1,S_2,\ldots,S_p$*, such that*

*(i)* *each* $S_i$ $(1 \le i \le p)$ *contains* $r = r_1 + r_2$ *blocks of* $B_2$*;*

*(ii)* *each* $S_i$ *contains each of the* V *elements precisely* $\lambda$ *times;*

*(iii)* *any two of the* $S_i$ *intersect in precisely* $\lambda$ *blocks of* $B_2$*;*

*and* *(iv)* *any block of* $B_2$ *lies in* n *of the* $S_i$*, where* $n = k_1 - k_2$.

*Proof.* Suppose that the PB2-design is such that there exist p sets $S_1,S_2,\ldots,S_p$ satisfying *(i)* - *(iv)* above. Take p new elements $x_1,x_2,\ldots,x_p$, and adjoin $x_i$ to each of the r blocks in $S_i$, for $1 \le i \le p$. By *(i)*, each $x_i$ appears in r blocks; by *(ii)*, each $x_i$ is paired with each of the original V elements $\lambda$ times; *(iii)* ensures that each unordered pair $\{x_i,x_j\}$ of new elements occurs in $\lambda$ blocks, and finally *(iv)* ensures that each block of $B_2$ has exactly n new elements added to it. So the resulting design is a BIBD with the required parameters.

Conversely, suppose a PB2-design exists which can be extended to a BIBD with the given parameters. Write $r = r_1 + r_2$ and $n = k_1 - k_2$, and let the p new elements added during the extension be $x_1,x_2,\ldots,x_p$. Let $S_i$ denote the set of r blocks containing $x_i$, with the n new elements in each block deleted. Then $S_i$ contains r blocks of size $k_1 - n = k_2$. Hence *(i)* holds. And every element $x_i$ was paired $\lambda$ times with each of the original V elements, so $S_i$ contains $\lambda$ copies of each of those V elements, and *(ii)* holds. Also the unordered pair $\{x_i,x_j\}$ belonged to $\lambda$ blocks of the BIBD, so $S_i \cap S_j$ consists of $\lambda$ blocks, yielding *(iii)*, and finally since each block of $B_2$ had n new elements adjoined to it, it must belong to n of the sets $S_i$, so *(iv)* holds. □

It follows that necessary conditions for extendability of a PB2-design are

$$(\alpha)\ \lambda V = rk_2,$$

$$(\beta)\ nb_2 = pr,$$

and $$(\gamma)\ \lambda p(p - 1) = b_2 n(n - 1).$$

For *(i)* and *(ii)* of Theorem 1 imply $(\alpha)$, *(i)* and *(iv)* imply $(\beta)$ and *(iii)* yields $(\gamma)$ by counting pairs of new elements. However conditions $(\alpha)$, $(\beta)$ and $(\gamma)$ are not sufficient for extendability: Consider a PB2-design with parameters (21; 7,35; 2,5; 6,3; 1). Since $b_1 = k_1(r_1 - 1) + 1$, a result in [5] tells us that $B_1$ and $B_2$ each form a PBIBD(2), with isomorphic association schemes. Designs T20 and T65 in [2] together form such a PB2-design. But the design cannot be extended because the parameters (4) in this case are (36,42,7,6,1), and no such BIBD exists.

Denniston [4] proved that any Desarguesian projective plane of order $2^\alpha$ contains a maximal $2^\beta$-arc, $\beta \le \alpha$. So there exists a PB2-design with parameters

$$((q+1)(q+1-n);\ t(q+1-n),(q+1)(q+1-t);\ t,q+1-t;\ q+1,q+1-n;\ 1)$$

whenever $q = 2^\alpha$, $n = 2^\beta$ and $t = q/n = 2^{\alpha-\beta}$. Bose and Shrikhande [1] have proved a converse result in the case of a maximal 2-arc: a PB2-design with parameters

$$\left(q^2 - 1;\ \frac{q(q-1)}{2},\ \frac{(q+1)(q+2)}{2};\ \frac{q}{2},\ \frac{q}{2} + 1;\ q + 1, q - 1;\ 1\right) \tag{6}$$

where q is even, can be extended to a projective plane of even order q, provided $q > 6$. The main part of their proof involves strongly regular graphs. Although their preliminary Lemmas 2.2 to 2.7 generalise as follows, there is no hope of a result for $n = 3$, say, because Thas [8] has shown that there is no maximal 3-arc in a projective plane of order $q > 3$.

The following five lemmas all refer to a PB2-design with parameters

$$((q+1)(q+1-n);\ t(q+1-n),(q+1)(q+1-t);\ t,q+1-t;\ q+1,q+1-n;\ 1) \tag{7}$$

where $q = tn$.

LEMMA 1. *Any two blocks of* $B_1$ *intersect in one element.*

*Proof.* Let E be a block of $B_1$ and suppose that $\alpha$ blocks of $B_1$ intersect E in no elements and $\beta$ blocks of $B_1$ intersect E in one element. Then $\alpha + \beta = b_1 - 1 = t(tn+1-n) - 1 = (tn+1)(t-1)$, and $\beta = k_1(r_1-1) = (q+1)(t-1)$. Hence $\alpha = 0$. □

Lemma 1 clearly holds for any PB2-design with parameters (1) that satisfies $\lambda = 1$ and $b_1 = k_1(r_1 - 1) + 1$.

LEMMA 2. *Any block of* $B_1$ *intersects any block of* $B_2$ *in one element.*

*Proof.* Let E be any block of $B_1$. Suppose that $\alpha$ blocks of $B_2$ do not intersect E and $\beta$ blocks of $B_2$ intersect E in one element. Then $\alpha+\beta = b_2 = (q+1)(q+1-t)$, and $\beta = k_1r_2 = (q+1)(q+1-t)$. Hence $\alpha = 0$. □

Lemma 2 also holds for any PB2-design with parameters (1) provided that both $\lambda = 1$ and $b_2 = k_1r_2$. So Lemmas 1 and 2 both hold in a PB2-design with parameters (1) that satisfies

$$\lambda = 1,\ b_1 = k_1(r_1 - 1) + 1 \text{ and } b_2 = k_1r_2.$$

But the parameters of such a PB2-design in fact reduce to (7) above. For $b_2 = k_1r_2$ and (2) imply that $V = k_1k_2$ and $b_1 = k_2r_1$. Then (3) becomes

$$k_1(k_2-1) = (r_1+r_2)(k_2-1) + \{k_1(r_1-1) + 1 - k_2r_1\},$$

$$\text{or}\quad k_1(k_2-1) = (r_1+r_2)(k_2-1),$$

since $b_1 = k_2r_1 = k_1(r_1-1) + 1$. Therefore $k_1 = r_1+r_2$, provided the PB2-design is a non-trivial one with $k_2 > 1$. This means that the parameters (4) are those of a symmetric design with $\lambda = 1$, that is, a projective plane, and letting $k_1 = q + 1$ and $k_2 = q + 1 - n$, the PB2-design parameters reduce to those of (7) above. So there is no point in applying Lemmas 1-5 to an arbitrary PB2-design with $\lambda = 1$, because the two requirements $b_1 = k_1(r_1-1) + 1$ and $b_2 = k_1 r_2$ mean that the PB2-design does in fact have parameters expressible as in (7) above.

LEMMA 3. *Any block of* $B_2$ *has empty intersection with* $n^2t$ *blocks of* $B_2$, *and intersects* $(q+1-n)(q-t)$ *blocks of* $B_2$ *in one element.*

*Proof.* Let F be any block of $B_2$, and suppose $\alpha$ blocks of $B_2$ do not intersect F, and $\beta$ blocks of $B_2$ intersect F in one element. Then $\alpha + \beta = b_2 - 1 = (q+1)(q+1-t) - 1$, and $\beta = k_2(r_2-1) = (q+1-n)(q-t)$. Hence $\alpha = nq = n^2t$. □

LEMMA 4. *Let F be any block of* $B_2$ *and say* x *is an element not in* F. *Then there are* n *blocks of* $B_2$ *containing* x *and not intersecting* F.

*Proof.* Let $F = \{x_1, x_2, \ldots, x_{q+1-n}\}$. Suppose $\alpha$ blocks of $B_2$ contain x but do not intersect F, and $\beta$ blocks of $B_2$ contain x and intersect F in one element. Then $\alpha+\beta = r_2 = q+1-t$. Each pair $\{x, x_i\}$ belongs to just one block of the PB2-design, either a block of $B_1$ or a block of $B_2$. But x belongs to $r_1 = t$ blocks of $B_1$, and each of these t blocks of $B_1$ intersects block F, by Lemma 2. So t of the pairs $\{x, x_i\}$ belong to blocks of $B_1$, leaving $(q+1-n) - t = (t-1)(n-1)$ pairs $\{x, x_i\}$ to occur in blocks of $B_2$. Hence $\beta = (t-1)(n-1)$ and $\alpha = (q+1-t) - (t-1)(n-1) = n$. □

LEMMA 5. *(i) Let* X *and* Y *be two non-intersecting blocks of* $B_2$. *Then there are*

| | |
|---|---|
| $n(n+t-2)$ | *other blocks of* $B_2$ *not intersecting* X *or* Y, |
| $(n-1)(tn-n+1)$ | *other blocks of* $B_2$ *intersecting* X *but not* Y, |
| $(n-1)(tn-n+1)$ | *other blocks of* $B_2$ *intersecting* Y *but not* X, |
| *and* $(t-1)(n-1)(tn-n+1)$ | *other blocks of* $B_2$ *intersecting* X *and* Y. |

*(ii) Let* X *and* Y *be two intersecting blocks of* $B_2$. *Then there are*

| | |
|---|---|
| $n^2$ | *other blocks of* $B_2$ *not intersecting* X *or* Y, |
| $n^2(t-1)$ | *other blocks of* $B_2$ *intersecting* X *but not* Y, |
| $n^2(t-1)$ | *other blocks of* $B_2$ *intersecting* Y *but not* X, |
| *and* $(n-1)\{n(t-1)^2+1\} - t$ | *other blocks of* $B_2$ *intersecting* X *and* Y. |

*Proof. (i)* Let $X = \{x_1, x_2, \ldots, x_{q+1-n}\}$ and $Y = \{y_1, y_2, \ldots, y_{q+1-n}\}$ be two non-intersecting blocks of $B_2$. From Lemma 3 there are $n^2t-1$ blocks of $B_2$, apart from Y, which do not intersect X. Also each element $y_i$, $1 \le i \le q+1-n$, belongs to exactly n-1 of these $n^2t-1$ blocks (from Lemma 4). Hence there are $(n^2t-1) - (n-1)(q+1-n) = n(n+t-2)$ blocks of $B_2$ which do not intersect X or Y, and $(n-1)(tn-n+1)$ blocks of $B_2$ intersecting Y but not X. By the symmetry of the situation, the same number of blocks of $B_2$ intersect X but not Y. So $b_2-2-2(n-1)(tn-n+1)-n(n+t-2) = (t-1)(n-1)(tn-n+1)$ blocks of $B_2$ intersect both X and Y.

*(ii)* Let $X = \{x,x_2,\dots,x_{q+1-n}\}$ and $Y = \{x,y_2,\dots,y_{q+1-n}\}$ be any two intersecting blocks of $B_2$. By Lemma 3, there are $n^2t$ blocks of $B_2$ which do not intersect X. Also from Lemma 4, each element $y_i$, $2 \le i \le q+1-n$, belongs to n of these $n^2t$ blocks. Hence there are $n(q-n) = n^2(t-1)$ blocks of $B_2$ intersecting Y but not X. So there are $n^2t - n^2(t-1) = n^2$ blocks which do not intersect X or Y. Similarly there are $n^2(t-1)$ blocks of $B_2$ intersecting X but not Y, and so $b_2-2-2n^2(t-1)-n^2 = (n-1)\{n(t-1)^2+1\} - t$ blocks of $B_2$ intersecting both X and Y. □

COROLLARY. *In a PB2-design with parameters (7), the blocks of* $B_2$ *(as elements) form a PBIBD(2) with parameters*

$$v^* = (tn+1)(tn+1-t)$$
$$b^* = (tn+1)(tn+1-n)$$
$$r^* = tn+1-n$$
$$k^* = tn+1-t$$
$$\lambda_1 = 0 \qquad n_1 = n^2t$$
$$\lambda_2 = 1 \qquad n_2 = t(tn+1-n)(n-1)$$
$$P^1 = \begin{bmatrix} n(n+t-2) & (n-1)(tn+1-n) \\ (n-1)(tn+1-n) & (n-1)(t-1)(tn+1-n) \end{bmatrix}$$
$$P^2 = \begin{bmatrix} n^2 & n^2(t-1) \\ n^2(t-1) & n(n-1)(t-1)^2+n-t-1 \end{bmatrix}.$$

*Proof.* The parameters all follow from the preceding five lemmas. □

Parameters (7) satisfy $b_1 = k_1(r_1-1)+1$ and $\lambda = 1$, and it was shown in [5] that in such a PB2-design, $B_1$ and $B_2$ each form a PBIBD(2). The above design in the corollary is the dual of the design which $B_2$ forms.

When n = 2 the design in the corollary is a PBIBD(2) with triangular association scheme. (See [2], page 14.) Two blocks of $B_2$ are first associates if they do not intersect. Connor [3] proves the following result for triangular association schemes.

THEOREM 2. (Connor). *The triangular association scheme for* $N(N-1)/2$ *treatments (elements),* $N > 8$, *exists if and only if there exist sets of treatments* $S_j$, $j = 1,\dots,N$, *such that:*

*(i) Each* $S_j$ *consists of* $N-1$ *treatments.*

*(ii) Any treatment is in precisely two sets* $S_j$.

*(iii) Any two distinct sets* $S_i$, $S_j$ *have exactly one treatment in common.* □

Now apply Theorem 2 to the triangular association scheme which arises from the corollary above when $n = 2$; $N = 2t+2$, and condition $N > 8$ then corresponds to $t > 3$, or $q = 2t > 6$. By Theorem 2 there exist sets of blocks of $B_2$, $S_1,\ldots,S_{2t+2}$ such that

(i) each $S_j$ consists of $2t+1$ blocks of $B_2$;

(ii) any block of $B_2$ belongs to two of the $S_j$;

(iii) any two of the $S_i$ intersect in one block of $B_2$.

These correspond respectively to conditions *(i)*, *(iv)* and *(iii)* of Theorem 1, with $n = 2$ and $\lambda = 1$. Connor also points out that treatments belong to the same set $S_j$ if and only if treatments are first associates; since in the corollary, blocks of $B_2$ are first associates if they do not intersect, it follows that $(2t+1)$ non-intersecting blocks of $B_2$ of size $2t-1$ in one $S_j$ must contain altogether $(2t+1)(2t-1)$ or $V$ elements once each, so *(ii)* of Theorem 1 is satisfied. So using the corollary and Theorems 1 and 2 we have

THEOREM 3. (Bose and Shrikhande). *If* $q$ *is even,* $q > 6$*, then a PB2-design with parameters*

$$\left(q^2-1;\ \frac{q(q-1)}{2},\ \frac{(q+1)(q+2)}{2};\ \frac{q}{2},\ \frac{q+2}{2};\ q+1,\ q-1;\ 1\right)$$

*extends to a projective plane of order* $q$. □

The cases $q = 4$ and $6$ are dealt with in [7] and [9] respectively.

*References.*


[1] R.C. Bose and S.S. Shrikhande, Embedding the complement of an oval in a projective plane of even order, *Discrete Math.* 6 (1973), 305-312.

[2] W.H. Clatworthy, *Tables of two-associate class partially balanced designs,* U.S. Dept. of Commerce, NBS, Washington, 1973.

[3] W.S. Connor, The uniqueness of the triangular association scheme, *Ann. Math. Stat.* 29 (1958), 262-266.

[4] R.H.F. Denniston, Some maximal arcs in finite projective planes, *J. Combinatorial Theory* 6 (1969), 317-319.

[5] Elizabeth J. Morgan, Arcs in block designs, *Ars Combinatoria* (to appear).

[6] D. Raghavarao, *Construction and combinatorial problems in design of experiments*, Wiley, New York, 1971.

[7] S.S. Shrikhande, On a characterization of the triangular association scheme, *Ann. Math. Stat.* 30 (1959), 39-47.

[8] J.A. Thas, Some results concerning $\{(q+1)(n-1);n\}$-arcs and $\{(q+1)(n-1)+1;n\}$-arcs in finite projective planes of order q, *J. Combinatorial Theory* (A) 19 (1975), 228-232.

[9] Paul de Witte, The exceptional case in a theorem of Bose and Shrikhande, *J. Australian Math. Soc.* (to appear).

Note added in proof: It has been pointed out to me that another proof of Bose and Shrikhande's result, also avoiding use of strongly regular graphs, is contained in

J.I. Hall, Bounds for equidistant codes and partial projective planes, Discrete Math. 17 (1977), 85-94.

05C35

# FURTHER RESULTS ON A PROBLEM IN THE DESIGN OF ELECTRICAL CIRCUITS

M.J. PELLING[1] and D.G. ROGERS[1],[2]

(1) Mathematical Institute, Oxford, England

(2) Department of Mathematics, University of Western Australia.

ABSTRACT

The problem of determining the minimum number $s(n,m)$ of make and break switches which may be used in an electrical circuit which allows a current to flow if $m$ or more voters of an $n$ member committee close their switches (vote yes) is considered. A limit result and a lower bound are established for $s(n,2)$ and an improved upper bound is obtained for $s(n,3)$. These results depend on dividing the committee into a number of subcommittees and, in this way, draw on the theory of subadditive functions.

## §1 STATEMENT OF THE PROBLEM AND RESULTS.

We consider the problem of designing an electrical circuit which will record the affirmative votes of a committee. Each member of the committee controls a number of make and break switches in the circuit which he closes if and only if he votes yes so that abstentions and negative votes are not recorded. A circuit of this sort which allows a current to flow if and only if $m$ members, $1 \le m \le n$, of an $n$ member committee $C$ vote yes is called an *m-threshold circuit* for the committee $C$ and we denote an arbitrary circuit of this sort by $A(n,m)$ or $A_C(m)$, labelling all the switches controlled by one voter with the same label. We are then especially interested in the minimum number $s(n,m)$ of switches which may be used in a circuit $A(n,m)$.

The problem may readily be rephrased as one in term of edge-colourings of graphs by representing switches controlled by one voter by edges coloured with the same colour. Let $G(n,m)$ be the set of graphs $\Gamma(n,m)$ having the following properties: (i) $\Gamma(n,m)$ has two distinguished vertices, $\alpha$ and $\omega$, say; (ii) the edges of $\Gamma(n,m)$ are coloured with any of $n$ colours; (iii) any path from $\alpha$ to $\omega$ in $\Gamma(n,m)$ uses at least $m$ distinctly coloured edges; and (iv) for any $m$ out of the $n$ available colours there is a path from $\alpha$ to $\omega$ in $\Gamma(n,m)$ whose edges are coloured using exactly these colours. Then, if $e(n,m)$ is the number of edges in $\Gamma(n,m)$

$$s(n,m) = \min\{e(n,m) : \Gamma(n,m) \in G(n,m)\}$$

This restatement suggests a variety of other similar problems and, in particular, by modifying the conditions (i-iv) it may be possible to obtain bounds for $s(n,m)$. But we do not pursue this avenue here.

Rather the terminology of committees suggests a recursive approach in terms of subcommittees which, as in [4], we adopt here. In [4], attention was restricted to the use of only two subcommittees at a time and the results obtained there were correspondingly limited. It was shown that

$$s(n,m) = O(n(\log n)^{m-1}) \qquad m \ll n \; ; \; n \to \infty \tag{1}$$

but no attempt was made to determine the correct asymptotic formula for $s(n,m)$ for fixed $m$ as $n \to \infty$, nor were any lower bounds on $s(n,m)$ obtained. It was not even proved that

$$\overline{\lim_{n\to\infty}} \frac{s(n,2)}{n\log n} > 0$$

although it is plausible to conjecture that

$$\lim_{n\to\infty} \frac{s(n,2)}{n\log n} = \frac{1}{\log 2} \tag{2}$$

Here we are able to go some way to remedying these deficiencies by using larger, variable numbers of subcommittees. In this we again use results from the theory of subadditive functions (see §5). We show, (§3), in particular, that

$$\frac{s(nm,2)}{nm} \le \frac{s(n,2)}{n} + \frac{s(m,2)}{m} \tag{3}$$

and deduce that

$$\lim_{n\to\infty} \frac{s(n,2)}{n\log n} \le \frac{1}{\log 2} \tag{4}$$

Further, for any constant $K > 1/2\log 2$, we have

$$s(n,3) \le K \frac{n(\log n)^2}{\log\log n} \qquad , \qquad n \to \infty . \tag{5}$$

For $n \ge 1$, $s(n,1) = n$. We begin, in §2, by proving that, for $m \ge 2$,

$$\lim_{n\to\infty} \frac{s(n,m)}{n} = +\infty \tag{6}$$

the existence of this limit following from the superadditive inequality

$$s(n_1,m) + s(n_2,m) \le s(n_1 + n_2,m) \tag{7}$$

A variant of the argument for (6) is used, in §4, to show that for some $\delta > 0$ and $n$ large enough

$$s(n,2) > n(\log\log n)^{1-\delta} \tag{8}$$

and we conclude with a more detailed discussion of $s(n,2)$ and some further conjectures.

§2 SUPERADDITIVITY.

If a committee $C$ is the disjoint union of two subcommittees $C_i$ , $i = 1,2$, with $n_i$ members each, then a circuit $A_C(m)$ contains disjoint circuits $A_{C_i}(m)$ , $i = 1,2$ , as may be seen by deleting inturn the switches controlled by each of the subcommittees. Hence (7) follows and then the limit in (6) exists (possibly + ∞) by a result on superadditive functions (see §5).

Now consider a 2-threshold circuit $A_m(n,2)$ , $m \geq 1$ , for an $n$ member committee in which each member controls at most $m$ switches so that the switches controlled by th $i$-th member may be labelled (in some order) $i_s$ , $1 \leq s \leq m(i) \leq m$ . We show that $n$ cannot be too large.

For each pair $\{i,j\}$ , $1 \leq i < j \leq n$ , there is a minimal path in $A_m(n,2)$ from the positive pole to the negative pole through only switches controlled by the $i$-th and $j$-th members and any such path may be identified by the ordered string of labels of the switches in it. An $r$-tuple $\underset{\sim}{x} = (x_1,\ldots,x_s,\ldots,x_r)$ , $2 \leq r \leq 2m$ , where for $1 \leq s \leq r$ , $x_s = \alpha_p$ or $\beta_q$ for some $p$ or $q$ with $1 \leq p,q, \leq m$ is a *path type* if there is a pair $\{i,j\}$ , $1 \leq i < j < n$, and some minimal path of $r$ switches in $A_m(n,2)$ controlled by the $i$-th and $j$-th members in which the $s$-th switch has $i_p$ or $j_q$ according as $x_s = \alpha_p$ or $\beta_q$ . For given $m$ , the number of path types is finite and bounded by $m^{(2+\epsilon)m}$ for some $\epsilon > 0$ , at least for large enough $m$ ; let $b(m)$ be the correct bound.

By Ramsey's theorem, (see [1, pp54-7]), there is a greatest integer $n$ , denoted $R(b)$, such that whenever the 2-element subsets of an $n$-element set $S$ are divided into $b$ mutually disjoint classes there is a 3-element subset of an $n$-element subset of $S$ not all of whose 2-element subsets are contained in one class. It follows that if $n > R(b(m))$ then there are three integers $i,j,k$, $1 \leq i < j < k \leq n$, such that the pairs $\{i,j\}$ , $\{i,k\}$ , and $\{j,k\}$ all have the same path type $\underset{\sim}{x}$ in $A_m(n,2)$ . Since the labelling of the switches is conventional, we may assume that the $\alpha$ suffixes in $\underset{\sim}{x}$ are in ascending order and likewise for the $\beta$ suffixes.

As an example, suppose that $\{i,j\}$ , $\{i,k\}$ , and $\{j,k\}$ all have paths of path type

$$\underset{\sim}{x} = (\alpha_1,\beta_1,\alpha_2,\alpha_3,\alpha_4,\beta_2,\beta_3,\alpha_5,\beta_4,\alpha_6,\beta_5,\beta_6) \tag{9}$$

so that the $\{i,j\}$ and $\{i,k\}$ paths appear as in Figure (1) (ignoring for the moment the dotted lines)

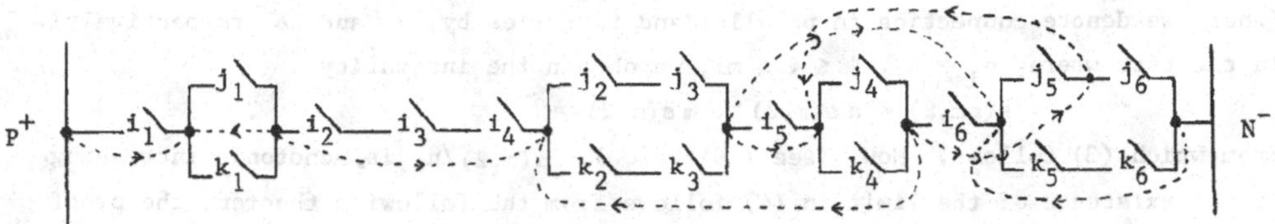

*Figure (1)*

*An illustration of paths with path type (9)*

Since $\{j,k\}$ also has a path of type (9), this $\{j,k\}$ path must be formed from the switches $\{j_1,\ldots,j_6,k_1\ldots,k_6\}$ by wiring them in the order $j_1,k_1,j_2,j_3,j_4,k_2,$ $k_3,j_5,k_4,j_6,k_5,k_6$ by additional connecting wires as shown by the dotted lines in Figure (1). But it is then apparent that these form a short circuit.

This example is typical since in the general case the dotted lines form the edges of a graph $\Gamma$ with vertices the poles and both ends of every unbroken run of $i$'s. Considering vertices other than the poles, at the start (resp. end) of a run of $i$'s there must be two dotted lines (wires) leading out from (resp. into) the preceding (resp. succeeding) $j$'s and $k$'s. Hence every vertex of $\Gamma$ other than the poles has valence two, the poles having valence one. Thus $\Gamma$ is Eulerian with the poles in the same connected component so that there is a short circuit between them.

Hence $A_m(n,2)$ can only exist for $n \le R(b(m))$. If $n > R(b(m))$, any circuit $A(n,2)$ must therefore have one voter controlling at least $(m + 1)$ switches, so, considering a minimal circuit and deleting all of this voters switches, we have

$$s(n,2) \ge s(n - 1,2) + m + 1 \qquad , \quad n > R(b(m))$$

$$s(n + 1,2) - s(n,2) \ge m \qquad , \quad R(b(m - 1)) \le n < R(b(m)) \qquad (10)$$

So (5) holds for $m = 2$ since

$$R(b(m)) \uparrow + \infty \qquad \text{as} \quad m \to + \infty$$

Further since, by [4,(2)], $s(n,m) \ge s(n - 1,m - 1)$, it follows that (5) holds for $m \ge 2$.

## §3 LOGARITHMIC SUBADDITIVITY.

Consider now a committee $C$ which is the disjoint union of $m$ subcommittees $C_i$, $1 \le i \le m$, of size $n_i$. Let $A^*(m,s)$ be the circuit obtained from a circuit $A_{C_i}(1)$ for single make and break switches with the $i$-th label. Then one $A_C(2)$ circuit is given by

$$A_C(2) \equiv A^*(m,2) \vee (\bigvee_{i=1}^{m} A_{C_i}(2)) \qquad (11)$$

(where we denote connection in parallel and in series by $\vee$ and $\wedge$ respectively). In the case where $n_i = n$, $1 \le i \le m$, we obtain the inequality

$$s(nm,2) \le n\,s(m,2) + m\,s(n,2)$$

from which (3) follows. Now, (see (19) below), $s(n,2)/n$ is monotonic increasing. So the existence of the limit in (4) follows from the following theorem, the proof of which we defer to §5:-

Theorem: If $f$ is a monotonic increasing function such that for integers $n$, $m \ge 1$

$$f(\log n + \log m) \le f(\log n) + f(\log m)$$

then the limit $\lim_{n\to\infty} \frac{f(\log n)}{\log n}$ exists and equals $\inf \frac{f(\log n)}{\log n}$.

The bound in (4) then follows since, from [4],

$$s(2^n + q,2) \le j(2^n + q,2) = n2^n + q(n+2), \quad 0 \le q \le 2^n \; ; n \ge 1 \tag{12}$$

Arguing in the same way as for (11), we obtain an $A_C(3)$ circuit given by

$$A_C(3) \equiv A^*(m,3) \vee \left( \bigvee_{i=1}^{m} (A_{C_i}(2) \wedge A_{C \setminus C_i}(1)) \vee \left(\bigvee_{i=1}^{m} A_{C_i}(3)\right)\right)$$

although the middle portion of this circuit does not appear to be very efficient.

If now $|C| = n$, so $\sum_{i=1}^{m} n_i = n$, then by splitting $C$ into $m = m(n) \le n$ subcommittees $C_i$ of size $n_i = \left\lceil \frac{n+i-1}{m} \right\rceil$, $1 \le i \le m$, that is as nearly equal as possible, we have

$$s(n,3) \le n_m s(m,3) + m s(n_m,2) + n(m-1) + m s(n_m,3)$$

$$\le m s(n_m,3) + m s(n_m,2) + nm K_m$$

where, by (1)

$$K_m = 1 - \frac{1}{m} + \frac{s(m,3)}{m}\left(\frac{1}{m} + \frac{1}{n}\right) \to 1 \quad , \quad \text{as } m \to \infty$$

Now, using (4)

$$s(n,3) \le m s(n_m,3) + n\,(1+\epsilon_n)(\log 2)^{-1} \log \frac{n}{m} \quad + nm K_m,$$

where $\epsilon_n \to 0$ as $n \to \infty$ provided that $\frac{n}{m} \to 0$ also. Thus we have

$$s(n,3) \le n\,g(n)$$

where
$$g(n) = g(n_m) + m K_m + (\log n - \log m)(\log 2)^{-1}(1+\epsilon_n) \quad . \tag{13}$$

So, for some constants A, B, we may take $m = A \log n$ and

$$g(n) \simeq B(\log n)^2 / \log\log n$$

provided that (13) holds as nearly as possible. This will be so if $B = \frac{1}{2}(A + (\log 2)^{-1})$ ; so (5) follows.

In [4], we conjectured that

$$s(n,m) \geq \frac{s(n,m)}{n} + s(n-1,m-1,) - s(n-2,m-1) \tag{14}$$

or

$$\frac{s(n,m)}{n} - \frac{s(n-1,m)}{n-1} \geq \frac{1}{n-1}(s(n-1,m-1) - s(n-2,m-1)) \tag{15}$$

For $m = 2$ , this leads to

$$s(n,2) > n \log n \quad , \quad n \to \infty$$

which is not in conflict with anything established so far. But then, from (15) for $m = 3$ , we have, for some $A > 0$ ,

$$s(n,3) > An(\log n)^2 \quad , \quad n \to \infty$$

in contradiction to (5). So the conjecture (14) is false in general.

§4 A LOWER BOUND AND SOME FURTHER RESULTS FOR $s(n,2)$

The argument in §2 may be used to derive the lower bound (8) for $s(n,2)$ as follows. Let the function $h$ be defined by

$$h(x) = m \quad , \quad R(b(m-1)) < x \leq R(b(m))$$

so that, from (10),

$$s(n,2) \geq \int_1^n h(x)dx \quad , \quad \text{as } n \to \infty$$

Now $b(m) \leq m^{(2+\epsilon)m}$ , for some $\epsilon > 0$ and, from [3, p255], $R(b) \leq b! \leq b^b$ . So, for some $\delta > 0$ ,

$$h(x) > (\log\log x)^{1-\delta}$$

and (8) follows. It is doubtful whether (8) is at all exact and in [4] we conjecture that, (see(12)),

$$s(1) = 0 \ ; \ s(n,2) = s(\left[\frac{n}{2}\right],2) + s(\left[\frac{n+1}{2}\right],2) + n = j(n,2) \quad , \quad n \geq 2 \tag{16}$$

leading to (2). An analysis by path length, as opposed to that by path type in §2, appears complicated but we conjecture that to achieve $s(n,2)$ it is sufficient to consider circuits $A(n,2)$ in which the poles are only two switches apart.

We now write $s(n) = s(n,2)$ and $A_n = A(n,2)$ . Let $S_n$ be a circuit $A_n$ achieving the minimum $s(n)$ and let $\lambda(S_n)$ and $\gamma(S_n)$ be respectively the least and greatest number of switches in $S_n$ having the same label $i$ , say $i_\lambda$ and $i_\gamma$ respectively, so

$$\lambda(S_n) \leq \frac{s(n)}{n} \leq \gamma(S_n) \tag{18}$$

From any circuit $S_n$ we may obtain a circuit $A_{n+1}$ by putting $\lambda(S_n)$ new switches labelled $(n+1)$ in parallel with those labelled $i_\lambda$ and introducing in parallel a pair of switches in series labelled $i_\lambda$ and $(n+1)$. On the other hand by deleting the switches labelled $i_\gamma$ in a circuit $S_{n+1}$ we obtain a circuit $A_n$. Hence

$$s(n) + \gamma(S_{n+1}) \le s(n+1) \le s(n) + \lambda(S_n) + 2 \qquad (18)$$

and combining (18) with (17) we have

$$0 \le \frac{s(n+1)}{n+1} - \frac{s(n)}{n} \le \frac{2}{n+1} \qquad (19)$$

We conjecture that, in fact, equality holds in the second inequality in (18).

If in a circuit $S_n$ there are $s_i$ switches labelled $i$, $1 \le i \le n$, then we may obtain another circuit $A_n$ by deleting switches labelled $j$ in $S_n$, putting new switches labelled $j$ in parallel with those labelled $k$ and introducing in parallel a pair of switches in series labelled $j$ and $k$. Since $A_n$ cannot have fewer switches than $S_n$, we must have

$$s_j + s_k \le 2 s_k + 2$$

Hence $\gamma(S_n) \le \lambda(S_n) + 2$ and for some circuit $S_n^+$, say, $\gamma(S_n^+) \le \lambda(S_n^+) + 1$; so relabelling the switches if necessary, the number $s_i^+$ of switches with label $i$ in $S_n^+$ is given by

$$s_i^+ = \left[\frac{s(n) + i}{n}\right].$$

Now using $S_m^+$ as the circuit $A(m,2)$ in (11), we have

$$s(n) \le \min \left\{ \sum_{i=1}^{m} s(n_i) + \sum_{i=1}^{m} \left[\frac{s(m) + i}{m}\right] n_{m-i+1} \right\} \qquad (20)$$

where the minimum is taken over all m-tuples $(n_1, n_2, \ldots, n_m)$ with $2 \le m$; $n_i \le n_j$, $i \le j$; and $n = \sum_{i=1}^{m} n_i$. We conjecture that equality holds in (20) where, in order to give a recursion, $m < n$. Indeed we conjecture that it suffices to take $m = 2$ so that, from [4], (16) holds.

§5 APPENDIX : SUBADDITIVITY

The fundamental result on subadditive functions is, (see [2, p244]

Theorem A: If $f$ is a subadditive function on the positive integers, that is

$$f(n+m) \le f(n) + f(m) \qquad ,n,m \ge 1,$$

then the limit $\lim_{n\to\infty} \frac{f(n)}{n}$ exists and equals $\inf_{n \ge 1} \frac{f(n)}{n}$

If $f$ is superadditive then $-f$ is subadditive; so we may apply Theorem A in §2. As a modification of Theorem A we have the theorem of §3. For if the assumptions in §3 hold we may extend $f$ to a function $\tilde{f}$ on $[0,\infty]$ by putting

$$\tilde{f}(x) = \inf\{f(\log n) : n \geq e^x\}$$

so that (i) $\tilde{f}(\log n) = f(\log n)$; (ii) $\tilde{f}$ is subadditive; and (iii) $\tilde{f}$ is bounded on bounded sets. Conditions (ii) and (iii) ensure that $\lim_{x\to\infty} \frac{\tilde{f}(x)}{x}$ exists and then taking $x = \log n$, the theorem follows. If $f$ is not increasing the limit $\lim_{n\to\infty} \frac{f(\log n)}{\log n}$ need not exist as is shown by the example

$$f(\log n) = 2\log n\ ,\ n \text{ a power of } 2;\ = \log n\ , \text{ otherwise.}$$

We are pleased to acknowledge the collaboration of C.J.K. Batty in our work on the theory of subadditive functions. We hope collectively to report on other related aspects of this theory, such as recursive minimization, (compare (20)), elsewhere.

REFERENCES.

[1] M. Hall. *Combinatorial Theory*. Blaisdell, Waltham, (1967).

[2] E. Hille and R. Phillips. *Functional Analysis and Semigroups*, revised edition. American Math. Soc., Providence, (1957).

[3] R.W. Irving. "Generalized Ramsey numbers for small graphs," Discrete Math. 9(1974), 251-264.

[4] M.J. Pelling and D.G. Rogers. "A problem in the design of electrical circuits, a generalized subadditive inequality and the recurrence relation $j(n,m) = j\left(\left[\frac{n}{2}\right],m\right) + j\left(\left[\frac{n+1}{2}\right],m\right) + j(n, m-1)$". *Combinatorial Mathematics* V : *Proceedings of the Fifth Australian Conference*. Lecture Notes in Mathematics, Springer-Verlag, Berlin, (to appear).

05A99, 05B30

# TRANSVERSALS AND FINITE TOPOLOGIES

R. A. Razen

School of Mathematics,
University of New South Wales,
P.O. Box 1, Kensington, New South Wales, 2033
Australia

ABSTRACT

Steffens [10] has shown that a family A of finite sets has a transversal if and only if the collection of all 'critical subfamilies' is a topology on A. In [6] these 'transversal topologies' have been characterized as well as families whose transversal topologies satisfy separation axioms. The purpose of this paper is to apply these results to enumerating certain finite topologies.

## 1. DEFINITIONS FROM TOPOLOGY

Let S be a topology on a set X. For all $x \in X$ we write

$$\{x\}' = \{y \neq x \mid \forall\, G \in S \quad y \in G \Rightarrow x \in G\},$$

and

$$\{x\}^{\sim} = \bigcap_{x \in G \in S} G \setminus \{x\}.$$

DEFINITION 1. S is a $T_o$-topology $\Leftrightarrow$ $\forall\, x,y \in X$, $x \neq y$ either

$$\exists\, G_x \in S \quad x \in G_x \quad y \notin G_x$$

or

$$\exists\, G_y \in S \quad y \in G_y \quad x \notin G_y.$$

DEFINITION 2. S is a $T_{DD}$-topology $\Leftrightarrow$
$\forall\, x \in X \quad \exists\, G_1, G_2 \in S$ such that $\{x\} = G_1 \cap (X \setminus G_2)$, and $\forall\, x \in X \quad |\{x\}^{\sim}| \leq 1$.

DEFINITION 3. S is a $T_{FF}$-topology $\Leftrightarrow$ either of the following conditions hold:

(i) $\{x\}' = \phi$ for all but at most one $x \in X$, or

(ii) $\{x\}^{\sim} = \phi$ for all but at most one $x \in X$.

*Remark* 1. Both $T_{DD}$- and $T_{FF}$-topologies also satisfy the $T_o$-property [1].

DEFINITION 4. S is regular $\Leftrightarrow$ $\forall\, x \in X$ and for all closed sets C with $x \notin C$ there exist disjoint $G_x, G_C \in S$ with $x \in G_x$ and $C \subset G_C$.

DEFINITION 5. S is symmetric $\Leftrightarrow$ every open set is closed.

DEFINITION 6. S is a partition topology $\Leftrightarrow$ S possesses a base which is a partition of X.

## 2. REQUIREMENTS FROM TRANSVERSAL THEORY

Let $\mathbf{A} = (A_i | i \in I)$ be an arbitrary family of finite non-empty subsets of a ground set $E = \bigcup_{i\in I} A_i$. We write

$$A(J) = \bigcup_{i\in J} A_i \quad \text{and} \quad \mathbf{A}_J = (A_i | i \in J)$$

for $J \subset I$. A set $T \subset E$ is called a *transversal* of $\mathbf{A}$ if there is a bijection $\Phi: T \to I$ such that $x \in A_{\Phi(x)}$ for all $x \in T$.

DEFINITION 7. A subfamily $\mathbf{A}_J$ is called *critical* iff it has a unique transversal.

*Remark* 2. If $\mathbf{A}_J$ is a critical subfamily with transversal T, then $T = A(J)$. Therefore, for a finite critical subfamily $\mathbf{A}_J$ we have $|A(J)| = |J|$.

The following result is due to Steffens [10].

THEOREM A. $\mathbf{A}$ *has a transversal* $\iff$ $\mathbf{T}(\mathbf{A}) = \{\mathbf{A}_J \subset \mathbf{A} | \mathbf{A}_J$ *critical*$\} \cup \{\phi, \mathbf{A}\}$ *is a topology on* $\mathbf{A}$.

DEFINITION 8. We call this topology the *transversal topology of the family* $\mathbf{A}$ and refer to its elements as the *open subfamilies of* $\mathbf{A}$.

DEFINITION 9. Let $\mathbf{S}_i$ be topologies on $X_i$ $(i = 1,2)$. Then $\mathbf{S}_1$ and $\mathbf{S}_2$ are said to be *base-homeomorphic* if the topologies arising after identifying all elements which occur in no other open set but $X_i$, are homeomorphic.

Using this notion, the transversal topologies can be characterized as follows [6]:

THEOREM B. *Let* $\mathbf{S}$ *be a topology on* X. *Then* $\mathbf{S}$ *is base-homeomorphic to a transversal topology* $\mathbf{T}(\mathbf{A})$ *if and only if* $\{G \in \mathbf{S} \,|\, |G| < \aleph_0\} \cup \{X\}$ *is a base for* $\mathbf{S}$.

The proof has shown that the transversal topology of the family essentially defined by $\mathbf{A} = (G_x | x \in X)$ where $G_x$ denotes the smallest (with respect to inclusion) set of $\mathbf{S}$ containing x satisfies the theorem. Note that if X is finite, the theorem remains true if we replace base-homeomorphic by homeomorphic.

DEFINITION 10. An open subfamily is called *minimal* if it contains no proper open subfamily.

We write $I^* = \{i \in I \,|\, \exists\, \mathbf{A}_J \in \mathbf{T}(\mathbf{A}) \setminus (\mathbf{A}) \text{ with } A_i \in \mathbf{A}_J\}$.

*Remark* 3. If $\mathbf{T}(\mathbf{A})$ is a $T_0$-topology, then $|I \setminus I^*| \leq 1$.

The following characterizations of families whose transversal topologies satisfy different separation axioms have been obtained [6]:

THEOREM C. $T(A)$ *is a* $T_o$*-topology* $\Leftrightarrow$ *∃ ordinal* $\alpha$ *and a bijection* $\psi:\alpha \to I^*$ *such that*

$$|A_{\psi(\beta)} \setminus \bigcup_{\gamma<\beta} A_{\psi(\gamma)}| = 1 \quad \forall \beta < \alpha.$$

THEOREM D. $T(A)$ *is a* $T_{DD}$*-topology* $\Leftrightarrow$ $|A_i| \leqslant 2$ *and* $|A_i \setminus \bigcup_{\substack{|A_j|=1 \\ j \neq i}} A_j| = 1 \quad \forall i \in I.$

THEOREM E. $T(A)$ *is a* $T_{FF}$*-topology* $\Leftrightarrow$ *either of the following conditions hold:*

(i) $|A_i| = 1$ *for all but at most one* $i \in I$,

(ii) $\exists A_i \in A$ *with* $|A_i| = 1$ *such that* $|A_j \setminus A_i| = 1 \quad \forall j \in I \setminus \{i\}$.

THEOREM F. *The following conditions are equivalent:*

(a) $T(A)$ *is regular*,

(b) $T(A)$ *is symmetric*,

(c) $T(A)$ *is a partition topology*.

## 3. APPLICATIONS TO TOPOLOGIES ON FINITE SETS

For the following let $X = \{x_1,\ldots,x_n\}$ be a finite set and $S$ a topology on X. Then Theorem B reads as follows:

THEOREM B'. *There is a family* $A$ *of* $n$ *subsets of a set with* $n$ *elements such that* $T(A)$ *is homeomorphic to* $S$.

This theorem enables us to answer several questions for finite topologies using the previous results.

Let $A = (A_1,\ldots,A_n)$ be critical and $J_o \subset \{1,\ldots,n\}$ such that $A_{J_o}$ is the smallest critical subfamily of $A$ containing $A_n$. ($J_o$ is uniquely determined.) Let $x \in E \setminus A(J_o)$. We define $B = (B_1,\ldots,B_n)$, where

$$B_i = A_i \qquad (i \notin J_o),$$

$$B_i \supset A_i \cup \{x\} \qquad (i \in J_o).$$

Then the following Lemma holds:

LEMMA. $|T(B)| < |T(A)|$.

*Proof.* We have $|B(J_o)| > |A(J_o)| = |J_o|$, but for all $J \neq J_o$ we have $|B(J)| \geqslant |A(J)| \geqslant |J|$, therefore $A$ has at least one critical subfamily more than $B$.

THEOREM 1 (Stephen [11], Sharp [8]). *There exists no topology* S *on* n *elements with* $(3/4)2^n < |S| < 2^n$.

*Proof.* Consider the family $A = (\{x_1\},\dots,\{x_{n-1}\},\{x_1,x_n\})$. The critical subfamilies of A consist of all subfamilies of $(\{x_1\},\dots,\{x_{n-1}\})$ and all subfamilies of A containing both $\{x_1\}$ and $\{x_1,x_n\}$. Therefore,

$$|T(A)| = 2^{n-1} + 2^{n-2} = (3/4)2^n.$$

Since every non-discrete topology not homeomorphic to T(A) is obtained by adding elements to the sets of A, the theorem follows from the lemma.

Repetition of this process of adding elements yields a generalization of this result:

| A = | \|T(A)\| = |
|---|---|
| $(\{x_1\},\dots,\{x_{n-2}\},\{x_1,x_{n-1}\},\{x_1,x_n\})$ | $(5/8)2^n$ |
| $(\{x_1\},\dots,\{x_{n-2}\},\{x_2,x_{n-1}\},\{x_1,x_n\})$ | $(9/16)2^n$ |
| $(\{x_1\},\dots,\{x_{n-2}\},\{x_{n-1}\},\{x_1,x_2,x_n\})$ | $(5/8)2^n$ |
| $(\{x_1\},\dots,\{x_{n-3}\},\{x_1,x_{n-2}\},\{x_1,x_{n-1}\},\{x_1,x_n\})$ | $(9/16)2^n$ |
| $(\{x_1\},\dots,\{x_{n-3}\},\{x_2,x_{n-2}\},\{x_1,x_{n-1}\},\{x_1,x_n\})$ | $(15/32)2^n$ |
| $(\{x_1\},\dots,\{x_{n-3}\},\{x_{n-2}\},\{x_1,x_{n-1}\},\{x_1,x_2,x_n\})$ | $(1/2)2^n$ |
| $(\{x_1\},\dots,\{x_{n-3}\},\{x_{n-2}\},\{x_1,x_{n-1}\},\{x_1,x_{n-1},x_n\})$ | $(1/2)2^n$ |
| $(\{x_1\},\dots,\{x_{n-3}\},\{x_3,x_{n-2}\},\{x_2,x_{n-1}\},\{x_1,x_n\})$ | $(27/64)2^n$ |
| $(\{x_1\},\dots,\{x_{n-3}\},\{x_{n-2}\},\{x_2,x_3,x_{n-1}\},\{x_1,x_n\})$ | $(15/32)2^n$ |
| $(\{x_1\},\dots,\{x_{n-3}\},\{x_{n-2}\},\{x_{n-1}\},\{x_1,x_2,x_3,x_n\})$ | $(9/16)2^n$ |

After one more step (adding at most 4 elements), noticing that for

$$A_1 = (\{x_1\},\dots,\{x_{n-i}\},\{x_1,x_{n-i+1}\},\dots,\{x_1,x_n\})$$

and

$$A_2 = (\{x_1\},\dots,\{x_{n-1}\},\{x_1,\dots,x_i,x_n\})$$

one has

$$|T(A_1)| = |T(A_2)| = (2^i + 1)2^{n-i-1} \quad (1 \leqslant i \leqslant n-1),$$

and using the lemma, one obtains the following theorem:

THEOREM 2 (Stanley [9]). *If* $n \geq 5$, *then up to homeomorphism there is one* $T_o$*-space with* $n$ *points and* $2^n$ *open sets, one with* $(3/4)2^n$ *open sets, two with* $(5/8)2^n$, *three with* $(9/16)2^n$, *two with* $(17/32)2^n$, *two with* $(1/2)2^n$, *two with* $(15/32)2^n$, *five with* $(7/16)2^n$, *and for each* $m = 6,7,\dots,n$, *two with* $(2^{m-1}+1)2^{n-m}$. *All other* $T_o$*-spaces with* $n$ *points have* $< (7/16)2^n$ *open sets.*

The same method applies to 'non-$T_o$' topologies:

| A = | $\lvert T(A)\rvert$ = |
|---|---|
| $(\{x_1\},\dots,\{x_{n-2}\},\{x_{n-1},x_n\},\{x_{n-1},x_n\})$ | $(1/2)2^n$ |
| $(\{x_1\},\dots,\{x_{n-3}\},\{x_1,x_{n-2}\},\{x_{n-1},x_n\},\{x_{n-1},x_n\})$ | $(3/8)2^n$ |
| $(\{x_1\},\dots,\{x_{n-3}\},\{x_{n-2}\},\{x_1,x_{n-1},x_n\},\{x_1,x_{n-1},x_n\})$ | $(3/8)2^n$ |
| $(\{x_1\},\dots,\{x_{n-3}\},\{x_{n-2},x_{n-1},x_n\},\{x_{n-1},x_n\},\{x_{n-1},x_n\})$ | $(3/8)2^n$ |

After two more steps and using the lemma again, we obtain a refinement of another theorem of Stanley's:

THEOREM 3. *If* $n \geq 6$, *then up to homeomorphism there is one* 'non-$T_o$' *space with* $n$ *points and* $(1/2)2^n$ *open sets, three with* $(3/8)2^n$, *six with* $(5/16)2^n$, *nine with* $(9/32)2^n$, *and all the rest have* $< (9/32)2^n$ *open sets.*

Let $\omega$ denote a topological property of a finite topological space. Then we denote by $t_\omega(n)$ and $\tau_\omega(n)$ respectively the number of non-homeomorphic topologies and the total number of topologies on a set with $n$ elements which have property $\omega$. ($\omega$ = DD, FF, R, S, P shall mean $T_{DD}$-, $T_{FF}$-, regular, symmetric and partition topologies respectively.) Furthermore, we write $p(n)$ for the number of partitions of $n$ into a sum of positive integers and $p_k(n)$ for the number of partitions of $n$ into a sum of $k$ positive integers and we recall the following formula:

$$p_1(n-k) + p_2(n-k) + \dots + p_k(n-k) = p_k(n) \qquad (*)$$

THEOREM 4. (a) $t_{DD}(n) = p(n)$;

(b) $\tau_{DD}(n) = \sum_{k=0}^{n-1} (n-k)^k \binom{n}{k}$.

*Proof* (a). In view of Theorem D, we have to determine the number of families of the form

$$A = (\{x_1\},\ldots,\{x_k\},$$
$$\{x_1,x_{k+1}\},\ldots,\{x_1,x_{k+k_1}\},\ldots\ldots,$$
$$\{x_i,x_{k+k_1+\ldots+k_{i-1}+1}\},\ldots,\{x_i,x_{k+k_1+\ldots+k_i}\}),$$

where $1 \leqslant k \leqslant n$, $0 \leqslant i \leqslant k$, $k_1 \geqslant \ldots \geqslant k_i$ and $k + k_1 + \ldots + k_i = n$.

For fixed k the required number is equal to the number of partitions of n-k into at most k summands. By (*), this is $p_k(n)$. Hence,

$$t_{DD}(n) = \sum_{k=1}^{n} p_k(n) = p(n).$$

(b). For $1 \leqslant k \leqslant n-1$ there are $\binom{n}{k}$ ways of choosing the k singletons of A, and each choice appears as often as the remaining n-k elements can be mapped into the set of these k elements, i.e. $(n-k)^k$ times. Since for k = n there is exactly one family, we obtain $\tau_{DD}(n)$ as

$$\tau_{DD}(n) = 1 + \sum_{k=1}^{n-1} (n-k)^k\binom{n}{k} = \sum_{k=0}^{n-1} (n-k)^k\binom{n}{k}.$$

*Remark* 4. It is easy to see that $\tau_{DD}(n)$ is also

(i) the number of idempotents of the symmetric semigroup on an n-set [2];
(ii) the number of forests of rooted labeled trees of height at most one [7].

THEOREM 5. (a) $t_{FF}(n) = 2(n-1)$ $(n \geqslant 2)$;
(b) $\tau_{FF}(n) = n(2^n-n-1)+1$ $(n \geqslant 1)$.

*Proof* (a). Using Theorem E, we have to determine the number of families of the form

$$A_1 = (\{x_1\},\ldots,\{x_k\},\{x_1,x_{k+1}\},\ldots,\{x_1,x_n\}) \qquad 1 \leqslant k \leqslant n, \text{ or}$$
$$A_2 = (\{x_1\},\ldots,\{x_{n-1}\},\{x_1,\ldots,x_k,x_n\}) \qquad 2 \leqslant k \leqslant n-1.$$

Now (a) is obvious.

(b). There are $\binom{n}{k}$ choices for the k singletons of $A_1$ and for $k < n$ each of the elements $x_1,\ldots,x_k$ can be combined with the remaining elements. In $A_2$ we have n choices for the element $x_n$ and $\binom{n-1}{k}$ for the elements that are combined with $x_n$.

Therefore,

$$\tau_{FF}(n) = 1 + \sum_{k=1}^{n-1} k\binom{n}{k} + \sum_{k=2}^{n-1} n\binom{n-1}{k} = 1 + n2^{n-1} - n + n(2^{n-1} - 1 - n + 1) = n(2^n - n - 1) + 1.$$

THEOREM 6. (a) $t_R(n) = t_S(n) = t_P(n) = p(n)$ (Knopfmacher [4]);

(b) $\tau_R(n) = \tau_S(n) = \tau_P(n) = B(n)$ (see e.g. [5]),

*where* B(n) *denote the Bell numbers.*

*Proof.* In view of Theorem F it suffices to consider partition topologies. Now any partition of a set of n elements gives rise to a partition topology, which proves (b); to obtain $t_P(n)$ we have to partition n like elements, which yields (a).

*Remark* 5. For finite topologies, (a) ⟺ (b) of Theorem F has been proved by Knopfmacher [4], whilst (a) ⟺ (c) was shown by Huebener [3].

## REFERENCES


1. C.E. Aull, and W.J. Thron, "Separation axioms between $T_o$ and $T_1$", *Indag. Math.* **24** (1962), 26-37.
2. B. Harris, and L. Schoenfeld, "The number of idempotent elements in symmetric semigroups", *J. Comb. Th.* **3** (1967), 122-135.
3. M.J. Huebener, "Complementation in the lattice of regular topologies", Ph.D. dissertation, Univ. of Cincinnati, 1970.
4. J. Knopfmacher, "Note on finite topological spaces", *J. Austral. Math. Soc.* **9** (1969), 252-256.
5. R.E. Larson, and S.J. Andima, "The lattice of topologies: a survey", *Rocky Mt. J. of Math.* **5** (1975), 177-198.
6. R.A. Razen, "A characterization of transversal topologies", to appear.
7. J. Riordan, "Forests of labeled trees", *J. Comb. Th.* **5** (1968), 90-103.
8. H. Sharp, Jr., "Cardinality of finite topologies", *J. Comb. Th.* **5** (1968), 82-86.
9. R.P. Stanley, "On the number of open sets of finite topologies", *J. Comb. Th.* **10** (1971), 74-79.
10. K. Steffens, "Injektive Auswahlfunktionen", Schriften aus dem Gebiet der Angewandten Mathematik Nr. 2, Aachen 1972.
11. D. Stephen, "Topology on finite sets", *Amer. Math. Monthly* **75** (1968), 739-741.

05 C 30

# ASYMPTOTIC NUMBER OF SELF-CONVERSE ORIENTED GRAPHS

R.W. Robinson

Department of Mathematics,
University of Newcastle,
N.S.W., Australia 2308.

ABSTRACT

The exact number $\overline{O}_p$ of self-converse oriented graphs on p points has been known for some time; we give the first asymptotic analysis of $\overline{O}_p$. The formula for $\overline{O}_p$ was found by an application of Burnside's Lemma, and consists of a sum of terms corresponding to the partitions of p. The usual pattern in such cases is for one term to dominate the others when p is large. For $\overline{O}_p$ there is no single term which is dominant. Instead an infinite family of terms must all be considered in the asymptotic treatment. This leads to a much more complicated asymptotic form for $\overline{O}_p$ than the ones obtaining for graphs, digraphs, self-complementary graphs, self-complementary digraphs, oriented graphs, tournaments, or self-complementary tournaments. The values of $\overline{O}_p$ are tabulated for p up to 27. A sample is then compared with the corresponding sums of dominant terms and final asymptotic expressions.

## 1. INTRODUCTION

An *oriented graph* is a finite digraph with no symmetric pairs of arcs. Thus an oriented graph has no loops and at most one arc joining any unordered pair of points. The *converse* of any digraph is obtained by reversing the directions of all of its arcs. A digraph is *self-converse* just if it is isomorphic to its converse.

The number $\overline{O}_p$ of nonisomorphic self-converse oriented graphs on p points was found by Sridharan [9] using Burnside's Lemma. His result can be put in the form

$$\overline{O}_p = \sum 3^{c\langle h\rangle} \Big/ \prod_i i^{h_i} h_i!, \qquad (1)$$

the sum being over all sequences $\langle h\rangle = (h_1, h_2, \ldots)$ of non-negative integers with $\sum_{1\leq i} ih_i = p$. The exponent $c\langle h\rangle$ is given by

$$c\langle h\rangle = \sum_{1\leq i} (ih_{2i}^2 - h_{4i}) + \sum_{\substack{1\leq i<\ell \\ i\ell \text{ even}}} (i,\ell)h_i h_\ell . \qquad (2)$$

Here as usual $(i,\ell)$ denotes the greatest common divisor of i and $\ell$.

In this paper we give the first asymptotic analysis of $\overline{O}_p$. It is found that

$$(3) \quad \overline{O}_p \sim 3^{p^2/4} p^{-p/2+(\log_3 p)/4-\log_3\log_3 p-1/2+\log_3 e} e^{p/2} (\log_3 p)^{\log_3\log_3 p-1/2} C_p$$

where the factor $C_p$ is of constant order. The latter may be expressed in terms of the function $\Phi$ defined by

$$\Phi(x) = \frac{1}{\sqrt{\pi \log_3 e}} \sum_{k=-\infty}^{\infty} 3^{-(x-k)^2} .$$

Then (3) is completed by

$$(4) \qquad C_p = \sqrt{\frac{\log_3 e}{2\pi}}\, \Phi\left(\tfrac{1}{2}\log_3 p - \log_3\log_3 p - \tfrac{1}{2}\rho(p)\right)$$

where $\rho(p)$ denotes the residue of p modulo 2.

An unusual feature of the asymptotic behaviour of $\overline{O}_p$ is that as $p \to \infty$ infinitely many terms from the right side of (1) make contributions which cannot be neglected. The normal pattern is that a single term dominates the asymptotic behaviour in unlabelled graph counting problems to which the exact answer is obtained fairly directly using Burnside's Lemma. This has been shown to be the case for graphs by Oberschelp [8], for relational systems of uniform non-monodic type by Oberschelp [7], for all relational systems of non-monodoc type by Fagin [4], for digraphs by Harary and Palmer [2, p.200], for self-complementary graphs and digraphs by Palmer [6], for oriented graphs by Wille [10], for tournaments by Moon [5, p.88], and for self-complementary tournaments in [2, exercise 9.6].

The pattern of asymptotic behaviour shown here to hold for self-converse oriented graphs was first observed in a graphical context in the joint work [3] with Harary, Palmer and Schwenk. There similar asymptotic expressions are reported for the numbers of self-dual signed graphs and line-self-dual nets, but no details of the derivation are given. In fact they follow closely the development outlined herein for self-converse oriented graphs. The asymptotic number of self-converse digraphs also involves an infinite series of dominant terms, but in a sufficiently different fashion that a separate description is necessary; this will appear elsewhere.

We begin the proof of (3) and (4) in the next section by selecting the dominant terms in the right side of (1) and showing that the remainder can be neglected asymptotically. Then the sum of the dominant terms is evaluated asymptotically in Section 3. It is found that the cases p even and p odd must be treated separately. In the last section a table of the numbers $\overline{O}_p$ is presented for $p \leq 27$.

These are compared numerically with the corresponding sums of dominant terms and their asymptotic approximations.

## 2. DOMINANT TERMS

In this section we show that asymptotically it is sufficient to restrict attention to terms on the right side of (1) contributed by sequences <j> of the form $(j_1,j_2)$, that is, those with $j_3 = j_4 = \ldots = 0$. We call these the *dominant terms*. In showing that the contribution of the other terms in (1) can be neglected, we compare the term due to an arbitrary $\langle h\rangle = (h_1,h_2,\ldots)$ with the dominant term given by

$$(5) \qquad j_1 = \begin{cases} h_1+1 & \text{if } p-h_1 \text{ is odd,} \\ h_1 & \text{if } p-h_1 \text{ is even} \end{cases} .$$

Here p refers to the *order* $p = \sum_{1\leq i} ih_i$ of <h>. The dominant term to which we compare <h> is understood to be of the same order, so that

$$(6) \qquad j_2 = \frac{p-j_1}{2}.$$

For each sequence <h> let t<h> be the corresponding term in the expression (1) for $\overline{O}_p$ where p is the order of <h>; that is,

$$(7) \qquad t\langle h\rangle = 3^{c\langle h\rangle}/\prod i^{h_i}h_i! \,.$$

The power c<h> is simply the total number of ordinary (non-diagonal) line cycles of even length and diagonal line cycles of odd length induced by a point permutation which is the product of $h_i$ disjoint cycles of length i over $i \geq 1$. The fraction $1/\prod i^{h_i}h_i!$ is the proportion of permutations on $\{1,2,\ldots,p\}$ which can be written as a product of this sort, out of the p! possible permutations.

For the time being we fix a dominant term $t(j_1,j_2)$ and consider all terms which are to be compared with it. Any such term t<h> corresponds to a sequence <h> satisfying one of two conditions for some integer $m \geq 0$;

$$(a) \qquad h_1 = j_1, \quad h_2 = j_2-m \quad \text{and} \quad \sum_{2<i} ih_i = 2m, \quad \text{or}$$

$$(b) \qquad h_1 = j_1-1, h_2 = j_2-m \quad \text{and} \quad \sum_{2<i} ih_i = 2m+1.$$

In case (a) $m \neq 1$ and in case (b) $m \neq 0$, but otherwise all values of m up to $j_2$ are possible.

We now show how to calculate an upper bound on the sum of terms obtained from all <h> which are compared to <j> and satisfy (a) for a fixed $m \geq 0$. First we will show that

$$(8) \qquad c\langle h\rangle \leq j_2(j_2 - m) + \tfrac{m}{2}(m-1) + j_1(j_2 - \tfrac{m}{2}) .$$

The right side of (8) is the value of $c(j_1, j_2-m, 0, \frac{m}{2})$ computed from (2) regardless of whether $\frac{m}{2}$ is integral. The left side of (8) is given by (2); with the values $j_1$ for $h_1$ and $j_2-m$ for $h_2$ from (a) and the terms in $j_1$ and $j_2-m$ separated from the sums, (2) takes the form

$$(9) \qquad c\langle h\rangle = (j_2-m)^2 + \sum_{\substack{2<i \\ i \text{ even}}} \tfrac{i}{2} h_i^2 - \sum_{1\leq i} h_{4i} + j_1(j_2-m) + j_1 \sum_{\substack{2<i \\ i \text{ even}}} h_i$$

$$+ (j_2-m) \sum_{\substack{1<i \\ i \text{ odd}}} h_i + 2(j_2-m) \sum_{\substack{2<i \\ i \text{ even}}} h_i + \sum_{\substack{2<i<\ell \\ i\ell \text{ even}}} (i,\ell) h_i h_\ell .$$

Selecting two terms each from (9) and (8) we obtain the difference

$$j_2(j_2-m) + j_1(j_2-\tfrac{m}{2}) - (j_2-m)^2 - j_1(j_2-m),$$

which simplifies to

$$(j_1+2j_2-2m)\tfrac{m}{2} .$$

We make the replacement

$$\frac{m}{2} = \frac{1}{4} \sum_{\substack{2<i \\ i \text{ even}}} ih_i + \frac{1}{4} \sum_{\substack{1<i \\ i \text{ odd}}} ih_i$$

in view of (a). Now note that the result is greater than or equal to the sum

$$j_1 \sum_{\substack{2<i \\ i \text{ even}}} h_i + 2(j-m) \sum_{\substack{2<i \\ i \text{ even}}} h_i + (j_2-m) \sum_{\substack{1<i \\ i \text{ odd}}} h_i$$

contained in (9). This is because $\frac{i}{4} \geq 1$ for even $i > 2$ and $\frac{i}{2} > 1$ for odd $i > 1$.

Therefore to verify (8) it will suffice to show the inequality

$$\sum_{\substack{2<i\\ i\text{ even}}} \frac{i}{2} h_i^2 - \sum_{1\leq i} h_{4i} + \sum_{\substack{2<i<\ell\\ i\ell\text{ even}}} (i,\ell)h_i h_\ell \leq \frac{m^2}{2} - \frac{m}{2}$$

among the remaining terms from (8) and (9). By (a) again we may replace $\frac{m^2}{2} - \frac{m}{2}$ by

$$\frac{1}{8}\sum_{2<i} i^2 h_i^2 + \frac{1}{4}\sum_{2<i<\ell} i\ell h_i h_\ell - \frac{1}{4}\sum_{2<i} ih_i,$$

and upon rearranging obtain the equivalent inequality

$$\text{(10)}\qquad \sum_{\substack{2<i\\ i\text{ even}}} \frac{i}{2} h_i^2 + \frac{1}{4}\sum_{2<i} ih_i + \sum_{\substack{2<i<\ell\\ i\ell\text{ even}}} (i,\ell)h_i h_\ell \leq \frac{1}{8}\sum_{2<i} i^2 h_i^2 + \sum_{1\leq i} h_{4i} + \frac{1}{4}\sum_{2<i<\ell} i\ell h_i h_\ell .$$

Now (10) is verified by observing that $(i,\ell) < \frac{i\ell}{4}$ when $2 < i < \ell$ and one of $i,\ell$ is even, $\frac{i}{2} = \frac{i^2}{8}$ and $\frac{i}{4} = 1$ when $i = 4$, $h_i \leq h_i^2$ for all $i$ (since $h_i$ must be a non-negative integer), $\frac{i}{4} < \frac{i^2}{8}$ for odd $i > 2$, and $\frac{3}{4}i \leq \frac{i^2}{8}$ for even $i > 4$.

Next, the number of all $\langle h\rangle$ which are compared to $\langle j\rangle$ and satisfy (a) for a particular $m \geq 0$ is bounded above by

$$\text{(11)}\qquad \binom{p}{j_1}\binom{p-j_1}{2j_2-2m}\frac{(2j_2-2m)!}{(j_2-m)!2^{j_2-m}}(p-j_1-2j_2+2m)! \quad .$$

This is derived by considering the number $\binom{p}{j_1}$ of ways to choose a set of $j_1$ fixed points from the object set $1,2,\ldots,p$ followed by the number $\binom{p-j_1}{2j_2-2m}$ of ways to choose a set of $2j_2-2m$ elements from the remainder. The latter can be arranged into $j_2-m$ transpositions in exactly $(2j_2-2m)!/(j_2-m)!2^{j_2-m}$ ways, since the transpositions can be rearranged among themselves in $(j_2-m)!$ ways and have starting points chosen in 2 ways each for a total of $2^{j_2-m}$ ways altogether. There are then $p-j_1-2j_2+2m$ elements to be permuted among themselves with no fixed points or transpositions, and $(p-j_1-2j_2+2m)!$ gives a crude upper bound for the number of ways this can be done.

Finally, multiplying together the upper bound for $3^{c\langle h\rangle}$ from (8) with the upper bound (11) for the number of terms and then dividing by $p!$ gives

$$\text{(12)}\qquad \frac{3^{j_2(j_2-m)+(m-1)m/2+j_1(j_2-m/2)}}{j_1!(j_2-m)!2^{j_2-m}} .$$

This is the desired upper bound for the total contribution of the terms $t\langle h\rangle$ to $\overline{O}_p$ in the right side of (1), for all $\langle h\rangle$ satisfying (a) for fixed $m \geq 0$.

The terms bounded by (12) are all to be compared with the dominant term $t\langle j\rangle$, where

$$t\langle j\rangle = \frac{3^{j_2(j_1+j_2)}}{j_1!\,j_2!\,2^{j_2}} \quad . \tag{13}$$

Using the fact that $j_2!/(j_2-m)! \leq j_2^m$, we see that the ratio of the upper bound (12) to the dominant term (13) is at most

$$\left(\frac{2j_2}{3^{(j_1+2j_2-m+1)/2}}\right)^m \quad . \tag{14}$$

Then, from $j_1+2j_2 = p$ and $m \leq p/2$ it follows that the latter is less than

$$\left(p3^{-p/4}\right)^m \quad . \tag{15}$$

The next stage is to establish a similar upper bound for the sum of the terms $t\langle h\rangle$ contributed by sequences $\langle h\rangle$ which satisfy (b). If $\langle h\rangle$ satisfies (b) for some m, then $m \geq 1$ and the exponent $c\langle h\rangle$ obeys the inequality

$$c\langle h\rangle \leq j_1\left(j_2-\frac{m+1}{2}\right)+(j_2-m)(j_2-1)+\frac{(m-1)(m-2)}{2} \tag{16}$$

when $p \geq 12$. The right side of (16) is the value of $c(j_1-1,\ j_2-m,\ 1,\ \frac{m-1}{2})$ computed from (2) whether or not $\frac{m-1}{2}$ is an integer. A proof of (16) can be carried out along the same lines as the proof of the similar inequality (8). It is not quite as neat in this case, for a consideration of $\langle h\rangle = (i,0,1,0,0,1)$ for $i = 0,1$ or $2$ will verify that the condition $p \geq 12$ is essential.

The number of sequences $\langle h\rangle$ which are compared to $\langle j\rangle$ and satisfy (b) for a particular $m \geq 1$ is bounded by

$$\binom{p}{j_1-1}\binom{p-j_1+1}{2j_2-2m}\frac{(2j_2-2m)!}{(j_2-m)!\,2^{j_2-m}}\,(p-j_1-2j_2+2m+1)! \quad . \tag{17}$$

The reasoning is exactly the same as for the similar bound (11), the only difference being that now $h_1 = j_1-1$, so that $j_1$ has been replaced by $j_1-1$. Multiplying this together with the upper bound for $3^{c\langle h\rangle}$ obtained from (16) and dividing by $p!$ gives

$$(18) \qquad \frac{3^{j_1(j_2-(m+1)/2)+(j_2-m)(j_2-1)+(m-1)(m-2)/2}}{(j_1-1)!(j_2-m)!2^{j_2-m}} .$$

This is an upper bound for the total of the terms t<h> for all <h> satisfying (b) for some $m \geq 1$ and fixed $\langle j\rangle = (j_1, j_2)$. Dividing by the value of t<j> given in (13) and again using the fact that $j_2!/(j_2-m)! \leq j_2^m$, we find that the ratio is less than or equal to

$$(19) \qquad j_1 3^{-j_1/2-j_2+1}\left(2j_2 3^{-j_1/2-j_2+(m-1)/2}\right)^m .$$

From the relations $j_1+2j_2 = p$ and $m < p/2$ it follows that (19) is less than

$$(20) \qquad \left(p3^{-p/4}\right)^{m+1}$$

for sufficiently large p.

We now collect each dominant term in the equation (1) for $\overline{O}_p$ together with all the other terms which are compared to it. These will satisfy (a) for $m \geq 2$ or (b) for $m \geq 1$; so by summing the upper bounds (15) and (20) over these values of m we find that the total contribution of the terms compared to $t(j_1, j_2)$ is

$$t(j_1, j_2)\left(1 + O(p^2 3^{-p/2})\right) .$$

Since all of these terms are positive and the upper bounds used are independent of the particular dominant term, we may sum over all dominant terms to obtain

$$(21) \qquad \overline{O}_p = \left(\sum_{m \leq p/2} t(p-2m, m)\right)\left(1 + O(p^2 3^{-p/2})\right) .$$

Thus we may neglect the nondominant terms of $\overline{O}_p$ asymptotically. The rapid convergence of the sum of the dominant terms to $\overline{O}_p$ is illustrated later in Table 2 for selected values of p up to 27.

## 3. ASYMPTOTIC FORMULAE

In this section we discuss the asymptotic behaviour of the sum of the dominant terms t(p-2m,m). A general approach to the asymptotic evaluation of sums with positive terms is described by Bender [1, Section 3]. Two atypical features appear in the process of following the usual approach. One is that the result depends on whether p is even or odd. The other is that the factor of constant order contains a periodic function of $\frac{1}{2}\log_3 p - \log_3\log_3 p$.

Consider first the case of even p, and let p = 2n. Denote by t(k) the dominant term t(2k,n-k). Then from (13) we have

$$t(0) = 3^{n^2}/n!2^n \tag{22}$$

and

$$t(k)/t(0) = n!2^k/(2k)!(n-k)!3^{k^2}. \tag{23}$$

The ratio of successive terms is

$$t(k+1)/t(k) = (n-k)/(k+1)(2k+1)3^{2k+1}. \tag{24}$$

From this it is not hard to calculate that t(k) should attain its maximum value for k near to f(n), where

$$f(n) = \tfrac{1}{2}\log_3 n - \log_3\log_3 n + \tfrac{1}{2}\log_3 2. \tag{25}$$

To study t(k) in the vicinity of its maximum value, let x = k-f(n) and restrict attention to the region $|x| \leq (\log_3 n)^{\frac{1}{4}}$. Using Stirling's formula for the factorial functions in (23) it is straightforward to calculate

$$t(k)/t(0) = \frac{(2n)^{f(n)}3^{-x^2}}{\{2f(n)\}!3^{f(n)^2}} \{1 + O(\log\log n/\sqrt{\log n})\} . \tag{26}$$

Here the implied constant in the error term can be taken independent of x and n, and $\{2f(n)\}!$ is defined in terms of the $\Gamma$ function.

From (24) it can be seen that the ratio t(k+1)/t(k) is a unimodal function of k. In the region $|x| \leq (\log_3 n)^{\frac{1}{4}}$ it is easy to calculate that t(k+1)/t(k) is of order $3^{-2x}$. Thus we have

$$\sum_{k=0}^{n} t(k) = \{\textstyle\sum' t(k)\}\{1 + O(\exp(-\sqrt{\log n}))\} , \tag{27}$$

where $\Sigma'$ denotes summation over the integers k such that $|k-f(n)| \leq (\log_3 n)^{\frac{1}{4}}$. On the other hand

$$\sum_{k=-\infty}^{\infty} 3^{-(f(n)-k)^2} = \{\textstyle\sum' 3^{-(f(n)-k)^2}\}\{1 + O(\exp(-\sqrt{\log n}))\} \tag{28}$$

also, so that in approximating the right side of (27) using (26) the sum can be extended over all k. The resulting expression for the sum of the dominant terms is

$$\sum_{k=0}^{n} t(k) = \frac{(2n)^{f(n)}\,t(0)}{(2f(n))!\,3^{f(n)^2}} \sum_{k=-\infty}^{\infty} 3^{-(f(n)-k)^2}\,(1+E(n)) \tag{29}$$

where $E(n) = O(\log\log n/\sqrt{\log n})$.

In the terminology introduced for equation (4) the summation on the right side of (29) is

$$\sqrt{\pi \log_3 e}\ \Phi(f(n)).$$

The asymptotic formula for $\overline{O}_p$ given in (3) and (4) for even p can now be calculated by combining (21), (22), (25) and (29), eliminating the factorials with Stirling's formula and expressing the result in terms of p. The order of the error term obtained is $\log\log p/\log p$.

The derivation of the asymptotic behaviour of $\overline{O}_p$ for odd p is quite similar to that for even p. If $t^*(k)$ denotes the dominant term $t(2k+1,n-k)$ for $p = 2n+1$, the maximum is attained for k near to

$$f^*(n) = \tfrac{1}{2}\log_3 n - \log_3\log_3 n + \tfrac{1}{2}\log_3 2 - \tfrac{1}{2}. \tag{30}$$

There are a number of minor differences between the expressions for $\overline{O}_{2n}$ and $\overline{O}_{2n+1}$ in terms of n. These cancel out asymptotically when the formulae are put in terms of p, all except for the respective periodic factors $\Phi(f(p/2))$ and $\Phi(f^*((p-1)/2))$.

From the definition it is evident that $\Phi$ is a periodic function, with period 1 and average 1. This is one of a class of functions considered by Wright [11] in connection with the asymptotic number of labelled k-coloured graphs. With 3 replaced by 2 as the base, Wright showed that this function varies very little from unity. The same method of proof would apply to our function $\Phi(x)$ in which 3 is the base. Numerical calculation indicates that the extreme values of $\Phi(x)$ differ by no more than $0.5 \times 10^{-3}$.

In the next section numerical values of the asymptotic expression are compared with the exact values of $\overline{O}_p$ for selected $p \leq 27$. No tendency toward convergence of the two is evident. As will be seen, much higher values of p would have to be used in order to observe the convergence which must eventually occur.

| $\overline{O}_p$ | p |
|---:|---:|
| 1 | 1 |
| 2 | 2 |
| 5 | 3 |
| 18 | 4 |
| 102 | 5 |
| 848 | 6 |
| 12 452 | 7 |
| 265 759 | 8 |
| 10 454 008 | 9 |
| 598 047 612 | 10 |
| 63 620 448 978 | 11 |
| 9 974 635 937 844 | 12 |
| 2 905 660 724 913 768 | 13 |
| 1 268 590 412 128 132 389 | 14 |
| 1 023 130 650 177 394 611 897 | 15 |
| 1 258 149 993 547 327 488 275 562 | 16 |
| 2 834 863 110 716 120 144 290 954 314 | 17 |
| 9 900 859 865 505 110 360 978 721 901 778 | 18 |
| 62 789 966 700 541 818 490 820 660 260 219 085 | 19 |
| 626 754 770 688 026 263 598 465 802 369 258 636 679 | 20 |
| 11 256 824 990 452 447 883 980 190 391 781 422 490 449 643 | 21 |
| 322 709 197 224 240 210 184 021 932 827 114 034 505 695 268 296 | 22 |
| 16 500 229 648 971 657 958 448 757 393 252 063 495 966 875 945 342 103 | 23 |
| 1 363 854 423 811 897 643 093 117 996 577 197 078 873 109 190 304 984 874 334 | 24 |
| 199 403 043 497 277 732 688 200 817 094 827 202 121 041 725 841 248 825 841 470<br>489 | 25 |
| 47 673 458 041 008 308 339 959 667 197 303 459 424 735 715 919 978 766 546 406<br>524 143 862 | 26 |
| 20 006 994 424 095 895 824 681 312 306 066 615 139 962 829 403 824 584 051 878<br>040 528 717 145 556 | 27 |

TABLE 1.

Exact numbers of self-converse oriented graphs for $p \leq 27$

## 4. NUMERICAL RESULTS*

The exact values of the numbers $\overline{O}_p$ of self-converse oriented graphs for $p \leq 27$ are presented in Table 1. These were produced by computer computation using a program which implemented equations (1) and (2) as directly as possible. In Table 2 are shown the results of comparing these numbers to the sum $D_p$ of dominant terms and the final asymptotic formula $A_p$ obtained by combining the right sides of equations (3) and (4). These are shown for sample values of p.

It will be seen that $D_p$ approaches $\overline{O}_p$ rapidly and evenly as p increases. The same certainly cannot be said of $A_p$ for the values of p considered. Since it was impractical to carry the exact computation of $\overline{O}_p$ significantly further, numerical testing was conducted on the various steps taken in deriving $A_p$ as an asymptotic expression for $D_p$. One particular step with a $\log\log p/\log p$ error term was observed to contribute the major part of the apparently divergent behaviour for small p. Computation showed that convergent behaviour for this step set in at values of $\log_3 p$ between 30 and 40. Thus $A_p/\overline{O}_p$ cannot be expected to start decreasing as p increases, until p reaches $10^{15}$ or so. By that time $\overline{O}_p$ would have almost $10^{30}$ digits in its decimal representation.

| $(\overline{O}_p/D_p)-1$ | $A_p/\overline{O}_p$ | p |
|---|---|---|
| $0.538\times10^{-2}$ | 1.624 | 8 |
| $0.227\times10^{-2}$ | 1.546 | 9 |
| $0.245\times10^{-4}$ | 1.701 | 14 |
| $0.956\times10^{-5}$ | 1.631 | 15 |
| $0.732\times10^{-7}$ | 1.741 | 20 |
| $0.272\times10^{-7}$ | 1.697 | 21 |
| $0.176\times10^{-9}$ | 1.768 | 26 |
| $0.636\times10^{-10}$ | 1.746 | 27 |

TABLE 2.

Comparison of exact numbers of self-converse oriented graphs with the sum of the dominant terms and the asymptotic formula.

* The computer programming for the reported data was performed by A. Nymeyer. The author is grateful to the Australian Research Grants Committee for its financial support, which provided technical assistance and some of the computing equipment used in this research.

REFERENCES


(1) Edward A. Bender, "Asymptotic methods in enumeration", *SIAM Review* 16 (1974), 485-515.

(2) Frank Harary and Edgar M. Palmer, *Graphical Enumeration*. Academic Press, N.Y., 1973.

(3) Frank Harary, Edgar M. Palmer, Robert W. Robinson, and Allen J. Schwenk, "Enumeration of graphs with signed points and lines", *J. Graph Theory*, to appear.

(4) Ronald Fagin, "The number of finite relational structures", *Discrete Math*, to appear

(5) John Moon, *Topics on Tournaments*. Holt, N.Y., 1968.

(6) Edgar M. Palmer, "Asymptotic formulas for the number of self-complementary graphs and digraphs", *Mathematika* 17 (1970), 85-90.

(7) W. Oberschelp, "Structurzahlen in endlichen relationssystemen", *Contributions to Mathematical Logic* (Proceedings of Logic Colloquium), Hanover, 1966, 199-213.

(8) W. Oberschelp, "Kombinatorische Anzahlbestimmungen in Relationen", *Math. Ann.* 174 (1967), 53-58.

(9) M.R. Sridharan, "Self-complementary and self-converse oriented graphs", *Indag. Math.* 32 (1970), 441-447.

(10) Detlef Wille, "Asymptotic formulas for the number of oriented graphs", *J. Combinatorial Theory* (B) 21 (1976), 270-274.

(11) Edward M. Wright, "Counting coloured graphs", *Canad. J. Math.* 13 (1961), 683-693.

05A15

# SOME CORRESPONDENCES INVOLVING THE SCHRÖDER NUMBERS AND RELATIONS

D.G. ROGERS[(1)] and L.W. SHAPIRO[(2)]

[(1)]Mathematical Institute, Oxford, England and
Department of Mathematics, University of Western Australia

[(2)]Department of Mathematics, Howard University, U.S.A.

ABSTRACT

The Schröder relations are put into correspondence with a variety of other objects of combinatorial interest, including bushes, foliated trees, lattice paths with diagonal steps and certain sorts of two-coloured objects. A distinction between left and right Schröder relations is used in obtaining some of these correspondenced; others involve the use of ternary codes. By attending also to subsidiary features of the objects considered, interpretations of the Schröder numbers are obtained and some comparable results for the Motzkin numbers are then deduced.

## §1 INTRODUCTION

The *Schröder numbers* $r_n$ and $s_n$, [12, sequence 1163, also 1170 (correcting a misprint)], are given by

$$r_n = \sum_{i=0}^{n} \binom{2n-i}{i} C_{n-i} \quad , \quad n \geq 0 \tag{1}$$

$$s_0 = r_0 = 1 \; ; \; 2s_n = r_n \quad , \quad n \geq 0 \tag{2}$$

where $C_n$ is the n-th *Catalan number*, [12, sequence 577], given by

$$C_n = \frac{1}{n+1}\binom{2n}{n} \quad , \quad n \geq 0 \tag{3}$$

The Schröder numbers, like the Catalan numbers, occur in the enumeration of a variety of objects and, in particular, in the enumeration of the Schröder relations, [8, 10]. A preliminary account of some of these enumerative problems (concentrating, however, on arithmetic properties of the Schröder numbers) is given, together with further references, in [8]. The purpose of this paper is to exhibit the equivalence of these problems by establishing correspondences among the Schröder relations and these other combinatorial objects. (For similar accounts of the Catalan numbers see [4, 5, 6, 12].) A reflexive, symmetric binary relation R on a totally ordered set $X_n = \{x_i : 1 \leq i \leq n\}$, with ordering $x_i < x_j$ for $i < j$, is a *left* (resp. *right*) *Schröder relation* if and only if $x_s \not R x_t$ for $i \leq s < j < t$ (resp. $i \leq s < j < t$) whenever $x_i R x_j$, $i < j$. Such relations are considered in detail, as part of a general study of ladder graphs, in [10] and we refer to it and to [9] for definitions and notation used here. The graphs $\Gamma(R)$ of Schröder relations $R$ on $X_3$ are illustrated in Figures (1) and (2)

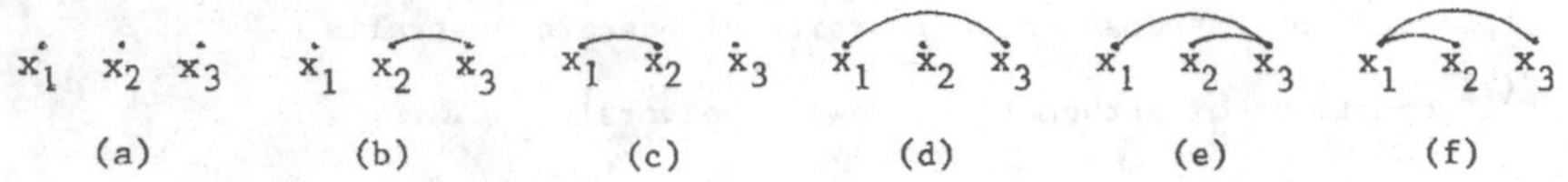

*Figure (1) $\Gamma(R)$ for left Schröder relations on $X_3$*

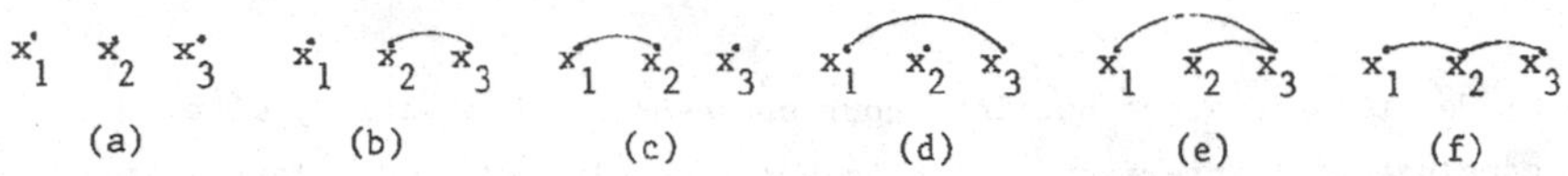

*Figure (2) $\Gamma(R)$ for right Schröder relations on $X_3$*

In [10], a bijection between the sets $S_\ell(n)$ and $S_r(n)$ of respectively left and right Schröder relations on $X_n$ was established (the graphs for corresponding relations have the same label in Figures (1) and (2)) and it was shown that the number $r(n,m)$ of relations $R$ in $S_\ell(n)$ or, equivalently, in $S_r(n)$, whose graphs $\Gamma(R)$ have $i$ edges is given by

$$r(n+1\ ,\ n-i) = \binom{2n-i}{i} c_{n-i} \qquad , \quad 0 \le i \le n \tag{4}$$

So the number $r(n)$ of relations in $S_\ell(n)$ $(S_r(n))$ is given by

$$r(n+1) = \sum_{i=0}^{n} r(n+1\ ,\ n-i) = r_n \tag{5}$$

If $R$ is a relation on $X_n$, then a *quasi-connected component* $X$ of $X_n$ under $R$ is a subset $X = \{x_i : s \le i \le t\}$ of $X_n$ such that $x_s R x_t$ and that $X$ maximal with respect to this property. $R$ is *qyasi-connected* on $X_n$ if $X_n$ is the (unique) quasi-connected component under $R$, that is $x_1 R x_n$; and then $\check{R}$ is the relation on $X_n$ agreeing with $R$ except that $x_1 \not R x_n$. The sets $S_\ell(n)$ and $S_r(n)$ have different structures as regards the property of being quasi-connected and it was also shown in [10] that the numbers $\tilde{r}(n)$ and $\hat{r}(n)$ of quasi connected relations in $S_\ell(n)$ and $S_r(n)$ respectively are given by

$$\tilde{r}(n+1) = s_n \qquad , \quad n \ge 0 \tag{6}$$

$$\hat{r}(n+1) = r_{n-1} \qquad , \quad n \ge 0 \tag{7}$$

In establishing correspondences between Schröder relations we utilize this structural difference (6, 7) between $S_\ell(n)$ and $S_r(n)$, (see §§2, 3). Ternary codes, generalizing the binary codes used in connection with the Catalan numbers (see, for example, [2, 13]) are also employed in setting up correspondences, (see §§4, 5). Moreover, by taking account of subsidiary features of the objects considered in these correspondences we are able to give (see, for example §4) further interpretations of (4, 5) as well as to deduce, from these, occurrences of the Motzkin numbers $m_n$, [12, sequence 456; 3], given by

$$m_n = \sum_{i=0}^{[n/2]} \binom{n}{2i} C_i \quad , \quad n \geq 0 \tag{8}$$

§2 BUSHES, BRACKETINGS, AND DISSECTIONS OF POLYGONS.

The number of bracketings of a product of $n$ non-associative, non-commutative terms in which a bracket may contain any number $i$, $2 \leq i \leq n$, of sub-products is $s_{n-1}$, $n \geq 2$, [1, pp. 56-7; 8]. A quasi-connected relation $R$ in $S_\ell(n)$ may be interpreted as such a bracketing of the product of the $n$ terms $x_i$, $1 \leq i \leq n$, by introducing a pair of brackets enclosing $x_i$ and $x_j$ whenever $x_i R x_j$, $i < j$; and then $\check{R}$ may be interpreted as the same bracketing but with the outermost pair of brackets removed.

A *bush* or *series reduced tree* is a rooted planar tree with at least one edge in which every vertex, other than the endpoints, has valence at least three : the number of bushes with $n$ endpoints (other than the root) is $r_{n-1}$, $n \geq 1$, and the number of these which are planted is $s_{n-1}$, $n \geq 1$, (compare (5, 6)). Bushes may be put into correspondence with the bracketings of the previous paragraph by the rule that, reading from the root upwards, the number of upward edges at a vertex (that is, the valence of the vertes less one) is the number of terms in a bracket, reading in from the outermost brackets. The bracketings and corresponding bushes associated with the relations of Figure (1) are shown (in the same order) in Figure (3).

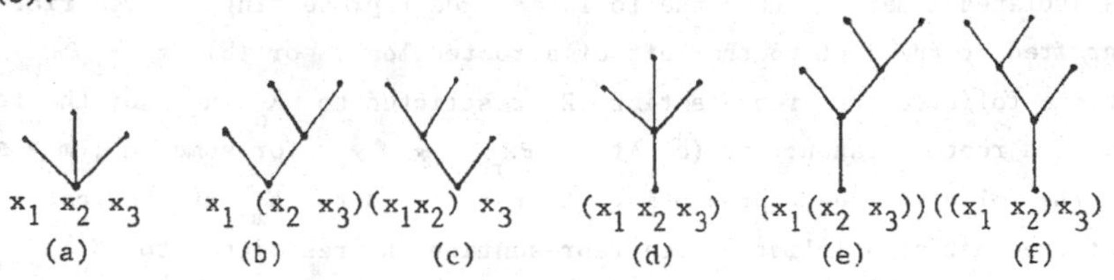


*Figure (3) Bushes and bracketings corresponding to the relations of Figure (1).*

More directly, combining the previous two correspondences, the relations in $S_\ell(n)$ may be put inductively into one-to-one correspondence with bushes with $n$ endpoints. For $n = 1$, the unique objects in these sets correspond. If $R$ is a relation in $S_\ell(n)$ with more than one quasi-connected component, then the trees representing $R$ restricted to each of its connected components are grafted together at their roots in the same order as the components which they represent. If R is a quasi-connected relation $S_\ell(n)$ then $\check{R}$ is a relation in $S_\ell(n)$ with more than one quasi-connected component and the tree representing $\check{R}$ is grafted to the tip of a rooted branch. Thus, under this correspondence, quasi-connected relations correspond to planted bushes. Further, relations in $S_\ell(n)$ whose graphs have $i$ edges, $0 \le i \le n-1$, correspond to bushes with $n$ endpoints (other than the root) and $n + i$ edges, so that the number of such bushes is $r(n,i)$.

As noted in [8], the bracketing of $n$ terms (without the outermost pair of brackets removed) may be put into correspondence with the dissections by non-crossing diagonals of a planar convex $(n + 1)$-gon into planar convex polygons, by extending a correspondence given in [4].

## §3 FOLIATED TREES AND SOME TWO-COLOURED OBJECTS

A *foliated tree* is a rooted planar tree with loops. The loops and edges may be given various weights and the weight of a foliated tree is then the total weight of its edges and loops. If both edges and loops have unit weight, then the number of foliated trees of weight $n$ is $r_n$ ; $n \ge 0$, while the number of these which are planted is $r_{n-1}$, $n \ge 1$, (compare (5, 7)). Throughout this section, a foliated tree is one with unit weights on loops and edges. We may establish inductively a correspondence between the set $S_r(n)$ and the set of foliated trees of weight $n - 1$, $n \ge 1$, in such a way that quasi-connected relations correspond to planted foliated trees. Moreover, relations in $S_r(n)$ whose graphs have $n - 1 - i$ edges, $0 \le i \le n-1$, correspond to foliated trees of weight $n - 1$ with $i$ loops, so that the number of these is $r(n,n - 1 - i)$.

To start this correspondence, the unique relation on $X_1$ corresponds to the unique tree of weight zero. For $n \ge 2$ and $R$ a relation in $S_r(n)$ either: (a) if $x_n$ is isolated under $R$ then the foliated tree representing $R$ restricted to $X_{n-1}$ is grafted at the root to the left of a rooted loop; or (b) if $x_1 R x_n$ then graft the foliated tree representing $R$ restricted to $X_n \backslash \{x_1\}$ by the root to the tip of a rooted branch; or (c) if $x_n R x_m$, $x_n \not R x_s$ for some $m$, $m > s \ge 1$, then graft the foliated tree representing $R$ restricted to $X_m$ at the root to the left of the root of a foliated tree representing $R$ restricted to $X_n \backslash X_{m-1}$. This process may be reversed to recover relations in $S_r(n)$ from foliated trees of weight $n - 1$, $n \ge 1$. The foliated trees corresponding to the relations of Figure (2) are shown in Figure (4).

*Figure (4) Foliated trees corresponding to the relations of Figure (2)*

The number of rooted planar trees with $n$ edges in which the endpoints (other than the roots) may be coloured independently with any one of two colours is also $r_n$, $n \geq 0$. This may be seen by taking foliated trees of weight $n$ and colouring their endpoints (other than the root) with one colour, converting the loops to edges whose endpoints are coloured with the other colour. Conversely, a two-coloured tree may be reconverted to a foliated tree by replacing all the edges adjacent to endpoints coloured in one specified colour by loops. From this, in turn, via the correspondences established in [9, 11], we obtain further correspondences with a variety of two-coloured objects. For example, the number of two-coloured connective relations on $X_n$ is $s_n$, $n \geq 1$, [10] and this is also the number of two-coloured non-decreasing bipartite graphs, the two disjoint subsets of whose vertex set both contain $n$ vertices [11].

§4 LATTICE PATHS WITH DIAGONAL STEPS.

A *restricted walk* is an outward directed walk on the non-negative quadrant of the two dimensional integral square lattice from the origin to some point $(n,m)$, $0 \leq m \leq n$, which never rises above the diagonal $y = x$. The number of restricted walks from the origin to $(n,n)$ is $C_n$, $m \geq 0$, [13], and the number of such walks when diagonal steps from $(s,t)$ to $(s+1, t+1)$ are permitted for $s,t \geq 0$ is $r_n$, $n \geq 0$, [1, pp. 80-1; 8]. The correspondence, given in [2, 13], between rooted planar trees and restricted walks without diagonal steps may be extended to one between foliated trees as in §3 and restricted walks in which diagonal steps are allowed. As in [2, 14], we trace around the tree introducing, in addition to U's and D's for respectively upward and downward steps, L's for when we encounter a loop. For a foliated tree of weight $n$ with $i$ loops we obtain a string of U's, D's and L's with $2(n-i)+i = 2n-i$ symbols in all, $i$ of which are L's. By interpreting the letters U, D and L respectively as steps to the right, upwards and diagonally, this string encodes a restricted walk from the origin to $(n,n)$ having $i$ diagonal steps. This construction is illustrated in Figure (5)

*Figure (5) An illustration of the contruction of lattice paths.*

Now there are $C_{n-i}$ restricted walks from the origin to $(n - i , n - i)$, $0 \le i \le n$, without diagonal steps and each of these walks passes through $2(n - i) + 1$ lattice points. To construct a restricted walk from the origin to $(n,n)$ with $i$ diagonal steps we may, using the translational invariance of the lattice, insert, possibly repeated, diagonal steps at any of these $2(n - i) + 1$ points in

$$\binom{2(n - i) + 1 + i - 1}{i} = \binom{2n - i}{i}$$

ways. Since this construction yields a bijection, the number of restricted walks from the origin to $(n,n)$ with $i$ diagonal steps is, (compare(4)),

$$\binom{2n - i}{i} C_{n-i}$$

giving another combinatorial interpretation of $r_n$. (This argument is more direct than that in [8].)

§5 SOME TERNARY CODES.

Codes of the kind used in §4 provide an alternative method to that of proof by induction as used in §§2,3 and, to illustrate their use further, we give another example here. The left - most upward branch (if any) at a vertex of a rooted planar trees is called the *eldest branch* at that vertex. The number of rooted planar trees with $n$ edges in which the eldest branches may be independently marked or unmarked is $r_n$, $n \ge 0$. One way of proving this is to put these trees inductively into correspondence with foliated trees, following the pattern of §3, so that marked edges become loops. Another proof is to note that a rooted planar tree with $n$ edges and $k$ eldest branches has $n + 1 - k$ endpoints and the number of such trees is the same as the number of rooted planar trees with $n$ edges and $k$ endpoints [9], so that the result then follows from that for two-coloured trees in §3. Here we give a proof using ternary codes.

Again we trace around the tree using the "up-down" code, except that we use an M instead of a U when we go up a marked branch. In this way we obtain a string of U's , D's and M's having the properties: (i) that at each stage the combined number of U's and M's is never less that the number of D's , with equality between these numbers for the string as a whole; and (ii) that M never directly follows D . Any such ternary string arises in this way; and any M in such a string occurs at the start of a unique minimal segment having the same properties which, in particular, ends with a D . Thus each M may be paired, in this way, with a unique D . We first replace each of these pairs by a pair of parentheses and then we delete all the left parentheses and replace the right parentheses by L's to obtain a sequence of U's , D's and L's which encodes a foliated tree or a lattice path as in §4. Conversely, results on bracketing ensure that the original sequence of U's , D's and M's is uniquely recoverable from the sequence of U's , D's and L's . Moreover, the numbers of M's and L's in corresponding sequences are the same. An illustration of these correspondences is given in Figure (6)

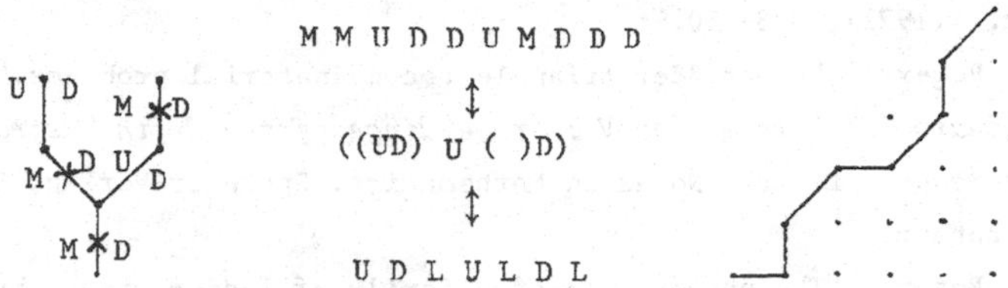


*Figure (6) An illustration of ternary codes.*

§6 SOME OCCURRENCES OF THE MOTZKIN NUMBERS.

It follows from (4) that the number of left (right) Schröder relations R on $X_{n+1-i}$ whose graphs $\Gamma(R)$ have $i$ edges for some $i$ , $0 \le i \le \frac{n}{2}$ , is given

$$\sum_{i=0}^{[n/2]} r(n+1-i,i) = \sum_{i=0}^{[n/2]} \binom{n}{2i} C_i = m_n \quad , \quad n \ge 0 \tag{9}$$

where $m_n$ is the n-th Motzkin number as in (8). In view of (5) and (9), it follows similarly that, for example, the number of bushes with $n + 1$ edges is $m_n$ , $n \ge 0$ , and this is also the number of foliated trees of weight $n$ in which loops have unit weight and edges double weight, both examples discussed in [3].

Further examination of the above correspondences produces similar results for bushes or foliated trees in which the root has valence one in terms of the numbers $\gamma_n$ given by, (see [3, 9] and compare (2)),

$$\gamma_0 = 1 \; ; \; \gamma_n + \gamma_{n+1} = m_n \quad , \quad n \ge 0$$

This connection between the numbers $\gamma_n$ and $s_n$, and hence between the numbers $m_n$ and $r_n$, may also be seen in Table 1, p.344, of [7], where the row sums are $s_{h-2}$, $h \geq 2$, and the column sums are $\gamma_n$, $n \geq 2$.

REFERENCES.


[1] L. Comtet. *Advanced Combinatorics*. D. Reidel, Dordrecht, 1974.

[2] N.G. de Bruijn and B.J.M. Morselt. "A note on plane trees." *J. Combinatorial Theory* 2(1967), 27-34.

[3] R. Donaghey and L.W. Shapiro. "Motzkin numbers." *J. Combinatorial Theory Ser. A*, (to appear).

[4] H.G. Forder. "Some problems in combinatorics." *Math. Gaz.* 45(1961), 199-201.

[5] M. Gardner. Mathematical Games. *Sci. Amer.* 234:6 (June 1976), 120-125.

[6] H.W. Gould. *Research bibliography of two special number sequences*. Mathematicae Monongaliae. Dept. of Math., W. Va. Univ., 1971; revised 1976.

[7] G. Kreweras. "Sur les partitions non croisées d'un cycle." *Discrete Math.* 4(1972), 333-350.

[8] D.G. Rogers. "A Schröder triangle : combinatorial problems." *Combinatorial Mathematics V : Proceedings of the Fifth Australian Conference.* Lecture Notes in Mathematics. Springer-Verlag, Berlin, (to appear).

[9] D.G. Rogers. "The enumeration of a family of ladder graphs by edges. Part I : Connective relations." *Quart. J. Math. Oxford (2)*, (to appear)

[10] D.G. Rogers. "The enumeration of a family of ladder graphs by edges. Part II : Schröder relations," (submitted).

[11] D.G. Rogers. "Rhyming Schemes : crossings and coverings", (submitted).

[12] N.J.A. Sloane. *A handbook of integer sequences.* Academic Press, New York, (1973).

[13] J.H. van Lint. *Combinatorial Theory Seminar, Eindhoven University of Technology,* Lecture Notes in Mathematics, vol 382, Springer-Verlag, Berlin, (1974).

05B20

# A COMPUTER LISTING OF HADAMARD MATRICES

Jennifer Seberry

Applied Mathematics Department,
University of Sydney,
N.S.W., 2006

ABSTRACT.

A computer has been used to list all known Hadamard matrices of order less than 40,000. If an Hadamard matrix is not known of order $4q$ ($q$ odd) then the smallest $t$ so that there is an Hadamard matrix of order $2^t q$ is given.

Hadamard matrices are not yet known for orders 268, 412, 428.

INTRODUCTION.

An *Hadamard matrix* of order $n$ has every entry $+1$ or $-1$ and its distinct row vectors are orthogonal.

These were discussed by Sylvester [16] in 1867 and Hadamard [8] conjectured in 1893 that they exist for orders 1, 2 and $4t$, for every natural number $t$. Hadamard proved that any complex $n \times n$ matrix $A = (a_{ij})$ with entries in the unit disc satisfies

$$(\det A)^2 \leq \prod_{i=1}^{n} \sum_{j=1}^{n} |a_{ij}|^2 ,$$

and Hadamard matrices satisfy the equality of this inequality.

In 1933 Paley [10] produced a list showing that Hadamard matrices of orders 92, 116, 156, 172, 184 and 188 where the only unsolved cases of order $\leq 200$.

The existence of the matrix of order 172 was settled by Williamson [27] in 1944.

This list induced L.D. Baumert, S.W. Golomb and Marshall Hall Jr to use sophisticated and exciting computer techniques with

Williamson's method to find the Hadamard matrices of orders 92 and 184 (in 1962 [2]), 116 (in 1966 [1]) and 156 (in 1965 [3]).

The case for 188 has been settled by R.J. Turyn using a technique of Goethals and Seidel [7] which generalized that of Williamson.

These results, and those of E. Spence, J. Cooper, J.S. Wallis and A.L. Whiteman have largely given Hadamard matrices of "low" order.

Four matrices $W_1, W_2, W_3, W_4$ of order $w$ with entries $+1$ or $-1$ which satisfy

$$W_iW_j^T = W_jW_i^T \qquad i, j \in \{1, 2, 3, 4\},$$

$$\sum_{i=1}^{4} W_iW_i^T = 4wI_w,$$

are called *Williamson* matrices.

A square matrix $A$ of order $n$ with entries from the set of commuting variables $\{0, \pm x_1, \pm x_2, \ldots, \pm x_s\}$ will be called an *orthogonal design of order* $n$ *and type* $(u_1, u_2, \ldots, u_s)$ if $x_i$ occurs $u_i$ times in each row and column of $A$ and if the rows of $A$ are formally orthogonal.

Then we can express the highly important result of Baumert and Hall (1965 [3]) as

BAUMERT-HALL THEOREM. *If there is an orthogonal design of order* $4t$ *and type* $(t, t, t, t)$ *and Williamson matrices of order* $w$, *then there exists an Hadamard matrix of order* $4wt$.

Orthogonal designs, introduced by Geramita, Geramita and Wallis [6], were used by Wallis [22] to prove

THEOREM. *Let* $q$ *be any odd natural number. Then there exists an integer* $t = [2 \log_2(q-3)]$ *such that there exists an Hadamard matrix of order* $2^sq$ *for every natural number* $s \geq t$.

| q | t | q | t | q | t | q | t | q | t | q | t | q | t | q | t | q | t | q | t |
|---|---|---|---|---|---|---|---|---|---|---|---|---|---|---|---|---|---|---|---|
| | | 103 | 3 | | | | | | | | | | | | | 803 | 3 | | |
| | | 107 | 10 | | | | | | | | | | | | | | | 907 | 5 |
| | | | | | | | | | | 509 | 8 | | | | | 809 | 12 | | |
| | | | | | | 311 | 14 | | | | | | | | | | | | |
| | | | | 213 | 3 | | | | | | | 613 | 3 | | | | | 913 | 3 |
| | | | | | | | | | | 515 | 4 | | | | | | | | |
| | | | | | | | | | | | | | | | | | | 917 | 3 |
| | | | | 219 | 3 | | | 419 | 4 | 519 | 3 | | | 719 | 4 | | | 919 | 3 |
| | | | | | | | | | | | | | | 721 | 4 | | | | |
| | | | | 223 | 3 | | | | | 523 | 3 | 623 | 3 | 723 | 3 | 823 | 3 | | |
| | | 127 | 3 | | | | | | | | | | | | | | | | |
| | | | | | | | | | | | | 631 | 13 | | | | | | |
| | | | | 233 | 4 | | | | | | | | | 733 | 18 | | | 933 | 10 |
| | | | | | | 335 | 6 | | | | | | | | | | | | |
| | | | | | | | | | | 537 | 5 | | | 737 | 6 | | | | |
| | | | | 239 | 4 | | | | | | | | | 739 | 16 | 839 | 8 | | |
| | | | | | | | | | | | | | | | | | | 941 | 6 |
| | | | | | | | | 443 | 6 | | | 643 | 3 | | | | | | |
| | | | | | | | | 445 | 3 | | | | | | | | | | |
| | | | | | | 347 | 15 | | | | | 647 | 10 | | | | | 947 | 6 |
| | | | | 249 | 3 | | | | | | | | | 749 | 11 | | | 949 | 3 |
| | | 151 | 3 | 251 | 6 | | | | | | | | | 751 | 3 | | | | |
| | | | | | | | | | | | | 653 | 10 | | | 853 | 3 | 953 | 14 |
| | | | | | | | | | | | | 655 | 3 | | | | | 955 | 3 |
| | | | | | | | | | | | | 657 | 3 | 757 | 5 | 857 | 4 | | |
| | | | | | | 359 | 4 | | | | | 659 | 17 | | | 859 | 3 | 959 | 3 |
| | | 163 | 3 | | | | | 463 | 3 | | | | | 763 | 3 | 863 | 4 | | |
| | | | | | | | | | | | | | | | | 865 | 3 | 965 | 3 |
| 67 | 5 | 167 | 4 | 267 | 3 | | | | | | | | | | | | | | |
| | | | | 269 | 8 | | | | | | | 669 | 3 | | | | | | |
| | | | | | | | | | | 571 | 3 | | | | | | | 971 | 6 |
| | | | | | | 373 | 7 | | | 573 | 3 | | | 773 | 8 | | | | |
| | | 179 | 8 | | | | | 479 | 12 | 579 | 3 | | | | | 879 | 3 | 979 | 3 |
| | | | | | | | | | | 581 | 3 | | | 781 | 3 | | | | |
| | | | | 283 | 3 | | | | | | | | | | | 883 | 3 | | |
| | | | | | | | | | | | | | | | | | | 985 | 3 |
| | | | | | | | | 487 | 5 | | | | | 787 | 5 | | | | |
| | | | | | | | | | | | | | | 789 | 3 | | | | |
| | | 191 | 13 | | | | | 491 | 15 | | | | | | | | | 991 | 3 |
| | | | | | | | | | | 599 | 8 | 699 | 3 | | | | | | |

If an odd number q does not appear in the list it means that there is an Hadamard matrix of order 4q . When q appears in the list the number t next to q indicates there is an Hadamard matrix of order $2^t q$ and no smaller power of 2.

| q | t | q | t | q | t | q | t | q | t | q | t | q | t | q | t | q | t | q | t |
|---|---|---|---|---|---|---|---|---|---|---|---|---|---|---|---|---|---|---|---|
| | | | | | | 1301 | 14 | | | | | | | | | | | | |
| | | 1103 | 17 | | | 1303 | 3 | | | | | | | 1703 | 3 | | | 1903 | 3 |
| | | | | 1205 | 3 | | | | | | | | | | | | | | |
| | | | | | | | | | | 1507 | 3 | | | | | | | 1907 | 10 |
| | | 1109 | 19 | | | | | | | 1509 | 3 | | | | | | | 1909 | 3 |
| | | | | 1211 | 3 | | | 1411 | 3 | | | | | | | | | | |
| | | | | 1213 | 16 | | | | | 1513 | 3 | | | 1713 | 4 | | | 1913 | 12 |
| | | 1115 | 4 | | | 1315 | 3 | | | | | | | | | | | 1915 | 3 |
| 1019 | 8 | | | | | 1319 | 12 | | | | | 1619 | 4 | 1719 | 3 | | | | |
| | | | | | | | | | | | | | | | | | | 1921 | 3 |
| | | 1123 | 7 | 1223 | 8 | | | 1423 | 3 | | | | | 1723 | 3 | 1823 | 4 | | |
| | | | | 1227 | 3 | 1327 | 5 | 1427 | 15 | 1527 | 9 | | | | | | | | |
| | | | | | | | | | | | | | | | | | | 1929 | 4 |
| 1031 | 6 | | | 1231 | 3 | | | | | | | 1631 | 5 | | | 1831 | 15 | | |
| | | 1133 | 4 | | | | | 1433 | 6 | | | 1633 | 3 | 1733 | 14 | | | | |
| | | | | | | 1335 | 3 | | | | | | | | | | | | |
| | | | | | | | | 1437 | 6 | | | | | | | | | 1937 | 3 |
| 1039 | 3 | | | | | | | 1439 | 12 | | | 1639 | 3 | | | | | | |
| | | | | 1241 | 3 | 1341 | 3 | 1441 | 3 | | | | | | | 1841 | 3 | | |
| 1043 | 3 | | | 1243 | 3 | | | | | 1543 | 3 | | | | | | | 1943 | 6 |
| | | | | | | | | | | | | | | 1745 | 3 | | | | |
| 1047 | 3 | | | | | | | 1447 | 19 | | | | | 1747 | 5 | 1847 | 8 | | |
| | | | | | | 1349 | 3 | | | | | | | | | | | 1949 | 4 |
| 1051 | 3 | | | | | 1351 | 3 | 1451 | 6 | | | | | 1751 | 4 | | | 1951 | 3 |
| | | | | | | | | | | | | | | 1753 | 11 | | | | |
| | | | | 1255 | 3 | | | | | | | | | | | | | | |
| | | 1157 | 3 | 1257 | 5 | | | | | 1557 | 3 | | | | | | | 1957 | 4 |
| 1059 | 3 | | | 1259 | 4 | 1359 | 3 | | | 1559 | 4 | | | | | | | | |
| | | | | | | | | | | | | 1661 | 4 | | | | | | |
| 1063 | 3 | | | | | | | | | | | 1663 | 11 | | | | | 1963 | 4 |
| | | 1165 | 3 | | | | | 1465 | 3 | | | | | | | | | | |
| | | | | | | 1367 | 4 | | | 1567 | 19 | 1667 | 19 | | | | | | |
| | | 1169 | 5 | | | | | 1469 | 3 | | | | | | | 1869 | 3 | 1969 | 9 |
| | | | | | | | | 1471 | 13 | 1571 | 10 | 1671 | 3 | | | 1871 | 20 | | |
| | | | | 1273 | 6 | 1373 | 19 | 1473 | 3 | | | | | | | | | 1973 | 4 |
| | | | | | | | | | | | | 1675 | 3 | | | | | | |
| | | 1177 | 5 | | | | | | | 1577 | 3 | | | | | | | | |
| 1079 | 3 | | | | | 1379 | 3 | | | 1579 | 5 | 1679 | 3 | | | 1879 | 3 | 1979 | 4 |
| | | | | | | 1381 | 3 | | | | | | | 1781 | 3 | | | 1981 | 4 |
| | | | | 1283 | 12 | | | 1483 | 3 | 1583 | 8 | | | 1783 | 7 | 1883 | 6 | | |
| | | | | | | 1385 | 3 | | | | | | | | | | | 1985 | 3 |
| 1087 | 18 | 1187 | 6 | | | 1387 | 3 | 1487 | 12 | | | | | 1787 | 4 | | | 1987 | 16 |
| | | | | | | | | | | 1589 | 3 | 1689 | 3 | | | 1889 | 4 | | |
| | | | | 1291 | 3 | | | 1491 | 3 | | | | | | | | | | |
| | | 1193 | 4 | | | | | 1493 | 4 | | | 1693 | 7 | 1793 | 4 | 1893 | 6 | 1993 | 15 |
| | | | | | | | | | | | | | | 1795 | 5 | | | | |
| | | | | | | 1397 | 4 | | | | | | | | | | | | |
| | | 1199 | 3 | 1299 | 3 | | | 1499 | 16 | | | 1699 | 3 | | | | | 1999 | 3 |

If an odd number $q$ does not appear in the list it means there is an Hadamard matrix of order $4q$. When $q$ appears in the list the number $t$ next to $q$ indicates there is an Hadamard matrix of order $2^t q$ and no smaller power of 2.

This result can be used to prove

THEOREM. *Let* $q$ *be any odd natural number. Then there exist an integer* $t$ *such that there exists a symmetric Hadamard matrix with constant diagonal of order* $2^{2s}q^2$ *for every natural number* $s \geq t$.

THE LIST.

As mentioned previously Paley [10] constructed a list of orders $\leq 200$ for which Hadamard matrices were known in 1933. Various other lists appeared (see, for example, Florek [5], Raghavarao [12], Wallis [23]).

The current listing, which is available for $q < 10,000$ although we only give $q < 2,000$ in this note, has no entry after $q$ if an Hadamard matrix of order $4q$ is *known*. If no Hadamard matrix of order $4q$ is known then the smallest $t$ such that an Hadamard matrix of order $2^tq$ is known is given. We note that Wallis' theorem gives an upper bound on $t$ and often Hadamard matrices are known for smaller powers of 2 (but greater than 2) than that theorem indicates.

The computer tape on which these results are stored in order to be easily updated also has an indication of whether a skew-Hadamard matrix is known.

ACKNOWLEDGEMENTS.

I wish to thank Emma Lehmer and D.H. Lehmer who suggested to me the idea of computerizing this list. The largest part of the programming was done by Dr. Ian S. Williams while a student in the Research School of Physical Sciences at A.N.U. on the School's DEC-10 system. The updating and printing programmes were written by Mr. N. Wormald.

REFERENCES.

(1) L.D. Baumert, "Hadamard matrices of orders 116 and 232". *Bull. Amer. Math. Soc.*, 72 (1966), 237.

(2) L.D. Baumert, S.W. Golomb and Marshall Hall, Jr., "Discovery of an Hadamard matrix of order 92", *Bull. Amer. Math. Soc.*, 68 (1962), 237-238.

(3) L.D. Baumert and Marshall Hall, Jr., "A new construction for Hadamard matrices", *Bull. Amer. Math. Soc.*, 71 (1965), 169-170.

(4) Joan Cooper and Jennifer Seberry Wallis, "A construction for Hadamard arrays", *Bull. Austral. Math. Soc.*, 7 (1972), 269-278.

(5) K. Florek, "On the evaluation from below of extremal determinants", *Coll. Math.*, 10 (1963), 111-131.

(6) Anthony V. Geramita, Joan Murphy Geramita and Jennifer Seberry Wallis, "Orthogonal designs", *Linear and Multilinear Algebra*, 3 (1975/76), 281-306.

(7) J.M. Goethals and J.J. Seidel, "A skew-Hadamard matrix of order 36", *J. Austral. Math. Soc.*, 11 (1970), 343-344.

(8) Jacques Hadamard, "Résolution d'une question relative aux déterminants" *Bull. des Sciences Math.*, 17 (1893), 240-246.

(9) Marshall Hall, Jr., *Combinatorial Theory*, Blaisdell, [Ginn and Co.], Waltham, Massachusetts, 1967 .

(10) R.E.A.C. Paley, "On orthogonal matrices", *J. Math. Phys.*, 12 (1933), 311-320.

(11) Stanley E. Payne, *The Construction of Hadamard matrices*, ARL73-0117, Aerospace Research Labs, United States Air Force, Wright-Patterson Air Force Base, Dayton, Ohio, 1973.

(12) D. Raghavarao, *Constructions and Combinatorial Problems in Design of Experiments*, Wiley Series in Probability and Mathematical Statistics, John Wiley and Sons, Inc., New York-London-Sydney-Toronto, 1971.

(13) Peter J. Robinson and Jennifer Seberry Wallis, "A note on using sequences to construct orthogonal designs", *Colloquia Mathematica Societatis Janos Bolyai*.

(14) Edward Spence, "Hadamard matrices from relative difference sets", *J. Combinatorial Theory*, Ser. A, 19 (1975), 287-300.

(15) Edward Spence, "Skew-Hadamard matrices of Goethals-Seidel type", *Canad. J. Math.*, 27 (1975), 555-560.

(16) J.J. Sylvester, "Thoughts on inverse orthogonal matrices, simultaneous sign successions, and tesselated payments in two or more colours, with applications to Newton's Rule, ornamental title work, and the theory of numbers", *Phil. Mag. No. 4*, 34 (1867), 461-475.

(17) Richard J. Turyn, "Hadamard matrices, Baumert-Hall units, four-symbol sequences, pulse compressions and surface wave encodings", *J. Combinatorial Theory*, Ser. A., 16 (1974), 313-333.

(18) Jennifer Wallis, "Some matrices of Williamson type", *Utilitas Math.*, 4 (1973), 147-154.

(19) Jennifer Wallis, "Hadamard matrices of order 28m, 36m, and 44m", *J. Combinatorial Theory*, Ser. A., 15 (1973), 323-328.

(20) Jennifer Seberry Wallis, "Williamson matrices of even order", *Combinatorial Mathematics: Proceedings of the Second Australian Conference* (editor D.A. Holton), Lecture Notes in Mathematics, Vol. 403, Springer-Verlag, Berlin-Heidelberg-New York, 1974, 132-142.

(21) Jennifer Seberry Wallis, "Construction of Williamson type matrices", *J. Linear and Multilinear Alg.*, 3 (1975) 197-207.

(22) Jennifer Seberry Wallis, "On the existence of Hadamard matrices", *J. Combinatorial Th.*, Ser. A., 21 (1976), 444-451.

(23) W.D. Wallis, Anne Penfold Street, Jennifer Seberry Wallis, *Combinatorics: Room Squares, sum-free sets, Hadamard Matrices*, Lecture Notes in Mathematics, 272 Springer-Verlag, Berlin-Heidelberg-New York, 1972 .

(24) Albert Leon Whiteman, "Hadamard matrices of order 4(2p+1)", *J. Number Th.*, 8 (1976), 1-11.

(25) Albert Leon Whiteman, "Hadamard matrices of Williamson type", *J. Austral. Math. Soc.*, 21 (1976), 481-486.

(26) Albert Leon Whiteman, "An infinite family of Hadamard matrices of Williamson type", *J. Combinatorial Theory*, Ser. A., 14 (1973), 334-340.

(27) John Williamson, "Hadamard's determinant theorem and the sum of four squares", *Duke Math. J.*, 11 (1944), 65-81.

# A CLASS OF CODES GENERATED BY CIRCULANT WEIGHING MATRICES

94A10
(05B20, 05B30)

Jennifer Seberry and K. Wehrhahn

Applied Mathematics Department and
Pure Mathematics Department,
University of Sydney,
N.S.W., 2006

ABSTRACT.

Some properties of a new class of codes constructed using circulant matrices over GF(3) will be discussed. In particular we determine the weight distributions of the (14, 7) and two inequivalent (26, 13)-codes arising from the incidence matrices of projective planes of orders 2 and 3.

## 1. INTRODUCTION.

In this paper *"code"* will mean a linear code over GF(3). An *(n, k)-code* C has length n, dimension k. An *(n, k, d)-code* is an (n, k)-code with minimum non-zero weight d. Our notation and definitions are consistent with those of Blake and Mullin [2].

Let Q be the circulant incidence matrix of a projective plane of order q (See Hall [6]). Then Q, of order $q^2 + q + 1$ satisfies

$$QQ^T = qI + J, \qquad QJ = (q+1)J$$

where J is the appropriate all 1's matrix. $W = Q^2 - J$ is a circulant (0, 1, -1) matrix of order $q^2 + q + 1$ satisfying

$$WW^T = q^2 I, \qquad WJ = qJ$$

i.e. W is a *circulant weighing matrix* of weight $q^2$. We write $W = W(q^2+q+1, q^2)$ to denote its order and weight. More details of W can be found in Hain [5] and Wallis and Whiteman [10].

We call codes with basis

$$[I\ W] \quad \text{for} \quad q \equiv 0 \pmod 3$$

$$[I\ qW] \quad \text{for} \quad q \equiv 1 \text{ or } 2 \pmod 3$$

over GF(3) *weighing codes*. The purpose of this paper is to establish some general properties of weighing codes and to determine the weight distributions and design properties of the codes corresponding to $q = 2$ and $q = 3$ .

Note that if

$$G = [I\ W]$$

is the basis of $C$ then for $q \equiv 1$ or $2 \pmod 3$

$$G^{\perp} = [I\ \ -W]$$

is the basis of the dual code $C^{\perp}$ . Hence $C$ is neither self-dual nor self-orthogonal. However we shall see that $C$ and $C^{\perp}$ always have the same weight distribution and hence the same minimum distance d. By a well known result, cf. Delsarte [3], weighing codes are orthogonal arrays of strength d-1. In this sense the weighing codes belong to a family of codes including the self-dual codes, see Mallows, et. al [7] and the symmetry codes, see Pless [8, 9] and Blake [1] .

We observe that the one's vector $\underset{\sim}{1}$ is in $C$ for $q \equiv 1$ or $2 \pmod 3$ and is the sum of the basis vectors. The vector $\underset{\sim}{k} = (1, 1, \ldots, 1, -, \ldots, -)$ (where $-$ represents $-1$) of $q^2 + q + 1$ ones and $q^2 + q + 1$ minuses occurs in the dual code for $q \equiv 1$ or $2 \pmod 3$.

If $q \equiv 0 \pmod 3$ then the sum of the basis vectors

$$[I\ W] \text{ is not } \underset{\sim}{1} ,$$

and so the code cannot contain $\underset{\sim}{1}$ . Moreover, in this case rank $W <$ order of $W$ since $WW^T \equiv 0 \pmod 3$.

## 2. GENERAL PROPERTIES OF THE CODES.

If $A_i$ is the number of codewords of weight $i$ in $C$, then we call the bivariate polynomial

$$WE(x, y) = \sum_{i=0}^{n} A_i x^{n-i} y^i$$

the *weight enumerator* of $C$ . If $A_{ijk}$ is the number of codewords of weight $j+k$ in $C$ containing $j$ ones and $k$ twos (minus ones over $GF(3)$) then we call the trivariate polynomial

$$CWE(x, y, z) = \sum_{i=0}^{n} A_{ijk} x^i y^j z^k$$

the *complete weight enumerator* of $C$ .

THEOREM.

*Let $C$ be the code over $GF(q)$ with basis $G = [I\ X]$ where $X$ is a circulant matrix of order $k$ and $I$ is the identity matrix of order $k$. Then $C$ and $C^{\perp}$ have the same weight enumerators.*

Proof :

First recall that if $X$ is a circulant matrix and $R$ the back diagonal permutation matrix then

$$(XR)^T = XR .$$

Now $C^{\perp}$ has basis

$$[-X^T\ I]$$

and the basis vectors of $C^{\perp}$ may be written as

$$R[-X^T\ I] = [-RX^T\ R] = [-XR^T\ R] = [-XR\ R]$$

since this merely involves rearranging the order of the basis vectors. Hence $C^{\perp}$ is equivalent to the code $\mathcal{D}^{\perp}$ with basis

$[-XR\ I]$ as this just rearranges the columns of $R$. Since $XR$ is symmetric we have that $(\mathcal{D}^{\perp})^{\perp} = \mathcal{D}$ has basis $[I\ XR]$ .

If $\underset{\sim}{b}$ is a q-ary vector of length $k$

then $\quad WE(\underset{\sim}{b}[I\ XR]) = WE(\underset{\sim}{b}) + WE(\underset{\sim}{b}XR)$

whereas $\quad WE(\underset{\sim}{b}[-XR\ I]) = WE(-\underset{\sim}{b}XR) + WE(\underset{\sim}{b})$

and hence $\mathcal{D}$ and $C^{\perp}$ have the same weight enumerators. But $\mathcal{D}$ is equivalent to $C$ and hence the theorem holds. ∎

In particular $A_i = A_i^{\perp}$ for weighing codes, and so $C$ and $C^{\perp}$ form orthogonal arrays of maximum strength $d-1$ where $d$ is the minimum distance of $C$ (and $C^{\perp}$).

Any two vectors from the basis of $C$ can be written as

```
100...0|1...11...11...1|-...--...---...-|0...00...00
010...0|1...1-...-0...0|1...1-...-0...0|1...1-...-0
\_____/ \___/ \___/ \___/ \___/ \___/ \___/ \___/ \___/
q^2+q+1   a     b     c     d     e     f     g     h  1
```

and we obtain the following equations

$$a + b + c = a + d + g = \tfrac{1}{2}(q^2 + q) = \text{number of ones.}$$

$$d + e + f = b + e + h = \tfrac{1}{2}(q^2 - q) = \text{number of minus ones.}$$

$$1 + g + h = c + f + 1 = q + 1 = \text{number of zeros.}$$

$$a + e = b + d \qquad\qquad \text{(orthogonality).}$$

These equations can be solved for c, d, e, f, g, h in terms of q, a, b. The CWE of the sum and difference of two vectors are

$$x^{\frac{1}{2}(3q^2+q)}\, y^{2+q^2+q-3a}\, z^{-\frac{1}{2}q^2+\frac{1}{2}q+3a}$$

and

$$x^{\frac{1}{2}(3q^2+q)}\, y^{1+q^2-3b}\, z^{-\frac{1}{2}q^2+\frac{3}{2}q+3b+1}$$

respectively.

Of course the negatives of these vectors are also in C and hence the weight of every two combination is $\frac{1}{2}(q^2 + 3q + 4)$ and consequently there are at least $4\binom{q^2+q+1}{2}$ vectors of this weight.

We may observe that

$$\frac{1}{2}(q^2 + 3q + 4) < q^2 + 1 \quad \text{for } q \geq 4$$

and hence $\frac{1}{2}(q^2 + 3q + 4)$ provides an upper bound on the minimum distance of C for $q \geq 4$.

## 3. THE (14, 7) CODE WITH MINIMUM DISTANCE 5.

This code is generated by W with first row

$$-110100 .$$

In order to ensure the $\underset{\sim}{1}$ vector is in C we use the basis vectors

$$G = [I \; qW] = [I \; -W]$$

where $q = 2$.

We observe that the linear combinations given by XG where $X = I + Q + J$ (Q as before the incidence matrix of the projective plane of order 2 and $W = Q^2 - J$) are

$$H = [X \; -XW] = [I+Q+J \;\; 2Q+2J] \pmod 3$$

and $K = 2H - 3J$ satisfies the equation $KK^T = 16I - 2J$ over the real numbers.

Since each row of K has eight + 1's and six - 1's and each column has four + 1's and three - 1's we have a (7, 14, 8, 4, 4) - BIBD . In fact the 16 vectors $\underset{\sim}{1}$, $\underset{\sim}{2}$, H, 2H contain a (14, 16, 6) - block code. The vectors

$$\begin{bmatrix} \underset{\sim}{1} & \underset{\sim}{1} & -\underset{\sim}{1} \\ \underset{\sim}{1}^T & \underset{\sim}{1}^T & -H \end{bmatrix},$$

where $\underset{\sim}{1}$ is the vector of seven ones, are the first eight rows of an Hadamard matrix of order 16 (See Wallis, et al [11]) .

We note that since every vector in the code $C$ is orthogonal to every vector in $C^\perp$ the remaining 8 rows of this Hadamard matrix of order 16 (and their negatives) will be obtained from the vectors of full weight in $C^\perp$ .

We found the weight distribution for this code, which is given in Figure 1, and that of the dual code, given in Figure 2. As expected, we see $C$ and $C^\perp$ have the same weight distribution but not the same complete weight enumerator.

The (14, 7)-code has minimum distance 5 and hence forms an orthogonal array of strength 4.

| | | | | | | | | | | | | | | |
|---|---|---|---|---|---|---|---|---|---|---|---|---|---|---|
| | | | | | | | $A_{1400}$ | | | | | | | |
| | | | | | | $A_{932}$ | | $A_{923}$ | | | | | | |
| | | | $A_{860}$ | | $A_{842}$ | | $A_{833}$ | | $A_{824}$ | | $A_{806}$ | | | |
| | | | | $A_{752}$ | | $A_{743}$ | | $A_{734}$ | | $A_{725}$ | | | | |
| | $A_{680}$ | | $A_{662}$ | | $A_{653}$ | | $A_{644}$ | | $A_{635}$ | | $A_{626}$ | | $A_{608}$ | |
| | | $A_{572}$ | | $A_{563}$ | | $A_{554}$ | | $A_{545}$ | | $A_{536}$ | | $A_{527}$ | | |
| | $A_{482}$ | | $A_{473}$ | | $A_{464}$ | | $A_{455}$ | | $A_{446}$ | | $A_{437}$ | | $A_{428}$ | |
| $A_{392}$ | | $A_{383}$ | | $A_{374}$ | | $A_{365}$ | | $A_{356}$ | | $A_{347}$ | | $A_{338}$ | | $A_{329}$ |
| | $A_{293}$ | | $A_{284}$ | | $A_{275}$ | | $A_{266}$ | | $A_{257}$ | | $A_{248}$ | | $A_{239}$ | |
| | | | | $A_{0140}$ | | $A_{086}$ | | $A_{068}$ | | $A_{0014}$ | | | | |

| | | | | | | | | | | | | | | |
|---|---|---|---|---|---|---|---|---|---|---|---|---|---|---|
| | | | | | | | 1 | | | | | | | |
| | | | | | | 14 | | 14 | | | | | | |
| | | | 7 | | 42 | | 14 | | 42 | | 7 | | | |
| | | | | 42 | | 42 | | 42 | | 42 | | | | |
| | 7 | | 84 | | 42 | | 168 | | 42 | | 84 | | 7 | |
| | | 42 | | 42 | | 84 | | 84 | | 42 | | 42 | | |
| | 42 | | 42 | | 168 | | 84 | | 168 | | 42 | | 42 | |
| 14 | | 14 | | 42 | | 42 | | 42 | | 42 | | 14 | | 14 |
| | 14 | | 42 | | 42 | | 84 | | 42 | | 42 | | 14 | |
| | | | 1 | | | 7 | | 7 | | 1 | | | | |

Figure 1.

| | | | | |
|---|---|---|---|---|
| | | $A_{1400}$ | | |
| | $A_{941}$ | | $A_{914}$ | |
| $A_{860}$ | | $A_{833}$ | | $A_{806}$ |
| | $A_{752}$ | | $A_{725}$ | |
| $A_{671}$ | | $A_{644}$ | | $A_{617}$ |
| | $A_{563}$ | | $A_{536}$ | |
| $A_{482}$ | | $A_{455}$ | | $A_{428}$ |
| | $A_{374}$ | | $A_{347}$ | |
| $A_{293}$ | | $A_{266}$ | | $A_{239}$ |
| | | $A_{077}$ | | |

| | | | | |
|---|---|---|---|---|
| | | 1 | | |
| | 14 | | 14 | |
| 7 | | 98 | | 7 |
| | 84 | | 84 | |
| 42 | | 350 | | 42 |
| | 168 | | 168 | |
| 84 | | 420 | | 84 |
| | 112 | | 112 | |
| 56 | | 168 | | 56 |
| | | 16 | | |

Figure 2.

## 4. TWO (26, 13)-CODES WITH DISTANCE 3 AND 4

Richard M. Hain [5] conjectured and Peter Eades [4] verified (by computer) that there are two equivalence classes of circulant $W(13, 9)$. They have first rows

0-0-10011-111

and

0101100--11-1 .

Call the circulant matrices with these first rows $W_1$ and $W_2$ .

The linear codes $C_1$, $C_2$ with bases

$$[\ I\ W_1],\quad [I\ W_2]$$

respectively, were studied via the computer at the University of Sydney and their CWE's obtained. We give here their WE's in Figures 3 and 4 respectively.

It is most interesting to note that the codes have different minimum distances 3 and 4 respectively. Also, as expected since $q = 3 \equiv 0 \pmod 3$ for these codes, neither $C_1$ nor $C_2$ contains $\underset{\sim}{1}$ (and neither does $C_1^{\perp}$ nor $C_2^{\perp}$ as $\underset{\sim}{1}$ is not orthogonal to their basis vectors). Also neither contains any full weight vectors.

Since the codes have minimum distance 3 and 4 they are orthogonal arrays of strength 2 and 3 respectively.

$A_0 = 1$
$A_1 = 0$
$A_2 = 0$
$A_3 = 104$
$A_4 = 468$
$A_5 = 1404$
$A_6 = 4056$
$A_7 = 8424$
$A_8 = 11934$
$A_9 = 13442$
$A_{10} = 11258$
$A_{11} = 5928$
$A_{12} = 4264$
$A_{13} = 11260$
$A_{14} = 39780$
$A_{15} = 105768$
$A_{16} = 211224$
$A_{17} = 317538$
$A_{18} = 352638$
$A_{19} = 281632$
$A_{20} = 154128$
$A_{21} = 51168$
$A_{22} = 7904$
$A_{23} = 0$
$A_{24} = 0$
$A_{25} = 0$
$A_{26} = 0$

Weight Distribution of $C_1$

Figure 3 .

$A_0 = 1$
$A_1 = 0$
$A_2 = 0$
$A_3 = 0$
$A_4 = 26$
$A_5 = 0$
$A_6 = 156$
$A_7 = 624$
$A_8 = 0$
$A_9 = 1118$
$A_{10} = 3458$
$A_{11} = 8736$
$A_{12} = 24830$
$A_{13} = 54264$
$A_{14} = 100152$
$A_{15} = 152568$
$A_{16} = 212862$
$A_{17} = 259974$
$A_{18} = 272766$
$A_{19} = 222976$
$A_{20} = 145002$
$A_{21} = 73996$
$A_{22} = 37180$
$A_{23} = 16848$
$A_{24} = 6006$
$A_{25} = 780$
$A_{26} = 0$

Weight Distribution of $C_2$

Figure 4.

REFERENCES.


(1) Ian F. Blake, "On a generalization of the Pless symmetry codes", *Information and Control*, 27(1975), 369-373.

(2) Ian F. Blake and Ronald C. Mullin, *An Introduction to Algebraic and Combinatorial Coding theory*, Academic Press, N.Y. -San Francisco-London, 1976.

(3) P. Delsarte, "Four Fundamental Parameters of a code and Their Combinatorial Significance", *Information and Control*, 23(1973) 407-438.

(4) P. Eades, *On the Existence of Orthogonal Designs*, Ph.D. Thesis, Australian National University, Canberra, 1977.

(5) Richard M. Hain, *Circulant Weighing matrices*, M.Sc. Thesis, Australian National University, Canberra, 1977.

(6) Marshall Hall Jr., *Combinatorial Theory*. Blaisdell, [Ginn Co.], Waltham, Mass, 1967.

(7) C.L. Mallows, V. Pless and N.J.A. Sloane, "Self-Dual codes over GF(3)", *SIAM J. Appl..Math*. Vol 31, (1976), 649-666.

(8) V. Pless, "On a new family of symmetry codes and related new five designs", *Bull. Amer. Math. Soc*. 75(1969), 1339-1342.

(9) V. Pless, "Symmetry codes over GF(3) and new-five designs", *J. Combinatorial Th. Ser*. A 12(1972), 119-142.

(10) Jennifer Seberry Wallis and Albert Leon Whiteman, "Some results on weighing matrices", *Bull. Austral. Math. Soc*. 12(1975), 433-447.

(11) W.D. Wallis, Anne Penfold Street, Jennifer Seberry Wallis, *Combinatorics : Room Squares, sum-free sets, Hadamard matrices*, Lecture Notes in Mathematics, Vol. 292, Springer-Verlag, Berlin-Heidelberg-New York, 1972.

# AN APPLICATION OF MAXIMUM-MINIMUM DISTANCE CIRCUITS ON HYPERCUBES*

05C99

Gustavus J. Simmons
Department of Applied Mathematics
Sandia Laboratories, Albuquerque, New Mexico 87115, USA

ABSTRACT

The related questions of finding Hamilton circuits in the n-dimensional cube with d points on an edge which maximize the minimum "taxicab" distance between successive vertices and/or which maximize the sum of such distances over the entire circuit is investigated. A "good" bound for the first quantity and an achievable limit for the second are developed and several optimal constructions found. Both of these circuits are central to designing pseudo-color graphics displays in which minimal grey scale differences become maximal color differences.

## 1. INTRODUCTION

Pseudo-color computer graphics displays identify the color of each pictel by a three-dimensional vector whose nonnegative elements specify the relative intensities of the three primary colors - magenta, cyan and yellow - required to realize the color. In many applications the information to be displayed is a scalar field such as the temperature, pressure, density, etc. over a complex surface or region, so that the display is essentially monochromatic. However, if the scalar field has at the same time wide extremes in scalar values and regions of major interest in which the variation is very small, a grey scale or monochrome presentation cannot simultaneously show the global and local features. The problem treated in this paper arose from the question of whether there existed a transformation from the grey scale into the CIE chromaticity diagram (essentially the color triangle) in which adjacent steps in the quantized grey scale would be mapped into colors that were maximally different from each other. This question is actually much more complex than this statement of it would suggest, since the CIE chromaticity diagram has no metric on it; i.e., pairs of points (colors) which are judged by most observers to be equally distinguishable often differ greatly in their separation in the diagram. An extensive literature [2,3,5] exists on the subject of transforming the CIE diagram so as to best approximate a metric space (usually for the Euclidean metric). In fact, in the application of the codes developed here the colors are preconditioned to approximate a metric space. These details, important though they are to a useful answer to the original question, will be ignored in this paper and only the simplified problem will be considered of how the $d^n$ n-tuples on the

* This work was supported by the U. S. Energy Research and Development Administration (ERDA) under Contract No. AT(29-1)-789.

symbols 0,1,···,d-1 can be cyclically ordered such that the minimum Lee distance between adjacent n-tuples is maximized. In fact we are only able to solve completely the related question of finding orderings which maximize the sum of the Lee distances between adjacent n-tuples over the entire cycle. A bound for the maximum-minimum distance is derived and a number of cases are shown to achieve this bound, but unfortunately a number of other cases are also shown not to.

The pseudo-color system (COMTAL) for which this question originally arose commonly uses only four intensity levels in each of the three primary colors, or sixty-four colors in total. For this case, the maximum sum of the Lee distances is shown to be 380 and the maximum-minimum distance is consequently at most five. Cyclic orderings of the 64 three-tuples which have every pair of adjacent three-tuples uniformly at distance five apart and others which have every adjacent pair at least five apart but which realize the maximum sum of the distances of 380 are exhibited. Both of these solutions have been used for the COMTAL system and have proven to be very effective in displaying both global and local variations which could not otherwise be displayed simultaneously.

## 2. PRELIMINARY DEVELOPMENT

Label the $d^n$ vertices of the complete graph $K_{d^n}$ with the n-tuples on the symbols 0,1,···,d-1 and assign a weight w(x,y) to the edge joining vertices x and y, where w(x,y) is the Lee distance between x and y: $w(x,y) = \sum_{i=1}^{n} |x_i - y_i|$. The resulting network we denote by $\mathfrak{W}_{d^n}$. Interchangeably - and most commonly - we shall also denote the $d^n \times d^n$ matrix whose rows and columns are indexed by the $d^n$ n-tuples in lexicographical order and in which $w_{i,j} = w(i,j)$ by $\mathfrak{W}_{d^n}$. We define the complement of the n-tuple x, denoted by x′, to be the n-tuple $(x_i' = d - 1 - x_i)$. Because of the way in which $\mathfrak{W}_{d^n}$ (matrix) was constructed the entries on the subdiagonal are w(x,x′). For the case n = 1, we have the following result.

Theorem 1: For any pair of vertices x and y in $\mathfrak{W}_{d^n}$, if

a) $x < \frac{d-1}{2}$ and $y < \frac{d-1}{2}$ or $x > \frac{d-1}{2}$ and $y > \frac{d-1}{2}$

$$\text{then} \quad 2w(x,y) < w(x,x') + w(y,y') \tag{1}$$

while if b) $x \le \frac{d-1}{2}$ and $y \ge \frac{d-1}{2}$ or $x \ge \frac{d-1}{2}$ and $y \le \frac{d-1}{2}$

$$\text{then} \quad 2w(x,y) = w(x,x') + w(y,y') \tag{2}$$

Proof: Since n = 1 and w is a metric, by the triangle inequality

$$w(x,y) = |x-y| \le |x - \tfrac{d-1}{2}| + |y - \tfrac{d-1}{2}| = \tfrac{1}{2}\{w(x,x') + w(y,y')\} \tag{3}$$

from which the absolute value signs can be removed for each of the specified conditions on x and y by observing the sign of the argument to obtain the conclusions of the theorem. ▌

A simple matrix recursion links the $\mathfrak{W}_d$ and the $\mathfrak{W}_{d^n}$ matrices. The Kronecker or tensor product $A \otimes B$ of the $m \times n$ matrix A with the $p \times q$ matrix B is the $mp \times nq$ block matrix $(a_{i,j}B)$. By analogy we define a tensor sum $A \oplus B$ to be the $mp \times nq$ block matrix $(a_{i,j} + B)$. Clearly

$$\mathfrak{W}_{d^n} = \mathfrak{W}_d \oplus \mathfrak{W}_d \oplus \cdots \oplus \mathfrak{W}_d \quad .$$

Based on this observation, we can state the following corollary to Theorem 1.

Corollary: For vertices x and y in $\mathfrak{W}_{d^n}$

$$2w(x,y) \leq w(x,x') + w(y,y') \tag{4}$$

where equality holds in (4) if and only if equality (2) is satisfied by each component.

Define $\mathfrak{C}_{d^n}$ to be the subnetwork of $\mathfrak{W}_{d^n}$ consisting of the edges whose weights satisfy equality (4). In the $\mathfrak{W}_{d^1}$ matrices in Figure 1 and the $\mathfrak{W}_{d^2}$ matrices in Figure 2 the edges belonging to $\mathfrak{C}_{d^n}$ are set in bold type. The following theorems characterize the $\mathfrak{C}_{d^n}$ graphs.

| | 0 | 1 |
|---|---|---|
| 0 | 0 | **1** |
| 1 | **1** | 0 |

$\mathfrak{W}_{2^1}$

| | 0 | 1 | 2 |
|---|---|---|---|
| 0 | 0 | **1** | **2** |
| 1 | **1** | **0** | **1** |
| 2 | **2** | **1** | 0 |

$\mathfrak{W}_{3^1}$

| | 0 | 1 | 2 | 3 |
|---|---|---|---|---|
| 0 | 0 | 1 | **2** | **3** |
| 1 | 1 | 0 | **1** | **2** |
| 2 | **2** | **1** | 0 | 1 |
| 3 | **3** | **2** | 1 | 0 |

$\mathfrak{W}_{4^1}$

| | 0 | 1 | 2 | 3 | 4 |
|---|---|---|---|---|---|
| 0 | 0 | 1 | **2** | **3** | **4** |
| 1 | 1 | 0 | **1** | **2** | **3** |
| 2 | **2** | **1** | **0** | **1** | **2** |
| 3 | **3** | **2** | **1** | 0 | 1 |
| 4 | **4** | **3** | **2** | 1 | 0 |

$\mathfrak{W}_{5^1}$

| | 0 | 1 | 2 | 3 | 4 | 5 |
|---|---|---|---|---|---|---|
| 0 | 0 | 1 | 2 | **3** | **4** | **5** |
| 1 | 1 | 0 | 1 | **2** | **3** | **4** |
| 2 | 2 | 1 | 0 | **1** | **2** | **3** |
| 3 | **3** | **2** | **1** | 0 | 1 | 2 |
| 4 | **4** | **3** | **2** | 1 | 0 | 1 |
| 5 | **5** | **4** | **3** | 2 | 1 | 0 |

$\mathfrak{W}_{6^1}$

Figure 1.

| | 00 | 01 | 10 | 11 |
|---|---|---|---|---|
| 00 | 0 | 1 | 1 | 2 |
| 01 | 1 | 0 | 2 | 1 |
| 10 | 1 | 2 | 0 | 1 |
| 11 | 2 | 1 | 1 | 0 |

$\mathfrak{M}_{2^2}$

| | 00 | 01 | 02 | 10 | 11 | 12 | 20 | 21 | 22 |
|---|---|---|---|---|---|---|---|---|---|
| 00 | 0 | 1 | 2 | 1 | 2 | 3 | 2 | 3 | 4 |
| 01 | 1 | 0 | 1 | 2 | 1 | 2 | 3 | 2 | 3 |
| 02 | 2 | 1 | 0 | 3 | 2 | 1 | 4 | 3 | 2 |
| 10 | 1 | 2 | 3 | 0 | 1 | 2 | 1 | 2 | 3 |
| 11 | 2 | 1 | 2 | 1 | 0 | 1 | 2 | 1 | 2 |
| 12 | 3 | 2 | 1 | 2 | 1 | 0 | 3 | 2 | 1 |
| 20 | 2 | 3 | 4 | 1 | 2 | 3 | 0 | 1 | 2 |
| 21 | 3 | 2 | 3 | 2 | 1 | 2 | 1 | 0 | 1 |
| 22 | 4 | 3 | 2 | 3 | 2 | 1 | 2 | 1 | 0 |

$\mathfrak{M}_{3^2}$

Figure 2.

Theorem 2: The graph defined by the edges of $\mathfrak{C}_{d^1}$ is isomorphic to

a) the complete bipartite graph $K_{m,m}$ if $d = 2m$

or b) the complete tripartite graph $K_{1,m-1,m-1}$ if $d = 2m-1$

in which a loop of weight 0 on vertex m-1 is introduced for consistency.

Proof: Immediate from Theorem 1 and the definition of $\mathfrak{C}_{d^n}$. ▮

The direct product of two graphs $G = (X,E)$ and $H = (Y,F)$ is defined to be the graph whose vertex set is the Cartesian product $X \times Y$ and in which two vertices $(x,y)$ and $(x',y')$ are joined by an edge if and only if;

$$[x,x'] \in E \quad \text{and} \quad [y,y'] \in F.$$

Note that if the factor edges are all loops, the product edge is also a loop.

The graphs defined by $\mathfrak{C}_{d^n}$ have a particularly simple characterization as a graph product.

Theorem 3: The graph defined by the edges of $\mathfrak{C}_{d^n}$ is isomorphic to

a) $K^n_{m,m}$ if $d = 2m$

or b) $K^n_{1,m-1,m-1}$ if $d = 2m-1$ .

Proof: The Corollary to Theorem 1 says that the conditions for the graph product are met for each component; i.e., for each successive iteration in forming $\mathfrak{C}_{d^n}$ from $\mathfrak{C}_d$. ▮

Corollary: If $d = 2m$, $\mathfrak{C}_{d^n}$ consists of $2^{n-1}$ disjoint isomorphs of $K_{m^n,m^n}$.

Unfortunately $K^n_{1,m-1,m-1}$ has no such simple characterization, although we may note that vertex $(m-1,m-1,\cdots,m-1)$ is connected to every vertex in $\mathfrak{C}_{d^n}$ since vertex (m-1) is connected to all of the 2m-2 other vertices in $\mathfrak{C}_{2m-1}$.

3. MAXIMUM-MINIMUM DISTANCE CIRCUITS ON HYPERCUBES

Any simple circuit on the $d \times d \times \cdots \times d$ hypercube will have a pair of edges on each n-tuple and would consequently have two entries in each row and each column of $\mathfrak{M}_{d^n}$. Define $\mathfrak{B}_{d^n}$ to be the sum of the entries on the subdiagonal in $\mathfrak{M}_{d^n}$. Let $\mathfrak{C}$ be the set of all Hamilton circuits on $\mathfrak{M}_{d^n}$, and for each such circuit $C_i \in \mathfrak{C}$, let $S_{C_i}$ be the sum over the circuit of the Lee distance between successive pairs of vertices. We denote the maximum such sum by $\mathfrak{S}_{d^n}$

$$\mathfrak{S}_{d^n} = \max_{C_i \in \mathfrak{C}} S_{C_i} .$$

Theorem 4:
$$\mathfrak{S}_{d^n} \leq \mathfrak{B}_{d^n} . \tag{5}$$

Proof: Corollary to Theorem 1. ▌

Corollary:
$$\mathfrak{S}_d = \mathfrak{B}_d . \tag{6}$$

Proof: To show that (6) holds we must exhibit a cycle in $\mathfrak{M}_d$ whose edges are all in $\mathfrak{C}_d$. Since $\mathfrak{C}_{2m}$ is isomorphic to the complete bipartite graph $K_{m,m}$ with vertex sets $(0,1,\cdots,m-1)$ and $(m,m+1,\cdots,2m-1)$, any Hamilton cycle in $K_{m,m}$ determines an ordering of the 2m integers for which (6) holds. $\mathfrak{C}_{2m-1}$ is isomorphic to the complete tripartite graph $K_{1,m-1,m-1}$ with vertex sets $(m-1)$, $(0,1,\cdots,m-2)$ and $(m,m+1,\cdots,2m-2)$. As in the case of $\mathfrak{C}_{2m}$, choose any Hamilton cycle on $K_{m-1,m-1}$ and delete an arbitrary edge $[x,y]$ and add edges $[m-1,x]$ and $[m-1,y]$ to form the desired cycle in $\mathfrak{C}_{2m-1}$. Every such construction determines an ordering of the 2m-1 integers for which (6) holds. ▌

Theorem 5: For d = 2m-1 and all n

$$\mathfrak{S}_{d^n} = \mathfrak{B}_{d^n} . \tag{7}$$

Proof: Start with the pair of edges on vertex $(m-1,m-1,\cdots,m-1,m)$:

$$\ell_1 = [0,0,\cdots,0,0;\ \ m-1,m-1,\cdots,m-1,m]$$

$$\ell_2 = [m-1,m-1,\cdots,m-1,m;\ \ 0,0,\cdots,0,1]$$

and increase each vertex label, considered as a d-ary number by 1 to form a succession of pairs of edges which closes on vertex $(0,0,\cdots,0,0)$ on the $\frac{1}{2}(d^n + 1)$-th step. The components in each pair of vertices generated by this procedure satisfy the conditions of the Corollary to Theorem 1 and hence the edge joining them is in $\mathfrak{C}_{d^n}$ as was to be shown. ▌

Obviously, for d = 2m and n > 1, $\mathfrak{S}_{d^n} \neq \mathfrak{B}_{d^n}$ since $\mathfrak{C}_{d^n}$ consists of $2^{n-1}$ disjoint components so that no Hamilton cycle is possible. Generalize the relationships (2) and (4) by introducing a deficit, $\delta = 0,1,\cdots$, which is the amount by which the equality fails. For all x,y in $\mathfrak{M}_{d^n}$ there exists a $\delta \geq 0$ such that

$$2(w(x,y) + \mathfrak{s}) = w(x,x') + w(y,y') \quad . \tag{8}$$

$\mathfrak{E}_{d^n}$ was defined to be those edges for which $\mathfrak{s} = 0$. It is easy to see that if $d = 2m-1$, there are no vertex pairs which have a deficit of 1; however, for $d = 2m$ we have the following result when $n = 1$.

Theorem 6: In $\mathfrak{W}_{2m}$ a vertex pair $x,y$ $(x \geq y)$ has deficit 1,

$$2(w(x,y) + 1) = w(x,x') + w(y,y') \tag{9}$$

if and only if

a) $x = m-1$ and $y \leq m-1$

or

b) $x \geq m$ and $y = m$ .

Proof: As in the proof of Theorem 1, lift the absolute value signs in (3) with due regard for the sign of the arguments. ▌

Corollary: For vertices x and y in $\mathfrak{W}_{d^n}$

$$2(w(x,y) + 1) = w(x,x') + w(y,y') \tag{10}$$

if and only if equality (2) is satisfied by all but one component for which (9) holds.

Identify the $2^{n-1}$ components of $\mathfrak{E}_{(2m)^n}$ each of which is isomorphic to $K_{m^n, m^n}$ as $\mathfrak{J}_1, \mathfrak{J}_2, \cdots, \mathfrak{J}_{2^{n-1}}$ with vertex sets $U_i$ and $V_i$. We then have the following lemma.

Lemma: The component graphs $\mathfrak{J}_i$ may be indexed in a natural way by the $2^{n-1}$ binary numbers $(0,0,\cdots,0)$ to $(0,1,\cdots,1)$ such that for every i and for any $x \in U_i \cup V_i$ there exists a j and a $y \in U_j \cup V_j$ for which

$$2(w(x,y) + 1) = w(x,x') + w(y,y')$$

if and only if $w(i,j) = 1$ or $w(i, 2^n - 1 - j) = 1$, where i and j are binary n-tuples with a leading zero.

Proof: The vertex sets in $\mathfrak{E}_{2m}$ are $U = (0,1,\cdots,m-1)$ and $V = (m,m+1,\cdots,2m-1)$ as determined in Theorem 1. Since by Theorem 3, $\mathfrak{E}_{(2m)^n}$ is isomorphic to $(\mathfrak{E}_{2m})^n$, the vertex sets of the component $\mathfrak{J}_i$ are just the $2^{n-1}$ possible Cartesian products of factors U and V - where with no loss of generality the first factor in one product is fixed to be U: $U^n$ and $V^n$, $U^{n-1}v$ and $V^{n-1}u$, etc. In other words the vertex sets for $\mathfrak{J}_i$, $0 \leq i \leq 2^n-1$, are formed from the n-1 digit binary representations of i and $2^n-1-i$ by introducing a factor of U in $U_i$ wherever a 0 occurs in i and a factor of V wherever a 1 appears and conversely for $V_i$. The lemma then follows by noting that (10) can hold if and only if it is possible to select a vertex y in $U_j \cup V_j$ in which all but one component is in the complementary set, $U^c = V$, to the one in which it occurs in x and in which one component is in the same set for both vertices. ▌

Theorem 7: For d = 2m and all n > 1

$$\mathfrak{S}_{d^n} = \mathfrak{B}_{d^n} - 2^{n-1} \quad . \tag{11}$$

Proof: Each of the $2^{n-1}$ disjoint component subgraphs $\mathfrak{F}_i$ of $\mathfrak{G}_{d^n}$ is isomorphic to $K_{m^n, m^n}$ by the Corollary to Theorem 3 and hence has a Hamilton cycle. The maximum possible value the cycle sum of distances could have would be realized if a single edge could be added with a deficit of 1 to link the cycles on the subgraphs. In other words, clearly

$$\mathfrak{S}_{d^n} \leq \mathfrak{B}_{d^n} - 2^{n-1} \quad .$$

It is well known that the binary n-tuples can be ordered so that the maximum Hamming distance (Hamming and Lee matrices are the same in this case) is 1 [1,4]. Such circuits are known as Gray codes. Index the $2^{n-1}$ subgraphs so that the indices form a Gray code. By the lemma it is then possible to delete an edge in each $\mathfrak{F}_i$ such that the endpoint vertices are joined to vertices in $\mathfrak{F}_{i+1}$ and $\mathfrak{F}_{i-1}$ with a deficit of 1 in each case. The resulting cycle in $\mathfrak{G}_{d^n}$ realizes the equality (11).▌

The final task is to determine the value of $\mathfrak{B}_{d^n}$.

Theorem 8:

$$\mathfrak{B}_d = \begin{cases} 2m^2 & \text{if} \quad d = 2m \\ 2m(m-1) & \text{if} \quad d = 2m-1 \end{cases} \quad . \tag{12}$$

Proof: If d is even the distances w(x,x′) on the subdiagonal are all odd, each such difference occurring twice. Hence

$$\mathfrak{B}_{2m} = 2\sum_{i=0}^{m-1} (2i+1) = 2m^2 \quad .$$

While if d is odd the distances are all even, each such difference also occurring twice. Hence in this case

$$\mathfrak{B}_{2m-1} = 2\sum_{i=0}^{m-1} 2i = 2m(m-1) \quad .$$

Theorem 9:

$$\mathfrak{B}_{d^n} = nd^{n-1}\,\mathfrak{B}_d \quad . \tag{13}$$

Proof: Recalling the recursive definition of $\mathfrak{W}_{d^n}$ in terms of $\mathfrak{W}_d$, the diagonal sum must satisfy the recursion

$$\mathfrak{B}_{d^n} = d\,\mathfrak{B}_{d^{n-1}} + d^{n-1}\,\mathfrak{B}_d$$

and (13) follows.▌

Corollary: For d = 2m

$$\mathfrak{S}_{d^n} = (2nm^{n+1} - 1)2^{n-1} \tag{14}$$

while for d = 2m-1

$$\mathfrak{S}_{d^n} = 2nm(m-1)(2m-1)^{n-1} \quad . \tag{15}$$

In the preceding development we have solved the question of finding circuits in $\mathfrak{M}_{d^n}$ which maximize the cycle sum; however, this global maximum may not insure that the minimum distance between adjacent pairs in the cycle is maximal. The maximum-minimum distance $L_{d^n}$ is bounded by

$$L_{d^n} \le \left[\frac{\mathfrak{S}_{d^n}}{d^n}\right] \quad . \tag{16}$$

For example

$$L_{3^2} \le \left[\frac{24}{9}\right] = 2$$

and the circuit described in the construction used in proving Theorem 5 realizes this maximum-minimum

(0,0)(1,2)(0,1)(2,0)(0,2)(2,1)(1,0)(2,2)(1,1)

while at the same time achieving the maximum possible cycle sum of 24. On the other hand the similar cycle

(0,0)(1,2)(0,1)(2,0)(0,2)(2,1)(1,0)(1,1)(2,2)

also achieves the maximum cycle sum but has a pair of adjacent vertices (1,0)(1,1) which are only one distant from each other.

There are not too many cases in which it can be shown that $L_{d^n}$ is achievable; d = 2 is one of these.

Theorem 10: For the binary hypercube, $2^n$, it is always possible to simultaneously realize the maximum cycle sum $\mathfrak{S}_{2^n} = (2n-1)2^{n-1}$ while at the same time achieving the maximum-minimum distance between successive n-tuples of $L_{2^n} = n-1$.

Proof: The weights on the subdiagonal of $\mathfrak{M}_{2^n}$ are all equal to n and $\mathfrak{C}_{2^n}$ consists of $2^{n-1}$ complementary pairs, i.e., isomorphs of $K_{1,1}$. By the construction used to prove the Lemma preceding Theorem 7 these pairs can be joined into a cycle by edges all of which have weight n-1. ∎

It is easy to exhibit infinitely many cases in which $L_{d^n}$ cannot be realized, however. For n > 1, $L_{d^n}$ is an integer if and only if d is odd and n = kd, in which case $L_{d^n} = 2km(m-1)$. But if a cycle were possible with this minimum, the distance between each adjacent pair of n-tuples would be uniformly $L_{d^n}$, i.e., a Hamilton cycle on $\mathfrak{M}_{d^n}$ with all edges of even length would have to exist. But this is impossible; hence, for all such cases, the maximum-minimum distance is less than $L_{d^n}$.

## 4. CONCLUSION

Table 1 summarizes the current status of our knowledge about orderings of n-tuples to maximize both the cyclic sum of distances and the minimum distance between adjacent pairs. Bold type indicates the cases in which $L_{d^n}$ is realized while entries in parentheses indicate the best construction known in those cases where it is still an open question whether $L_{d^n}$ or a closer value than known at present may be possible.

Table 1.

| n \ d | 2 | 3 | 4 | 5 | 6 | 7 | 8 |
|---|---|---|---|---|---|---|---|
| 1 | **1** | **1** | 1 | **2** | 2 | **3** | 3 |
| 2 | **1** | **2** | **3** | **4** | **5** | **6** | 7 |
| 3 | **2** | **3** | **5** | ≤7(6) | 7 | ≤10(9) | ≤11 |
| 4 | **3** | ≤5(4) | **7** | ≤9(8) | ≤11 | ≤13(12) | ≤15 |
| 5 | **4** | ≤6(5) | ≤9 | ≤11(10) | ≤14 | | |
| 6 | **5** | ≤7(6) | | | | | |
| 7 | **6** | ≤8(7) | | | | | |
| 8 | **7** | | | | | | |
| 9 | **8** | | | | | | |
| 10 | **9** | | | | | | |

The pseudo-color graphic display system whose application prompted this research codes colors with vertices of $\mathfrak{B}_{4^3}$. $\mathfrak{S}_{4^3} = 380$ and $L_{4^3} = 5$. Table 2 is a joining of four cycles in $\mathfrak{S}_{4^3}$ which realizes the maximum cycle sum and the maximum-minimum distance between adjacent three-tuples.

Table 2.

| | | | |
|---|---|---|---|
| 000 | 112 | 203 | 211 |
| 222 | 330 | 021 | 033 |
| 001 | 113 | 202 | 210 |
| 223 | 331 | 020 | 032 |
| 010 | 002 | 312 | 201 |
| 232 | 220 | 130 | 023 |
| 011 | 003 | 313 | 200 |
| 233 | 221 | 131 | 022 |
| 100 | 013 | 303 | 311 |
| 322 | 231 | 121 | 133 |
| 101 | 012 | 302 | 310 |
| 323 | 230 | 120 | 132 |
| 110 | 102 | 213 | 301 |
| 332 | 320 | 031 | 123 |
| 111 | 103 | 212 | 300 |
| 333 | 321 | 030 | 122 |

Table 3 is a more difficult construction - a cycle in which the distance between adjacent pairs of three-tuples is uniformly five. This construction is a Hamilton cycle in the graph on the vertices of $\mathfrak{W}_{4^3}$ in which an edge [x,y] is present if and only if w(x,y) = 5.

Table 3.

| | | | |
|---|---|---|---|
| 003 | 203 | 103 | 303 |
| 200 | 000 | 300 | 100 |
| 012 | 212 | 112 | 312 |
| 220 | 020 | 320 | 120 |
| 032 | 232 | 132 | 332 |
| 202 | 002 | 302 | 102 |
| 010 | 210 | 110 | 310 |
| 222 | 022 | 322 | 122 |
| 030 | 230 | 130 | 330 |
| 233 | 033 | 333 | 133 |
| 021 | 221 | 121 | 321 |
| 213 | 013 | 313 | 113 |
| 001 | 201 | 101 | 301 |
| 231 | 031 | 331 | 131 |
| 023 | 223 | 123 | 323 |
| 211 | 011 | 311 | 111 |

BIBLIOGRAPHY


(1) E. N. Gilbert, "Gray Codes and Paths on the n-Cube," Bell System Tech. Jour. (37) 3 (1958), 815-825.

(2) David L. MacAdam, "Specification of Small Chromaticity Differences," Jour. of the Optical Soc. of Amer. (33) 1 (1943), 18-26.

(3) Parry Moon, "A Metric Based on the Composite Color Stimulus," Jour. of the Optical Soc. of Amer. (33) 5 (1943), 270-277.

(4) D. H. Smith, "Hamiltonian Circuits on the n-Cube," Canadian Math. Bulletin (17) 5 (1975), 759-761.

(5) W. D. Wright, "The Graphical Representation of Small Color Differences," Jour. of the Optical Soc. of Amer. (33) 11 (1943), 632-636.

05C99

# DECOMPOSITIONS OF GRAPHS AND HYPERGRAPHS

T.P. Speed

Department of Mathematics,
The University of Western Australia,
Nedlands, W.A. 6009
Australia.

ABSTRACT

The notion of a decomposition of a class of hypergraphs is introduced. Stimulated by the requirements of certain problems in probability and statistics, the problem of describing all decompositions of such a hypergraph is attacked.

## 1. INTRODUCTION

This paper concerns systems $(C,\mathcal{C})$ consisting of a non-empty finite set C and a class $\mathcal{C}$ of pairwise incomparable subsets of C. A number of writers in the area of probability and statistics have made use of such systems in recent years, see, for example, Vorob'ev (7), (8), Kellerer (4), Haberman (2) and Suomela (6), and it is the purpose of this paper to discuss certain combinatorial problems which have arisen in this work. The system $(C,\mathcal{C})$ is clearly a *hypergraph* in the sense of Berge (1), indeed a rather special hypergraph in that no element of $\mathcal{C}$ may contain any other element of $\mathcal{C}$.

The main concept in this paper is a way of decomposing such a hypergraph $(C,\mathcal{C})$ into two or more components which generalises the *principle of separation into pieces* of a graph relative to a complete articulation set, (1, p.329), and we call this a *decomposition* of $(C,\mathcal{C})$. We begin by giving three problems from probability and statistics which have led naturally to such a notion of decomposition, and, after defining it, we show what a decomposition means in each of the three problems. Following this a combinatorial problem is extracted, and a first attempt at its solution is described. The paper closes with some remarks concerning further work.

All hypergraphs in this paper will be defined on *finite* base sets, and will consist of a class of *pairwise incomparable* subsets of the base set. This assumption will not be repeated.

## 2. THE MOTIVATING PROBLEMS

In this section we describe three problems in which our particular type of hypergraph arises.

*PROBLEM 1.* Let $\{(X_\gamma,\mathcal{X}_\gamma):\gamma \in C\}$ be a set of *measure spaces* indexed by C ; i.e. for each $\gamma \in C$ , $X_\gamma$ is a non-empty set and $\mathcal{X}_\gamma$ is a σ-field of subsets of $X_\gamma$ . For each subset $c \subseteq C$ write $(X_c,\mathcal{X}_c)$ for the *product* measure space $\underset{\gamma\in c}{\otimes}(X_\gamma,\mathcal{X}_\gamma)$ , and put $(X,\mathcal{X}) = (X_C,\mathcal{X}_C)$ . We will take as *given* a class $\mathcal{C}$ of subsets of $C$ , and for each $c \in \mathcal{C}$ , a probability measure $\mu_c$ on $(X_c,\mathcal{X}_c)$ , such that the system $\{\mu_c : c \in \mathcal{C}\}$ satisfies the following *consistency condition*: if $d \subseteq a \cap b$ for $a,b \in \mathcal{C}$ , then the *images* $\mu_{a,d} = \mu_a \circ \pi_{a,d}^{-1}$ and $\mu_{b,d} = \mu_b \circ \pi_{b,d}^{-1}$ of $\mu_a$ and $\mu_b$ under the canonical projections $\pi_{a,d}:X_a \to X_d$ and $\pi_{b,d}:X_b \to X_d$ , respectively, *coincide*. An obvious way to get such a system is to take a measure $\mu$ on $(X,\mathcal{X})$ and put $\mu_c = \mu\circ\pi_c^{-1}$ where $\pi_c:X \to X_c$ is the canonical projection. In this case the measure $\mu$ is said to be an *extension* of the system $\{\mu_c : c \in \mathcal{C}\}$ .

The problem considered by Vorob'ev (7) and Kellerer (4) is the following: *for which classes $\mathcal{C}$ of subsets of $C$ does every consistent system $\{\mu_c : c \in \mathcal{C}\}$ admit an extension?* It is not hard to show that for $\mathcal{C}_1 = \{\{1,2\},\{2,3\}\}$ , every consistent system of measures *does* admit an extension, whilst for $\mathcal{C}_2 = \{\{1,2\},\{2,3\},\{3,1\}\}$ , this is *not* the case.

*PROBLEM 2.* In this case we let $X_\gamma$ denote the finite set of *categories* associated with a *response* $\gamma$ from a set $C$ of responses. The product set $X = \prod_{\gamma\in C} X_\gamma$ indexes the combinations of categories of responses, and we can consider $|C|$-dimensional *contingency tables* $\{n(x) : x \in X\}$ over $X$ . An *hierarchical log-linear model* for such a contingency table is uniquely specified by the (generating) class $\mathcal{C}$ of pairwise incomparable subsets of $C$ , whose marginals $\{n_c : c \in \mathcal{C}\}$ constitute the minimal sufficient statistics for the model; see (2) for background and further details.

A problem considered by Haberman (2) is the following: *for which classes $\mathcal{C}$ does there exist an explicit formula for the maximum likelihood estimator $\hat{m}_{\mathcal{C}}$ , of $m = \mathbb{E}\{n\}$ , under the model defined by $\mathcal{C}$ ?* For example, an explicit formula exists for $\mathcal{C}_3 = \{\{1,2\},\{2,3\},\{3,4\}\}$, but not for $\mathcal{C}_4 = \{\{1,2\},\{2,3\},\{3,4\},\{4,1\}\}$ .

*PROBLEM 3.* Let $\underset{\sim}{C} = (C,E(\underset{\sim}{C}))$ be a simple graph with vertex set $C$ and edge set $E(\underset{\sim}{C})$ , and suppose (continuing the notation of Problem 1) that for each $\gamma \in C$ we have a measure space $(X_\gamma,\mathcal{X}_\gamma)$ . An *X-valued random field* (r.f.) over $\underset{\sim}{C}$ consists of a random variable $\xi:\Omega \to X$ defined over some probability space $(\Omega,\underline{\underline{A}},\mathbb{P})$ , and we write $\xi = (\xi_\gamma : \gamma \in C)$ . For each $c \subseteq C$ we let $\xi_c:\Omega \to X_c$ be the *c-marginal* r.f. and let $P_c$ be the *distribution* $\mathbb{P}\circ\xi_c^{-1}$ of $\xi_c$ on $(X_c,\mathcal{X}_c)$ .

The r.f. $\xi$ is said to be $\underset{\sim}{C}$-*Markov* if for any three pairwise disjoint subsets *a*, *b* and *d* of C with *d* separating *a* from *b*, we have $\xi_a$ and $\xi_b$ *conditionally independent* given $\xi_d$. A question of great interest, particularly for those who wish to *simulate* such Markov r.f.'s, is the following: *for which graphs* $\underset{\sim}{C}$ *does there exist a closed-form expression for the distribution* $P = \mathbb{P} \circ \xi^{-1}$ *of any* X-valued $\underset{\sim}{C}$-*Markov* r.f. $\xi$? It is not hard to show that this problem can be reduced to a discussion of the interrelations between the distributions $\{P_c : c \in \mathcal{C}_{\underset{\sim}{C}}\}$ of the r.f.'s $\{\xi_c : c \in \mathcal{C}_{\underset{\sim}{C}}\}$, where $\mathcal{C}_{\underset{\sim}{C}}$ is the hypergraph of all *(maximal) cliques* of the graph $\underset{\sim}{C}$; see Suomela (6) and Vorob'ev (8) for further details.

We remark at this point that there *does* exist a simple expression for the distribution P when we are considering $\underset{\sim}{C}_5$ below, but *not* for $\underset{\sim}{C}_6$, where

$\underset{\sim}{C}_5$ :  $\underset{\sim}{C}_6$ : 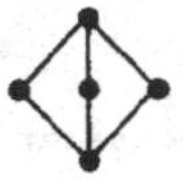 .

It will become apparent that all three problems have a common solution.

## 3. THE NOTION OF A DECOMPOSITION

For two hypergraphs $(A,\mathcal{A})$ and $(B,\mathcal{B})$ of the type we are considering, we write $(A,\mathcal{A}) \leq (B,\mathcal{B})$ if $A \subseteq B$ and for every $a \in \mathcal{A}$ there exists $b \in \mathcal{B}$ with $a \subseteq b$. It is easy to see that this relation is a partial ordering, and the partially ordered set it defines has a lattice structure: $(A,\mathcal{A}) \vee (B,\mathcal{B}) = (C,\mathcal{C})$ [resp. $(A,\mathcal{A}) \wedge (B,\mathcal{B}) = (D,\mathcal{D})$] iff $A \cup B = C$ [resp. $A \cap B = D$] and $\mathcal{C}$ [resp. $\mathcal{D}$] is the class of *maximal elements* of $\mathcal{A} \cup \mathcal{B}$ [resp. $\{a \cap b : a \in \mathcal{A}, b \in \mathcal{B}\}$]. For simplicity in what follows, we will denote the hypergraphs by their second element, i.e. $\mathcal{C}$ instead of $(C,\mathcal{C})$, where no confusion can result. The following definition is basic.

*DEFINITION 1* The hypergraph $\mathcal{C}$ is *decomposed* into $\{\mathcal{C}_i : i \in I\}$ relative to $d \subseteq C$, if (i) $\mathcal{C} = \bigvee_{i \in I} \mathcal{C}_i$; and (ii) $\mathcal{C}_i \wedge \mathcal{C}_j = \{d\}$ whenever $i \neq j$, $i,j \in I$.

*COROLLARY* For all pairs $i,j \in I$ with $i \neq j$ there exists $c_i^* \in \mathcal{C}_i$, $c_j^* \in \mathcal{C}_j$ such that $C_i \cap C_j = c_i^* \cap c_j^* = d$.

*Example* If $\mathcal{C} = \{\{1,2\},\{1,3\},\{1,4\}\}$, then $\mathcal{C}$ is decomposed into $\mathcal{C}_1 = \{\{1,2\}\}$, $\mathcal{C}_2 = \{\{1,3\}\}$, and $\mathcal{C}_3 = \{\{1,4\}\}$ relative to $d = \{1\}$.

If $\underset{\sim}{G} = (G, E(\underset{\sim}{G}))$ is a simple graph, then the (maximal) *clique hypergraph* $(G,\mathcal{C}_{\underset{\sim}{G}})$ is clearly a hypergraph of the type under discussion. Conversely, if $(C,\mathcal{C})$ is one of our hypergraphs, then we can define the graph $\underset{\sim}{C}_{\mathcal{C}} = (C, E(\underset{\sim}{C}_{\mathcal{C}}))$ over the set C, whose edges are just those unordered pairs $\{\alpha,\beta\}$ which are contained in an element of $\mathcal{C}$, the *2-section* of $\mathcal{C}$, see (4, p.390). With this notation we have the following:

*PROPOSITION 1* (i) If the clique hypergraph $\mathcal{C}_{\underset{\sim}{G}}$ of a connected graph $\underset{\sim}{G}$ is decomposed into hypergraphs $\{\mathcal{C}_i : i \in I\}$ relative to $d$, then $\{\underset{\sim}{G}_{\mathcal{C}_i} : i \in I\}$ is a separation of $\underset{\sim}{G}$ into pieces relative to the complete articulation set $d$.

(ii) Conversely, if a connected graph $\underset{\sim}{G}$ is separated into pieces $\{G_i : i \in I\}$ relative to the complete articulation set $d$, then $\{\mathcal{C}_{\underset{\sim}{G}_i} : i \in I\}$ is a decomposition of $\mathcal{C}_{\underset{\sim}{G}}$ relative to $d$.

The proof of this proposition is straightforward and so is omitted. Since not every hypergraph $\mathcal{C}$ under discussion is *conformal*, that is, arises as the clique hypergraph of a graph, (4), p.396, the notion of decomposition is a generalisation of the idea of separation of a graph into pieces relative to a complete articulation set.

## 4. ANOTHER LOOK AT THE MOTIVATING PROBLEMS

We now show how the notion of decomposition of a hypergraph provides the key to the solution of our three problems, for simplicity considering only two-element decompositions.

*PROBLEM 1. (Contd.)* Let $\{\mu_c : c \in \mathcal{C}\}$ be a consistent system of probability measures, and suppose that $\mathcal{C}$ is decomposed into $\mathcal{A}$ and $\mathcal{B}$ relative to $d$. *If* $\mu_{\mathcal{A}}$ is an extension of $\{\mu_a : a \in \mathcal{A}\}$ to $X_A$, $A = \cup\mathcal{A}$, and $\mu_{\mathcal{B}}$ an extension of $\{\mu_b : b \in \mathcal{B}\}$ to $X_B$, $B = \cup\mathcal{B}$, then

$$\mu_{\mathcal{C}}\left(\prod_{\gamma \in C} Y_\gamma\right) = \frac{\mu_{\mathcal{A}}\left(\prod_{\gamma \in A} Y_\gamma\right)\mu_{\mathcal{B}}\left(\prod_{\gamma \in B} Y_\gamma\right)}{\mu_d\left(\prod_{\gamma \in d} Y_\gamma\right)} \qquad Y_\gamma \in \mathcal{X}_\gamma, \gamma \in C,$$

defines an extension $\mu_{\mathcal{C}}$ of $\{\mu_c : c \in \mathcal{C}\}$ to $X$. Loosely speaking, if $\mathcal{C}$ is decomposed into $\mathcal{A}$ and $\mathcal{B}$ relative to $d$, then solutions $\mu_{\mathcal{A}}$ and $\mu_{\mathcal{B}}$ to the extension problems restricted to $\mathcal{A}$ and $\mathcal{B}$, respectively, can be "joined" across $d$, to give a solution to the original problem.

*PROBLEM 2. (Contd.)* A result similar to that just noted holds in the context of contingency tables. Let $M_{\mathcal{A}}, M_{\mathcal{B}}$ and $M_{\{d\}}$ be the (non-linear) maximum-likelihood operators, which convert the data $\{n(x) : x \in X\}$ into the maximum-likelihood estimates of $\{\mathbb{E}\{n(x)\} : x \in X\}$, evaluated under the models $\mathcal{A}, \mathcal{B}$ and $\{d\}$, respectively.

If $\mathcal{C}$ is decomposed into $\mathcal{A}$ and $\mathcal{B}$ relative to $d$, then

$$M_{\mathcal{C}}n = \frac{M_A n M_{\mathcal{B}} n}{M_{\{d\}} n} , \qquad n = \{n(x) : x \in X\} ,$$

is the maximum-likelihood operator for the model $\mathcal{C}$ . Thus any explicit formulae for $M_A$ and $M_{\mathcal{B}}$ combine with the one which always exists for $M_{\{d\}}$ , to give one for $M_{\mathcal{C}}$ .

*PROBLEM 3. (Contd.)* Let $\underset{\sim}{\xi}$ be an X-valued $\underset{\sim}{C}$-Markov random field and suppose that the graph $\underset{\sim}{C}$ , which we assume connected, is separated into two parts $\underset{\sim}{A}$ and $\underset{\sim}{B}$ relative to the complete articulation set d . Let the distribution P of $\underset{\sim}{\xi}$ have a density p with respect to a product measure $\lambda = \underset{\gamma \in C}{\otimes} \lambda_\gamma$ , and let $p_A , p_B$ and $p_d$ denote the corresponding densities of $\underset{\sim}{\xi}_A , \underset{\sim}{\xi}_B$ and $\underset{\sim}{\xi}_d$ , respectively. Then a consequence of the Markov assumption is the factorisation

$$p(\underset{\sim}{x}) = \frac{p_A(\underset{\sim}{x}_A) p_B(\underset{\sim}{x}_B)}{p_d(\underset{\sim}{x}_d)} , \qquad \underset{\sim}{x} \in X ,$$

where $\underset{\sim}{x}_A , \underset{\sim}{x}_B$ and $\underset{\sim}{x}_d$ are the projections of $\underset{\sim}{x}$ onto $X_A , X_B$ and $X_d$ respectively.

We can now refer to Proposition 1 and see that if there is an explicit formulae for the distributions $p_A$ and $p_B$ of the Markov field $\underset{\sim}{\xi}_A$ and $\underset{\sim}{\xi}_B$ in terms of $\{p_a : a \in \mathcal{C}_{\underset{\sim}{A}}\}$ and $\{p_b : b \in \mathcal{C}_{\underset{\sim}{B}}\}$ , then these combine in the above formula to give an explicit formula for p in terms of $\{p_c : c \in \mathcal{C}_{\underset{\sim}{C}}\}$ .

## 5. A COMBINATORIAL PROBLEM : DESCRIBE ALL DECOMPOSITIONS

It is clear from the three motivating problems that what would be very helpful is an adequate description of the end result of successively decomposing our hypergraph $(C\mathcal{L})$ . Equivalently, following the next proposition, we seek a description of *all* decompositions of $(C,\mathcal{C})$ , together with information concerning their inter-relationships. We now prove a sort of associativity or compatibility result which suggests that such an aim is well-founded.

*PROPOSITION 2* Suppose that the hypergraph $\mathcal{C}$ is decomposed into $\{\mathcal{C}_i : i \in I\}$ relative to $d \subseteq C$ , and that one of its components, $\mathcal{C}_1$ say, is itself decomposed into $\{\mathcal{C}'_k : k \in I_1\}$ relative to $d_1 \subseteq C_1$ . Then there exists a decomposition of $\mathcal{C}$ into $\{\tilde{\mathcal{C}}_k : k \in I_1\}$ relative to $d_1$ , and $\mathcal{C}'_k \leq \tilde{\mathcal{C}}_k$ for each $k \in I_1$ .

*PROOF* From the Corollary to Definition 1 we know that for every $i \neq 1$ there exists $c_i^* \in \mathcal{C}_i$ and $c_i' \in \mathcal{C}_1$ such that $c_i^* \cap c_i' = d$. Now each $c_i'$ belongs to at least one of the classes $\mathcal{C}_k'$, and, choosing any one, where choice is possible, we get a map $\varphi: I\backslash\{1\} \to I_1, i \mapsto k$. In terms of this map we define $\tilde{\mathcal{C}}_k$ by

$$\tilde{\mathcal{C}}_k = \mathcal{C}_k' \vee \bigvee_{i:\varphi(i)=k} \mathcal{C}_i .$$

Since $\bigvee_{k\in I_1} \mathcal{C}_k' = \mathcal{C}_1$ and $\bigvee_{i\in I} \mathcal{C}_i = \mathcal{C}$, we see that $\bigvee_{k\in I_1} \tilde{\mathcal{C}}_k = \mathcal{C}$, and it only remains for us to prove that if $k \neq \ell$, $\tilde{\mathcal{C}}_k \wedge \tilde{\mathcal{C}}_\ell = \{d_1\}$. But this will be achieved if we prove that for all $c_i \in \mathcal{C}_i$, $c_j \in \mathcal{C}_j$ where $\varphi(i) = k$, $\varphi(j) = \ell$, $c_{1,k} \in \mathcal{C}_k'$ and $c_{1,\ell} \in \mathcal{C}_\ell'$, we have $c_{1,k} \cap c_{1,\ell}$, $c_{1,k} \cap c_j$, $c_i \cap c_j$, and $c_i \cap c_{1,\ell}$ *all* contained in $d_1$.

Now $c_{1,k} \cap c_{1,\ell} \subseteq d_1$ from the definition of decomposition, and so this inclusion is proved. For a similar reason each of the other three intersections is contained in $d$, and our argument goes on from there. Writing $d = c_i^* \cap c_i'$, we find that $c_i \cap c_{1,\ell} \subseteq c_i' \cap c_{1,\ell} \subseteq d_1$. Similarly, writing $d = c_j^* \cap c_j'$, we get $c_{1,k} \cap c_j \subseteq c_{1,k} \cap c_j' \subseteq d_1$ and $c_i \cap c_j \subseteq c_i' \cap c_j' \subseteq d_1$, completing the proof.

## 6. A FIRST ATTACK : DECOMPOSABLE HYPERGRAPHS

The simplest sort of hypergraphs of the type under discussion are those of the form $(A,\{A\})$, the one-element hypergraphs. A first attack on our problem might well be an attempt to describe that class of hypergraphs for which the procedure of successive decomposition ultimately leads to one-element hypergraphs in all cases. Following Haberman (2), we call such hypergraphs *decomposable*, and have the following characterisation (2, p.181) : $\mathcal{C}$ is decomposable if and only if:

(S) $\mathcal{C}$ can be enumerated as $c_1, c_2, \ldots, c_n$ in such a way that for all $m, 1 < m \leq n$, there exists $m^* < m$ such that for all $\ell < m : c_m \cap c_\ell \subseteq c_m \cap c_{m^*}$.

Turning now to another type of characterisation of decomposability, we begin by noting the easily proved fact that *if $\mathcal{C}$ is decomposable, then $\mathcal{C}$ is conformal* (1, p.396), i.e. the class $\mathcal{C}$ coincides with the class $\mathcal{C}_{\underset{\sim}{G}}$ of all (maximal) cliques of the 2-section $\underset{\sim}{G} = \underset{\sim}{G}_{\mathcal{C}}$ of $\mathcal{C}$. Thus there is no loss of generality in supposing that every decomposable hypergraph is the *clique hypergraph* of a graph $\underset{\sim}{G}$. A standard type of argument, which we omit, allows us to restrict further to connected graphs. Within this framework, we define a notion of *adjacency* over the clique hypergraph $\mathcal{C}_{\underset{\sim}{C}}$ : cliques $a$ and $b$ are said to be adjacent if $a \cap b$ is an articulation set. Then decomposability of $\mathcal{C}_{\underset{\sim}{C}}$ is equivalent to the condition:

(C) $\mathcal{C}_{\underset{\sim}{C}}$ is a connected graph.

Next we analyse the rôle of the clique intersections $d$ more closely, letting $\mathcal{D} = \mathcal{D}_{\underset{\sim}{C}}$ denote the class of all such, by defining an important index $\nu$ on $\mathcal{D}$ as follows: for $d \in \mathcal{D}, \nu(d) + 1$ is the number of components of $\underset{\sim}{C}\backslash d$ which contain the difference $c\backslash d$ for some clique $c \supseteq d$ . Space prevents us developing the properties of this interesting index, but we note that for a connected graph $\underset{\sim}{C}$ , $\mathcal{C}_{\underset{\sim}{C}}$ is decomposable if and only if

(I) $$|\mathcal{C}_{\underset{\sim}{C}}| - \sum_{d \in \mathcal{D}_{\underset{\sim}{C}}} \nu(d) = 1 .$$

Finally we give a characterisation of the class of graphs $\underset{\sim}{C}$ for which $\mathcal{C}_{\underset{\sim}{C}}$ is decomposable, noting that it coincides with a class introduced by Hajnal and Suranyi (3), see also (1, p.368). $\mathcal{C}_{\underset{\sim}{C}}$ is decomposable if and only if

(P) No subset $s \subseteq C$ generates a subgraph $\langle s \rangle$ isomorphic to a cycle $Z_n$ with $n > 3$ .

Graphs with property (P) have been called *triangulated* by Berge, but this name is perhaps unsuitable because of the following examples:

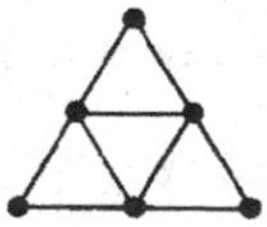

Triangulated

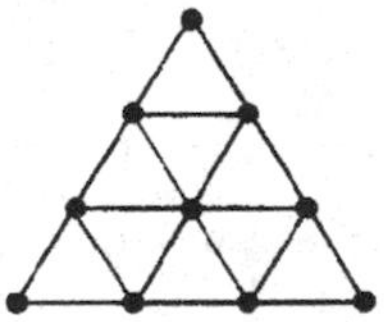

Not triangulated.

Proofs of all the above assertions can be found in (5), and we see that the results concerning decomposability are quite satisfactory. In particular our three problems may all be answered positively in the decomposable case, and separate arguments show that this is the only case in which such answers may be given.

## 7. FINAL COMMENTS

Although we have a good understanding of the situation when the decomposition process is complete, i.e. goes right down to one-element hypergraphs, the more general case still poses interesting problems. Restricting ourselves to the *conformal* hypergraphs, we see (by Proposition 1) that our question of §5 becomes: describe the end result of repeated separation into pieces of a connected graph $\underset{\sim}{G}$. Let us call a connected graph *indecomposable* if it *cannot* be separated into pieces relative to any complete articulation set. The following proposition characterises such graphs.

*PROPOSITION 2* For a connected graph $\underset{\sim}{G}$ the following are equivalent:

(i) $\underset{\sim}{G}$ has no complete articulation sets.

(ii) $\underset{\sim}{G}$ has no complete separating sets.

(iii) Every pair $\alpha,\beta$ of distinct vertices is contained in a subset $s$ which generates a cyclic subgraph of $\underset{\sim}{G}$.

By making use of this result, and arguments analogous to those in §6, a satisfactory answer can be given in this conformal case as well. The details will be published elsewhere, as will a discussion of the general case, which builds on the results just mentioned.

## 8. REFERENCES


(1) C. Berge. *Graphs and Hypergraphs.* North Holland/American Elsevier, 1973.

(2) Shelby J. Haberman. *The Analysis of Frequency Data.* The University of Chicago Press, Chicago and London, 1974.

(3) András Hajnal und János Surányi. Über die auflösung von graphen in vollständige teilgraphen. *Ann. Univ. Sci. Budapest.* 1 (1958) 113-121.

(4) Hans G. Kellerer. Verteilungsfunktionen mit gegebenen Marginalverteilungen. *Z. Warscheinlichkeitstheorie.* 3 (1964) 247-270.

(5) S.L. Lauritzen, T.P. Speed and K. Vijayan. *Decomposable graphs and Hypergraphs* (1976) Manuscript, 25 pp. To be submitted.

(6) P. Suomela. *Construction of Nearest-neighbour systems. Annales Academiae Scientarum Fennicae Series A. Mathematical Dissertationes.* 10 (1976) 1-56.

(7) N.N. Vorob'ev. Consistent families of measures and their extensions. *Theor. Prob. Appl.* 7 (1962) 147-163.

(8) N.N. Vorob'ev. Markov measures and Markov extensions. *Theor. Prob. Appl.* 8 (1963) 420-429.

# SOME EXTREMAL PROBLEMS IN COMBINATORIAL GEOMETRY

05A99

E.G. Straus

Department of Mathematics UCLA
California 90024, USA

1. INTRODUCTION. P. Erdös and G. Purdy raised the following question :

PROBLEM 1. *Given a set* S *of* n *points in the Euclidean plane not all on one line. Let* T *be the maximal area of a triangle determined by* 3 *of the points and* t *the minimal (positive) area of a triangle determined by* 3 *of the points. What is* $f(n) = \inf_S T/t$ *?*

A second, at first glance not closely related problem, was also posed by Erdös and Purdy :

PROBLEM 2. *For the* n*-tuples described in Problem 1, what is the minimal number,* g(n) *, of distinct areas of triangles determined by* 3 *of the points?*

We were able to answer Problem 1 essentially completely and we conjecture that the answer to Problem 2 is given by the same arrangements of the points and leads to the same value. Since the details of the proof will be given in a forthcoming paper by Erdös, Purdy and myself in Discrete Mathematics, I shall only outline the ideas and discuss some of the many additional geometric and combinatorial conjectures raised.

2. SOLUTION TO PROBLEM 1.

THEOREM. *For all* $n > 35$ *we have* $f(n) = \left[\frac{n-1}{2}\right]$ *and an* n*-tuple which yields this minimal value is given by the set*

$$S = \left\{(0,0),(0,1),\ldots,\left(0,\left[\frac{n-1}{2}\right]\right),(1,0),(1,1),\ldots,\left(1,\left[\frac{n-2}{2}\right]\right)\right\}$$

*For even* $n > 35$ *all sets which yield* f(n) *are affinely equivalent to* S .

That is, the extremal set S is divided into two parts of equal or nearly equal numbers of points placed at equal distances on two parallel lines.

The condition $n > 35$ is presumably a fault of the method of proof and would be replaceable by $n > 5$ . For $n = 5$ the extremal value is given by the vertices of the regular pentagon.

The set S shows that in Problem 2 we have $g(n) \leq \left[\frac{n-1}{2}\right]$ (if we do not count 0 as an area) for all $n \geq 3$ and we conjecture that equality holds with affine uniqueness of the sets for $n > 7$.

Outline of proof. We rely on the following Theorem of E. Sas :

LEMMA 1. *Let* $C$ *be a convex closed curve in* $E^2$ *and let* $A_n$ *be the maximal area of an inscribed* $n$-*gon* $(n \geq 3)$. *Then the ratio of the areas satisfies* $|A_n|/|C| \geq \frac{n}{2\pi} \sin \frac{2\pi}{n}$ *with equality only for the case where* $C$ *is an ellipse.* (Here $|C|$ denotes the area of the interior of $C$).

The generalization of Lemma 1 to higher dimensions is known for inscribed simplices of convex closed surfaces (Blaschke).

LEMMA 1'. *The ratio of the volume of the maximal inscribed simplex to the volume of the interior of a closed convex surface is minimal when the surface is an ellipsoid; and only when it is an ellipsoid.*

The generalization of Lemma 1' to the case of the convex hull of more than $n+1$ points on a closed convex surface does not appear to be known, probably because of the usual difficulty that one doesn't even know the maximal volumes and corresponding polyhedra for the sphere.

We needed a variant of Lemma 1 for the case $n = 3$.

LEMMA 2. *If* $C$ *is a convex closed curve in* $E^2$ *and one of the sides of a maximal inscribed triangle* $T$ *lies on* $C$*; then the ratio of the areas satisfies* $|T|/|C| \geq 1/\sqrt{5}$, *with equality if and only if* $C$ *is an (affinely) regular pentagon.*

It is not clear what the corresponding variant of Lemma 1' would be. For the case $n \geq 3$ in the plane it should not be too difficult to generalize Lemma 2 :

CONJECTURE 1. *If* $A_n$ *is an inscribed* $n$-*gon of maximal area in the convex closed curve* $C$ *and* $A_n$ *has one edge on* $C$, *then* $|A_n|/|C|$ *is minimal in case* $C$ *is a regular* $(2n-1)$-*gon.*

More generally we conjecture the following :

CONJECTURE 2. *If* $A_n$ *is an inscribed* $n$-*gon of maximal area in the convex closed curve* $C$ *and* $A_n$ *has* $k$ *of its edges on* $C$, $(0<k \leq n)$, *then* $|A_n|/|C|$ *is minimal in case* $C$ *is a regular* $(2n-k)$-*gon.*

Returning to the proof of the Theorem, we assume that $S$ is a set of $n$ points in $E^2$ which yields the minimal value, $f(n)$, of $T/t$ and we let $C$ be the boundary of the convex hull of $S$. By Lemma 1 we know that

$$(1) \qquad T > \frac{3\sqrt{3}}{4\pi} |C|.$$

we now triangulate $C$ using the points of $S$ in such a way that all triangles have their vertices in S and no point of S is interior to one of the triangles or to one of the edges in the triangulation.

LEMMA 3. *Let* S *contain* a *points on the boundary of* C *and* $b = n-a$ *points in the interior of* C *then any triangulation of the prescribed kind divides the interior of* C *into* $2(n-1) - a$ *triangles.*

The proof follows by simple induction, starting with $n = 3$ vertices on C . Now any additional point on C increases the number of triangles by one; either through the addition of a new triangle if the vertex is a vertex of C ; or through the bisection of a previous triangle if the vertex is interior to an edge of C . On the other hand, any additional point interior to C increases the number of triangles by two; either through the trisection of a previous triangle, in case the point is interior to that triangle; or through the bisections of two adjacent previous triangles, in case the point is interior to their common edge. Thus we get $a - 2 + 2b = a - 2 + 2(n-a) = 2(n-1) - a$ triangles.

Since each of the triangles in Lemma 3 has area $\geq t$ we have

$$|C| \geq (2(n-1)-a)t \ . \tag{2}$$

Comparing (1) and (2) and using the fact that $f(n) \leq \left[\frac{n-1}{2}\right]$ we have

$$\frac{n-1}{2}t > \frac{3\sqrt{3}}{4\pi}(2(n-1)-a)t$$

and hence

$$a > \left(2 - \frac{2\pi}{3\sqrt{3}}\right)(n-1) \ . \tag{3}$$

In other words a substantial fraction of the points of S lies on the boundary of its convex hull.

If we now assume that a maximal triangle T has none of its edges on C , then the deletion of the interior of T from the interior of C leaves three convex regions whose total area is less than $\left(\frac{3\sqrt{3}}{4\pi} - 1\right)T$ and the triangulations using only the $a + 3$ boundary points (the three vertices of T are counted twice) would lead to triangles whose average area is $< 2T/(n-1)$ (provided $n > 35$), a contradiction. Thus one of the edges of T must be on C and we can apply Lemma 2 to improve (3) to

$$a > \left(2 - \frac{\sqrt{5}}{2}\right)(n-1) \ . \tag{4}$$

If we again triangulate the two regions obtained by deleting the interior of T from the interior of C we find that we would get triangles that are too small in area, unless we assume that a substantial fraction of the points of S lies on the edge of T which is part of C. It is now a simple matter to show that all the points of S which do not lie on that edge must lie on a line parallel to that edge if we are to avoid triangles whose area is $< 2T/(n-1)$. Once we know that the points lie on two parallel lines, it becomes easy to verify that the set S described in the theorem yields a minimal $f(n)$ and that for even n any other set which yields the same $f(n)$ must be affinely equivalent to that S. For odd n there is some freedom in the distribution of the points on the line which contains $(n-1)/2$ points.

3. RELATED PROBLEMS. In trying to generalize Problem 1 to k-gons with vertices in S, $k > 3$, one is faced with the added complication that these k-gons are not necessarily convex. It is therefore more plausible to refer to the convex hull of a k-tuple.

CONJECTURE 3. *Given a set* S *of* n *points in* $E^2$ *not all on a straight line. Let* $T_k$ *be the maximal area of the convex hull of* k *points* $(k>3)$ *of* S *and* $t_k$ *the minimal nonzero area of the convex hull of* k *points of* S. *Then for all sufficiently large* n *the value* $f(k,n) = \inf_S T_k/t_k$ *is attained for the set* S *described in the Theorem.*

Thus we would have $f(k,n) = [(n+1)/2]/(k-2)$ for $k \geq 4$ and $n \geq 2k$.

Another generalization would involve higher dimensional analogues of Problem 1:

PROBLEM 3. *Given a set* S *of* n *points in* $E^m$ *not all in one hyperplane. Let* T *be the maximal volume of an* m-*simplex with vertices in* S, *and* t *the minimal nonzero volume of an* m-*simples with vertices in* S. *What is* $f_m(n) = \inf_S T/t$ ?

CONJECTURE 4. *We have* $f_m(n) = \left[\frac{n-1}{m}\right]$ *for all large* n. *This valued is attained for*

$$S = \left\{(0,\ldots,0,0),\ldots,(0,\ldots,0,\left[\frac{n-1}{m}\right]);\ (1,0,\ldots,0,0),\ldots,(1,0,\ldots,0\left[\frac{n-2}{m}\right]);\right.$$

$$\left.\ldots;(0,\ldots,0,1,0),\ldots,(0,\ldots,0,1,\left[\frac{n-m}{m}\right])\right\}.$$

*For* $n \equiv 0 \pmod m$ *all sets which yield the value* $f_m(n)$ *are affinely equivalent to* S.

In other words, we conjecture that the minimal $T/t$ is attained by picking the vertices of an $(m-1)$-simplex in a hyperplane and putting nearly equal numbers of equally spaced points on parallel lines which intersect this hyperplane at the vertices of the simplex. The difficulty in trying to follow the steps which lead to the proof in case $m = 2$ is that Lemma 1' yields a relatively small fraction of the convex body for the maximal volume of the inscribed simplex when $m > 2$ and that we have no useful available analogue to Lemma 2.

We should also add that Problems 1 and 3 have been considered under the assumption that the set $S$ of $n$ points is in general position (no 3 points collinear, no 4 points coplanar etc.). This leads to problems of a rather different nature where exact extremal solutions appear to be very difficult to attain and even the correct asymptotic values present formidable difficulties.

Finally we return briefly to Problem 2 and its generalization :

CONJECTURE 5. *Let* $S$ *be a set of* $n$ *points in* $E^m$ *not all in one hyperplane, and let* $g_m(n)$ *be the minimal number of nonzero volumes of* m-*simplices with vertices in such an* $S$. *Then* $g_m(n) = \left\lceil \frac{n-1}{m} \right\rceil$, *attained for the set* $S$ *described in Conjecture 4.*

Problem 2 and Conjecture 5 naturally lead to the following which seems a very interesting problem in its own right.

PROBLEM 5. *Characterize the sets* $S$ *of* $n$ *points in* $E^m$ *not all in one hyperplane, so that for every hyperplane* $H$ *determined by* $m$ *points of* $S$ *there are no more than* $\left\lceil \frac{n-1}{m} \right\rceil$ *parallel hyperplanes passing through points of* $S$ *on one side of* $H$.

In case $m = 2$ the conditions in Problem 5 are satisfies by (affine equivalents of) the set $S$ described in the Theorem and also by the vertices of (affinely) regular $n$-gons. In case $n$ is odd we also get the vertices of an (affinely) regular $(n-1)$-gon plus its center. Perhaps this list completes the solution of Problem 5 for $m = 2$. If so it would yield the conjectured answer to Problem 2.

# DISTANCE-REGULAR GRAPHS

05C25

D.E. Taylor and Richard Levingston

Department of Pure Mathematics, University of Sydney,
Sydney, N.S.W., 2006, Australia

ABSTRACT

Inequalities are obtained between the various parameters of a distance-regular graph. In particular, if $k_1$ is the valency and $k_2$ is the number of vertices at distance two from a given vertex, then in general $k_1 \leq k_2$. For distance-regular graphs of diameter at least four, $k_1 = k_2$ if and only if the graph is simply a circuit. However when the diameter is two or three there are distance-regular graphs other than circuits for which $k_1 = k_2$.

## 1. INTRODUCTION

Let $\Gamma$ be a connected graph of diameter d. We shall think of $\Gamma$ as a symmetric irreflexive relation (called *adjacency*) on a set V of *vertices*. For $\alpha \in V$ let $\Gamma_i(\alpha)$ denote the set of vertices at distance i from $\alpha$. Our initial assumption means that for each $\alpha$, V is the disjoint union of the sets $\Gamma_0(\alpha)$, $\Gamma_1(\alpha), \ldots, \Gamma_d(\alpha)$ and that for some $\alpha$, $\Gamma_d(\alpha)$ is not empty.

The graph $\Gamma$ is said to be *distance-regular* if for all $\alpha \in V$ and all $\beta \in \Gamma_i(\alpha)$, $i \geqslant 0$, the numbers $b_i = |\Gamma_{i+1}(\alpha) \cap \Gamma_1(\beta)|$ and $c_i = |\Gamma_{i-1}(\alpha) \cap \Gamma_1(\beta)|$ depend only on i.

In particular, $\Gamma$ is a regular graph of *valency* $k = b_0$. The non-zero numbers in the above collection comprise the *intersection array* of $\Gamma$ which we write in the form

$$\iota(\Gamma) = \{k, b_1, b_2, \ldots, b_{d-1}; 1, c_2, \ldots, c_d\} \tag{1}$$

As an example, consider the graph depicted in Figure 1; its intersection array is $\{3,2;1,1\}$.

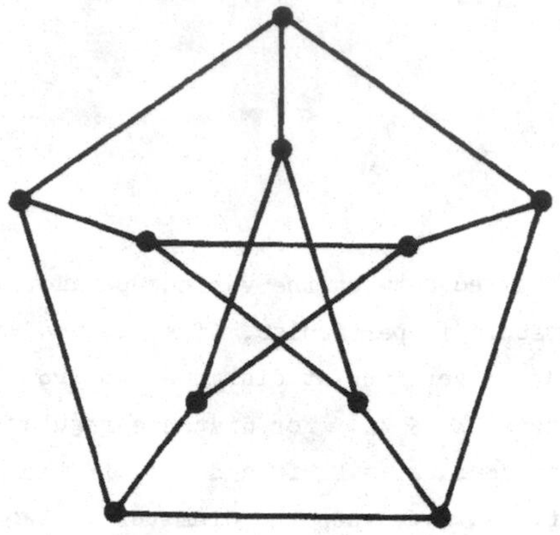

Figure 1. Petersen's graph

This concept was presented by Biggs in [1] as a generalization of the notion of a distance-transitive graph and it also occurs in the work of Higman [3] and Delsarte [2] under other names. A convenient reference for the general theory of distance-regular graphs is [1]. In this paper we shall investigate certain inequalities which hold between the entries of $\iota(\Gamma)$. We begin by introducing some auxiliary parameters.

For $\beta \in \Gamma_i(\alpha)$, let $a_i = |\Gamma_i(\alpha) \cap \Gamma_1(\beta)|$. Then $a_i$ depends only on i since $a_i + b_i + c_i = k$ and consequently we have $k \geq b_i + c_i$. Put $k_i = |\Gamma_i(\alpha)|$ and observe that the number of edges between $\Gamma_{i-1}(\alpha)$ and $\Gamma_i(\alpha)$ can be counted in two ways to yield

$$k_i c_i = k_{i-1} b_{i-1}, \quad \text{for } 1 \leq i \leq d . \tag{2}$$

The numbers $k_1, k_2, \ldots, k_d$ are called the *subdegrees* of $\Gamma$.

If $\beta \in \Gamma_h(\alpha)$ we put $a_{ijh} = |\Gamma_i(\alpha) \cap \Gamma_j(\beta)|$. It is not hard to show that these numbers can be computed from $\iota(\Gamma)$ and depend only on i, j and h (Biggs [1]). We have the following extension of (2):

$$a_{ijh}\, k_h = a_{ihj}\, k_j = a_{hji}\, k_i. \tag{3}$$

## 2. INEQUALITIES

Throughout this section let $\Gamma$ be a distance-regular graph with intersection array (1).

Proposition 1.

(i) $k > b_1 \geq b_2 \geq \ldots \geq b_{d-1}$.

(ii) $1 \leq c_2 \leq c_3 \leq \ldots \leq c_d$.

(iii) *If* $i + j \leq d$, *then* $c_i \leq b_j$.

Proof. Parts (i) and (ii) of the Proposition occur as Lemma 1 of [6]. For part (iii) choose vertices $\alpha$, $\beta$ and $\gamma$ so that $\beta \in \Gamma_i(\alpha)$ and $\gamma \in \Gamma_{i+j}(\alpha) \cap \Gamma_j(\beta)$. Then we have $\Gamma_{i-1}(\alpha) \cap \Gamma_1(\beta) \subseteq \Gamma_{j+1}(\gamma) \cap \Gamma_1(\beta)$ and hence $c_i \leq b_j$.

Remark. The antipodal graphs are characterized by the relations $b_i = c_{d-i}$, $0 \leq i \leq d-1$ and $i \neq [\frac{1}{2}d]$. (See Biggs [1] or Smith [6] for a definition of *antipodal*.)

Proposition 2.

(i) *If* $i + j \leq d$ *and* $i < j$, *then* $k_i \leq k_j$. *Moreover, if* $k_i = k_j$, *then* $k_{i+1} = k_{j-1}$ *and so on.*

(ii) *For some integer* $h \geq \frac{1}{2}d$ *we have* $k_1 \leq k_2 \leq \ldots \leq k_h$ *and* $k_h \geq k_{h+1} \geq \ldots \geq k_d$.

Proof. (i) We have $k_j = k_i \frac{b_i}{c_{i+1}} \cdot \frac{b_{i+1}}{c_{i+2}} \cdots \frac{b_{j-1}}{c_j}$ and therefore $k_j = k_i \frac{b_i}{c_j} \frac{b_{i+1}}{c_{j-1}} \cdots \frac{b_{j-1}}{c_{i+1}}$. By Proposition 1(iii) we have $k_j \geq k_i$ and equality holds if and only if $b_i = c_j$, $b_{i+1} = c_{j-1}, \ldots, b_{j-1} = c_{i+1}$. Thus $k_i = k_j$ implies $k_{i+1} = k_{j-1}$.

(ii) We have already observed that $k_{i+1} = k_i \frac{b_i}{c_{i+1}}$ and from Proposition 1(i) and 1(ii) we have $\frac{b_i}{c_{i+1}} \geq \frac{b_{i+1}}{c_{i+1}}$, whence (ii) holds for some h. It follows from (i) that it is possible to choose h so that $h \geq \frac{1}{2}d$.

Proposition 3.

$$k \leq b_1 + b_i + c_{i+1} - 1 \quad \text{for} \quad 1 \leq i \leq d - 1.$$

Proof. Choose vertices $\alpha$, $\beta$ and $\gamma$ so that $\beta \in \Gamma_i(\alpha)$ and $\gamma \in \Gamma_{i+1}(\alpha) \cap \Gamma_1(\beta)$. There are $c_{i+1}$ vertices of $\Gamma_i(\alpha)$ joined to $\gamma$. The remaining $k - c_{i+1}$ vertices joined to $\gamma$ lie in $\Gamma_{i+1}(\alpha)$ and $\Gamma_{i+2}(\alpha)$; at most $b_1$ of these are in $\Gamma_2(\beta)$, the others are in $\Gamma_1(\beta) \cap \Gamma_{i+1}(\alpha)$. It follows that $k - c_{i+1} \leq b_1 + (b_i - 1)$.

Remarks. (1) Proposition 1(iii) and Proposition 3 sharpen results of Biggs [1; p.139].

(2) The graphs for which $k = 2b_1 + c_2 - 1$ have been determined. Apart from the icosahedron the only possibilities are line graphs of certain graphs with no triangles. (See Figures 2, 3 and 4 for some small examples.)

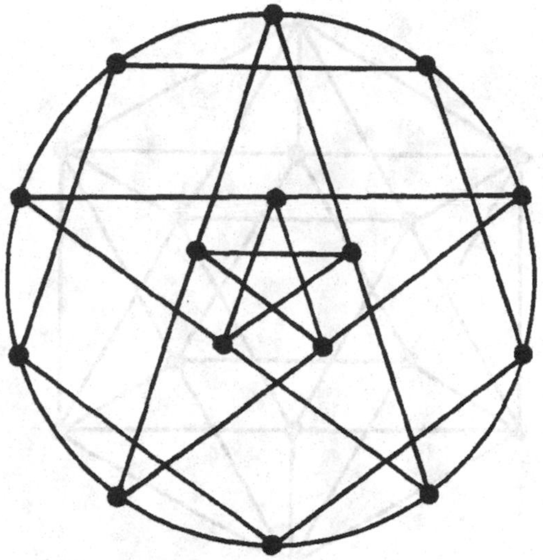

Figure 2. The line graph of Petersen's graph.

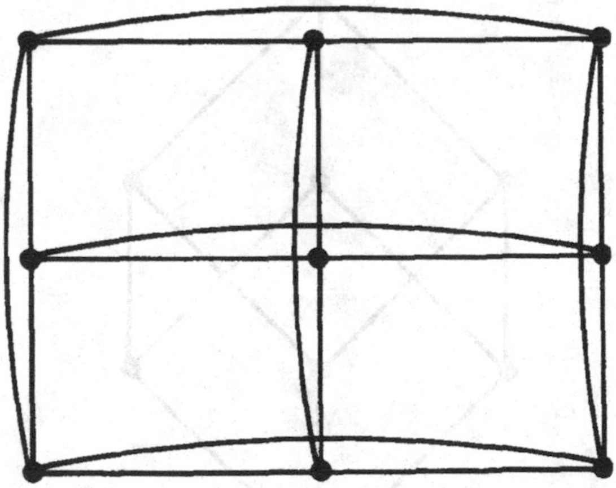

Figure 3. The line graph of $K_{3,3}$.

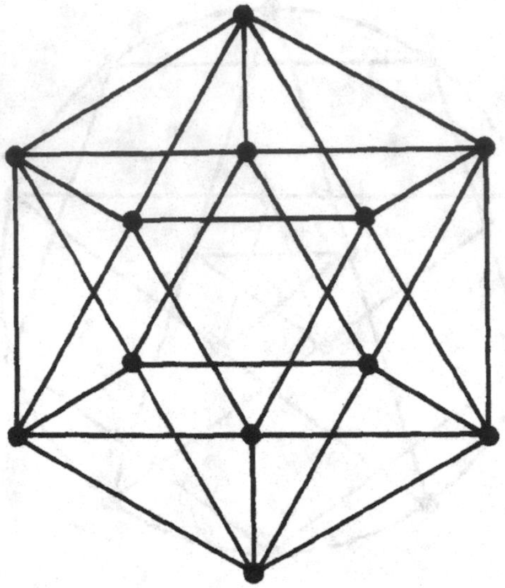

Figure 4. The icosahedron.

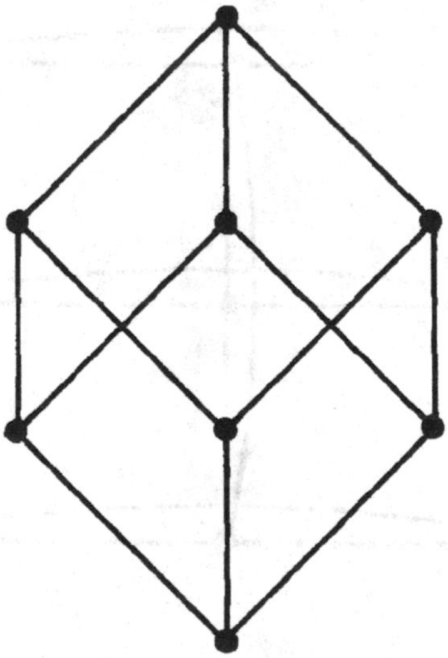

Figure 5. The cube.

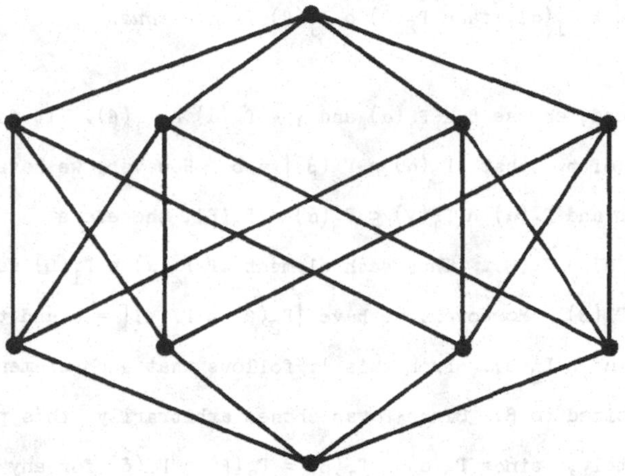

Figure 6. A double cover of $K_5$.

3. CHARACTERIZATION OF LARGE CIRCUITS

A distance-regular graph of valency two is simply a circuit. In this section we characterize circuits by means of their subdegrees provided the diameter of the graph is at least four. The reason for this restriction on the diameter will become apparent in the next section. In the proof the word *clique* denotes a complete subgraph.

<u>Theorem</u>. *If* $\Gamma$ *is a distance-regular graph of diameter at least four, then* $\Gamma$ *is a circuit if and only if* $k_1 = k_2$.

<u>Proof</u>. Suppose that the intersection array of $\Gamma$ is given by (1) and that $k_1 = k_2$. It follows from Proposition 1 that $k_1 = k_3$, hence $b_1$, $b_2$, $c_2$ and $c_3$ are all equal and we may denote their common value by b. The proof depends on establishing the following facts:

(a) *if* $\gamma \in \Gamma_2(\alpha)$, *then* $\Gamma_1(\alpha) \cap \Gamma_1(\gamma)$ *is a clique.*

(b) *if* $\beta \in \Gamma_1(\alpha)$, *then* $\Gamma_2(\alpha) \cap \Gamma_1(\beta)$ *is a clique.*

To this end, choose $\beta \in \Gamma_1(\alpha)$ and $\gamma \in \Gamma_2(\alpha) \cap \Gamma_1(\beta)$. It follows from (3) and the definition of $b_2$, that $|\Gamma_3(\alpha) \cap \Gamma_2(\beta)| = b$. However, we have $|\Gamma_3(\alpha) \cap \Gamma_1(\gamma)| = b$ and $\Gamma_3(\alpha) \cap \Gamma_1(\gamma) \subseteq \Gamma_3(\alpha) \cap \Gamma_2(\beta)$, therefore $\Gamma_3(\alpha) \cap \Gamma_1(\gamma) = \Gamma_3(\alpha) \cap \Gamma_2(\beta)$. Thus each element of $\Gamma_2(\alpha) \cap \Gamma_1(\beta)$ is joined to each element of $\Gamma_3(\alpha) \cap \Gamma_2(\beta)$. Moreover, we have $|\Gamma_2(\beta) \cap \Gamma_1(\gamma)| = b$ and therefore $\Gamma_2(\beta) \cap \Gamma_1(\gamma) = \Gamma_3(\alpha) \cap \Gamma_2(\beta)$. From this it follows that each element of $\Gamma_1(\alpha) \cap \Gamma_1(\gamma)$ is joined to $\beta$. Since $\beta$ was chosen arbitrarily, this proves (a). We deduce (b) immediately, since $\Gamma_2(\alpha) \cap \Gamma_1(\beta) = \Gamma_1(\beta) \cap \Gamma_1(\delta)$ for any $\delta \in \Gamma_3(\alpha) \cap \Gamma_1(\gamma)$.

For any $\delta \in \Gamma_3(\alpha) \cap \Gamma_2(\beta)$ we have $\Gamma_1(\alpha) \cap \Gamma_1(\gamma) \subseteq \Gamma_1(\alpha) \cap \Gamma_2(\delta)$ and equality holds since both these sets contain b elements. This shows that the clique $\Gamma_1(\alpha) \cap \Gamma_1(\gamma)$ does not depend on the choice of $\gamma \in \Gamma_2(\alpha) \cap \Gamma_1(\beta)$. It follows that $\Gamma_2(\alpha)$ is partitioned into cliques of the form $\Gamma_2(\alpha) \cap \Gamma_1(\beta)$, $\beta \in \Gamma_1(\alpha)$ and $\Gamma_1(\alpha)$ is similarly partitioned into cliques of the form $\Gamma_1(\alpha) \cap \Gamma_1(\gamma)$, $\gamma \in \Gamma_2(\alpha)$. In particular b divides k and therefore $k = 2b$ since from Proposition 3 we have $k \leq 3b - 1$. But now $a_2 = k - 2b = 0$ so no two vertices of $\Gamma_2(\alpha)$ are adjacent. This means that $b = 1$ and $k = 2$, hence $\Gamma$ is a circuit.

Conversely, if $\Gamma$ is a circuit of diameter at least four, then certainly $k_1 = k_2$.

## 4. GRAPHS OF DIAMETERS TWO AND THREE

If $\Gamma$ is a distance-regular graph of diameter *two* such that $k_1 = k_2$, then $b_1 = c_2 = \frac{1}{2} k_1$ and $1 + 2k_1$ must be a sum of two squares - see van Lint and Seidel [8]. In fact, there are infinitely many graphs apart from circuits which meet these conditions. A family of distance-transitive examples can be obtained by taking the finite field GF(q), q a prime power $\equiv 1 \pmod 4$, and joining two elements whenever their difference is a square. Examples with 25 and 45 vertices which are not distance-transitive can be found in Weisfeiler [9] and Mavron [4], respectively. The smallest example which is not a circuit is shown in Figure 3.

In the remainder of this section we consider distance-regular graphs of diameter *three* for which $k_1 = k_2$. Again there are infinitely many which are not circuits. In order to describe them it is useful to introduce a new concept.

A *regular 2-graph* $(\Omega,\Phi)$ consists of a set $\Omega$ and a set $\Phi$ of three-element subsets of $\Omega$ such that:

(i) every four-element subset of $\Omega$ contains an even number of elements of $\Phi$; and

(ii) every two-element subset of $\Omega$ is contained in the same number, say $a$, of elements of $\Phi$.

For the theory of 2-graphs see Seidel [5] and Taylor [7].

We construct a graph $\Gamma$ from a regular 2-graph $(\Omega,\Phi)$. The vertex set of $\Gamma$ is the disjoint union of two copies of $\Omega$. That is, for each $\alpha \in \Omega$ we adjoin a new vertex $\alpha'$. Now choose a vertex, say $\infty$, of $\Omega$ and join it to all other vertices of $\Omega$. For $\alpha, \beta \in \Omega - \{\infty\}$ join $\alpha$ to $\beta$ whenever $\{\infty,\alpha,\beta\} \in \Phi$, otherwise join $\alpha$ to $\beta'$ and $\alpha'$ to $\beta$. Finally, join $\alpha'$ to $\beta'$ whenever $\alpha$ is joined to $\beta$. Provided $\Phi$ is not the set of all three-element subsets of $\Omega$ this construction produces a distance-regular graph $\Gamma$ such that $k_1 = k_2 = n - 1$, $k_3 = 1$ and $b_1 = c_2 = k - a - 1$, where n is the number of elementsin $\Omega$. The graph $\Gamma$ does not depend on our choice of $\infty$.

As pointed out in [7] we can always obtain a regular 2-graph by adjoining a vertex to a distance-regular graph of diameter two for which $k_1 = 2c_2$. If we then perform the "doubling construction" just described we obtain a distance-regular graph $\Gamma$ in which the original graph appears as $\Gamma_1(\infty)$ and its complement appears as

$\Gamma_2(\infty)$. When this is applied to a pentagon, for example, we obtain the graph of the icosahedron (Figure 4). If we apply the "doubling construction" with $\Phi$ empty we obtain a bipartite graph (Figures 5 and 6).

<u>Theorem</u>. *If* $\Gamma$ *is a distance-regular graph of diameter three, then* $k_1 = k_2$ *if and only if* $\Gamma$ *is the "doubled graph" of a regular 2-graph or* $\Gamma$ *is a cycle of length seven.*

<u>Proof</u>. Suppose that $k_1 = k_2$. Then $b_1 = c_2 = b$, say and we may choose vertices $\alpha,\beta,\gamma$ and $\omega$ so that $\beta \in \Gamma_1(\alpha)$, $\gamma \in \Gamma_2(\alpha) \cap \Gamma_1(\beta)$ and $\omega \in \Gamma_3(\alpha) \cap \Gamma_1(\gamma)$. As in the proof of the previous theorem we have $\Gamma_3(\alpha) \cap \Gamma_2(\beta) = \Gamma_3(\alpha) \cap \Gamma_1(\gamma)$ and this set contains $b_2$ vertices. Thus every vertex of $\Gamma_2(\alpha) \cap \Gamma_1(\beta)$ is joined to every vertex of $\Gamma_3(\alpha) \cap \Gamma_2(\beta)$. In particular, $\Gamma_2(\alpha) \cap \Gamma_1(\beta) \subseteq \Gamma_1(\omega)$ and therefore $\Gamma_3(\omega) \cap \Gamma_1(\beta) \subseteq \Gamma_1(\alpha) \cup \{\alpha\}$. It follows that $\Gamma_3(\omega) \cap \Gamma_1(\beta)$ is a clique, hence $\Gamma_3(\alpha) \cap \Gamma_1(\gamma)$ is a clique as well. We aim to show that $b_2 = 1$ so suppose, by way of contradiction, that $b_2 \neq 1$ and let $\delta$ be a vertex of $\Gamma_3(\alpha) \cap \Gamma_2(\beta) - \{\omega\}$. If $\Delta$ is the set of those vertices of $\Gamma_2(\alpha)$ which are joined to all vertices of $\Gamma_3(\alpha) \cap \Gamma_2(\beta)$, then $\delta$ is joined to at least $b_2 - 2 + x$ vertices of $\Gamma_1(\omega)$, where $x = |\Delta|$. Now $|\Gamma_2(\omega) \cap \Gamma_2(\gamma)| = b$ and therefore $\Gamma_2(\omega) \cap \Gamma_1(\gamma) \subseteq \Gamma_1(\alpha)$ so that $\Gamma_2(\alpha) \cap \Gamma_1(\gamma) \subseteq \Gamma_1(\omega)$. It follows that $x \geq a_2 + 1$ and since we also have $x \leq a_1 - b_2 + 2 = a_2 + 1$ this proves that $x = a_2 + 1$. Thus $\Delta$ is a clique of size $a_2 + 1$ in $\Gamma_2(\alpha)$ and $\Delta \cup (\Gamma_3(\alpha) \cap \Gamma_2(\beta))$ is a clique of size $a_1 + 2$ in $\gamma_1(\omega) \cup \{\omega\}$. This means that $\Gamma_2(\ )$ is a disjoint union of cliques of size $a_2 + 1$ and $\Gamma_1(\alpha)$ is a disjoint union of cliques of size $a_1 + 1$. But $\Gamma_2(\omega) \cap \Gamma_1(\gamma)$ is a clique in $\Gamma_1(\alpha)$, hence $b \leq a_1 + 1$. We also have $a_1 + 1 \leq \frac{1}{2}k$ and so $k = 2b$. But now $\Gamma_2(\alpha)$ must be a union of exactly two disjoint cliques, whence $b_2 = 1$, contrary to our assumption. This proves that $b_2 = 1$. If $k = c_3$, then $k_3 = 1$ and the graph is easily seen to be the "doubled graph" of a regular 2-graph. Now suppose that $k \neq c_3$ and choose a vertex $\xi$ in $\Gamma_3(\alpha) \cap \Gamma_1(\omega)$. We know that for each vertex $\beta$ in $\Gamma_1(\alpha)$, the set $\Gamma_3(\alpha) \cap \Gamma_2(\beta)$ consists of a single vertex and for any $\gamma \in \Gamma_2(\alpha) \cap \Gamma_1(\omega)$ we have $\Gamma_2(\alpha) \cap \Gamma_1(\gamma) \subseteq \Gamma_1(\omega)$. Thus $\Gamma_1(\xi) \cap \Gamma_2(\alpha) \subseteq \Gamma_3(\gamma)$ and it follows that $c_3 = 1$. But we have $k \leq 3b - 1$ and $b \leq c_3$, hence $k = 2$ and $\Gamma$ is a cycle of length seven.

REFERENCES

[1] N. Biggs, *Algebraic Graph Theory* (Cambridge Math. Tracts. No. 67) Cambridge Univ. Press, London, 1974.

[2] P. Delsarte, "An algebraic approach to the association schemes of coding theory", *Philips Res. Repts. Suppl.* No. 10 (1973).

[3] D.G. Higman, "Coherent configurations, Part I, Ordinary representation theory", *Geometriae Dedicata 4* (1975), 1-32.

[4] Rudolf Mavron, "Symmetric conference matrices of order $pq^2 + 1$", *Canad. J. Math.* (to appear).

[5] J.J. Seidel, "A survey of two-graphs", *Proc. Intern. Coll. Teorie Combinatorie, Accad. Naz. Lincei, Roma,* (1976), 481-511.

[6] D.H. Smith, "Primitive and imprimitive graphs", *Quart. J. Math. Oxford (2) 22* (1971), 551-7.

[7] D.E. Taylor, "Regular 2-graphs", *Proc. London Math. Soc. (3) 34* (1977).

[8] J.H. van Lint and J.J. Seidel, "Equilateral point sets in elliptic geometry", *Nederl. Akad. Wetensch. Proc. Ser. A. 69 (Indag. Math. 28)*, (1966), 335-348

[9] Boris Weisfeiler, *On Construction and Identification of Graphs* (Lecture Notes in Mathematics 558), Springer-Verlag, Berlin, 1976.

# A NOTE ON BAXTER'S GENERALIZATION OF THE TEMPERLEY-LIEB OPERATORS


H.N.V. Temperley[(1)] and D.G. Rogers[(2)]



05A19, 82A05



(1) Department of Applied Mathematics, University College, Swansea, Wales, U.K.

(2) Mathematical Institute, Oxford, England, and Department of Mathematics, University of Western Australia



ABSTRACT

The number $b(n)$ of modes of connections of 2n points permissible under Baxter's generalization of the Temperley-Lieb operators is found to be

$$b(1)=1; nb(n) = \sum_{i=0}^{\left[\frac{n-1}{2}\right]} \binom{2n}{i}\binom{2n-i}{n-1-2i} + 2\sum_{i=0}^{\left[\frac{n-2}{2}\right]} \binom{2n}{i}\binom{2n-i}{n-2-2i}, \quad n \geqslant 2.$$

In particular $b(n)$ differs from the Schröder number $s_n$ for $n \geqslant 4$.


## 1. CONNECTIVE RELATIONS

A relation R on a set $X_n = \{x_i: 1 \leq i \leq n\}$ of, say points $x_i$ on a line, is a *connective relation* if and only if it is reflexive, symmetric and satisfies the 'planarity' condition that $x_s \not R x_t$ for $i \leqslant s \leqslant j < t$ whenever $x_i R x_j$, $i < j$, [4]. Temperley and Lieb considered, in [8], rows of lattice points $x_i$ connected according to a connective relation R, that is points $x_i, x_j$ are connected if and only if $x_i R x_j$, $i < j$, as representing "wave functions" arising under their operator calculus. In an alternative, but equivalent formalism, that of spin variables, these "wave functions" are represented by connecting 2n points (spin variables), $x_i$, $1 \leqslant i \leqslant 2n$, in n disjoint pairs again in a manner satisfying the planarity condition, that is by connecting pairs in accordance with a connective relation of valence exactly one (compare Errera's problem [3,4]). These are closely related to Bethe-Hulthen wave functions, see Temperley and Lieb [8].

The important feature of both reprepresentations is the planarity of the modes of connection of the points. This allows easy enumeration of the modes. In both formulations, if $c(n), n \geqslant 1$, is the number of modes of connection, with $c(0)=1$, then considering the point connected to $x_1$, we have

$$c(n) = \sum_{i=1}^{n-1} c(i)c(n-1-i) \qquad , n \geq 1 \tag{1}$$

or

$$C(x) = \sum_{n \geq 0} c(n)x^n = 1 + x(C(x))^2 \tag{2}$$

It follows that $c(n)$ is the n-th Catalan number $C_n$ [7 sequence 577] given by

$$c(n) = C_n = \frac{1}{n+1}\binom{2n}{n} \qquad , n \geq 0 \ . \tag{3}$$

## 2. BAXTER'S GENERALIZATION

Baxter [1] has proposed a generalization of the Temperley-Lieb operators (a replacement of them by "eight-vertex" operators) which introduces new "wave functions" or modes of connecting the points. More specifically, in the second of the above formulations, in addition to taking the points of $X_{2n}$ in disjoint pairs, Baxter's generalization allows us also to take the points in groups of four, so that all groups are pairwise diagonal and that the planarity condition is observed as regards any pair of points in a group. If $b(n)$, $n \geq 1$, is the number of modes of connection of $X_{2n}$ which are now permissible (taking $b(0)=1$), then $b(1)=1$, $b(2)=3$, $b(3)=11$, agreeing with the first few Schröder numbers $s_n$ [7, sequence 1163] given by

$$s_0 = 1, \ s_n = \tfrac{1}{2} \sum_{i=0}^{n} \binom{2n-i}{i} C_{n-i} \qquad , n \geq 1 \tag{4}$$

The Schröder numbers arise in the enumeration of the Schröder relations which are generalizations of connective relations [5,6]. The Baxter generalization, however, produces a different sequence, since, for example $b(4) = 46$ whereas $s_4 = 45$. The $b(n)$ may be determined as follows.

Consider a mode of connection of $X_{2n}$ arising under Baxter's generalization. Then $x_1$ is connected either (i) in a pair to, say, $x_{2i+2}$, $0 \leq i \leq n-1$; or (ii) in a quadruple to, say, $x_{2i+2}$, $x_{2(i+j)+3}$ and $x_{2(i+j+k)+4}$, $0 \leq i, j, k, i+j+k \leq n-2$. The number of possible modes of connection in the two cases is (compare (1)), (i) $b(i)\ b(n-1-i)$, $0 \leq i \leq n-1$; and (ii) $b(i)\ b(j)\ b(k)\ b(n-2-i-j-k)$, $0 \leq i, j, k, i+j+k \leq n-2$. Hence, in terms of generating functions (compare (2)),

$$B(x) = \sum_{n \geq 0} b_n x^n = 1 + x(B(x))^2 + x^2(B(x))^4 \tag{5}$$

Writing $A(x) = x(B(x))^2$, we have

$$B(x) = 1 + A(x) + (A(x))^2 = G(A(x))$$

with $$A(x) = x(1 + A(x) + (A(x))^2)^2 = xF(A(x))$$

where $$G(t) = 1 + t + t^2 \;;\; F(t) = (G(t))^2$$

By the Lagrange inversion formula [9, pp 132-3] , for $n \geq 1$

$$n\, b_{(n)} = \text{coefficient of } t^{n-1} \text{ in } G'(t)(F(t))^n$$

Now if the coefficient of $t^r$ in $(G(t))^n$ is $p(n,r)$, so that $p(n,r)$ is a trinomial coefficient [2, p.72], then $b(1) = 1$ and for $n \geq 2$,

$$n\, b(n) = p(2n,n-1) + p(2n,n-2)$$

More explicitly we find that

$$n\, b(n) = \sum_{i=0}^{\left[\frac{n-1}{2}\right]} \binom{2n}{i}\binom{2n-i}{n-1-2i} + 2\sum_{i=0}^{\left[\frac{n-2}{2}\right]} \binom{2n}{i}\binom{2n-i}{n-2-2i}, \quad n \geq 2.$$

For comparison we give the first six values of $b(n)$ and $s_n$. The sequence $\{b(n)\}$ does not appear in [7].

| $n$ | 0 | 1 | 2 | 3 | 4 | 5 | 6 |
|---|---|---|---|---|---|---|---|
| $s_n$ | 1 | 1 | 3 | 11 | 45 | 197 | 903 |
| $b(n)$ | 1 | 1 | 3 | 11 | 46 | 207 | 979 |

3. SOME FURTHER REMARKS

If $b_k(n)$ is the number of permissible modes of connection of $X_{2n}$ where now the points may be taken in disjoint groups of any even size less than 2k and the planarity condition holds for any pair of points in a group (so $b_1(n) = c(n)$, $b_2(n) = b(n)$) then the argument for [5] extends to show that, taking $b_k(0) = 1$

$$B_k(x) = \sum_{n \geq 0} b_k(n)x^n = \sum_{i=0}^{k} x^i (B_k(x))^{2i}.$$

More generally if $v(n)$ is the number of permissible modes of connection of $X_{2n}$ when the restriction on the size of the groups is lifted, with $v(0) = 1$, then

$$V(x) = \sum_{i \geq 0} x^i (V(x))^{2i}$$

or $$V(x) = 1 + x(V(x))^3$$

whence by the Lagrange inversion formulae, (compare (2)), [7, sequence 1174]

$$v(n) = \frac{1}{2n+1}\binom{3n}{n} \quad , n \geqslant 0$$

Although the Schröder numbers $s_n$ do not arise in Baxter's generalization we nevertheless obtain a representation of the Schröder numbers by allowing the points in $X_{2n}$ to be grouped in pairs or overlapping quadruples, subject to the planarity condition. More precisely we connect the points of $X_{2n}$ according to a connective relation R and then whenever $x_i \, R \, x_j$, $x_s \, R x_t$ with $i < j < s < t$ we allow points $x_j$ and $x_s$ to be connected or not subject always to the planarity condition, that is that if they are connected then no points $x_p, x_q$ are connected with $j \leqslant p \leqslant s < q$ unless $p=s$, $q=t$. If $s(n)$ is the number of these modes of connection, taking $s(0) = 1$, then compare (2), $s(1) = 1$

$$s(n) = 2\sum_{i=1}^{n-2} s(i)s(n-1-i) + s(n-1) \quad , \qquad n \geqslant 2$$

or $$S(x) = \sum s_n x^n = 1 + S(x)(2S(x)-1)$$

Writing $R(x) = 2S(x) - 1$,

$$R(x) = 1 + x\,R(x)(1 + R(x))$$

from which it follows, by the Lagrange inversion formula, that

$$s(n) = s_n \quad , \qquad n \geqslant 0 \qquad .$$

REFERENCES


[1] R. Baxter, private communication

[2] L. Comtet, Advanced Combinatories D. Reidel, Dovdrecht (1974)

[3] J. Riordan 'The distribution of crossings of chords joining pairs of 2n points on a circle', Mathematics of Computation, 29 (1975), 215-222

[4] D.G. Rogers, 'The enumeration of a family of ladder graphs Part I. Connective relations', Quart.J.Math. Oxford (2), (to appear)

[5] D.G. Rogers, 'The enumeration of a family of ladder graphs Part II. Schröder relations', (submitted).

[6] D.G. Rogers and L.W. Shapiro, 'Some correspondences involving the Schröder numbers and relations', Proceedings of International Conference on Combinatorial theory, Canberra (1977) (to appear).

[7] N.J.A. Sloane, 'A handbook of integer sequences' Academic Press, New York (1973)

[8] H.N.V. Temperley and E.H. Lieb, 'Relations between the 'percolation' and 'colouring' problem and other graph theoretical problems associated with regular planar lattices : Some exact results for the 'percolation' problem', Proc.Roy.Soc. Ser.A., 322 (1971), 251-280

[9] E.T. Whittaker and G.N. Watson, 'A course of modern analysis' 4th ed. C.U.P., Cambridge (1950).

05B99, 05B20, 05B30

# AUTOCORRELATION OF (+1,-1) SEQUENCES

Earl Glen Whitehead, Jr.

Department of Mathematics,
University of Pittsburgh,
Pittsburgh, Pennsylvania, 15260
U.S.A.

ABSTRACT

Nonperiodic autocorrelation functions of integer sequences have been studied in connection with Hadamard matrices and combinatorial designs. Here we study conditions under which distinct (+1,-1) sequences have the same nonperiodic autocorrelation function. These conditions involve the Hadamard (tensor) product of sequences and the concatenation of sequences. Generating functions for the nonperiodic autocorrelation functions are used to prove the main results of this paper.

## 1. INTRODUCTION

Turyn (5) uses (+1,-1) sequences satisfying a nonperiodic autocorrelation condition to construct new Hadamard matrices. His method involves the intermediate construction of Baumert-Hall units.

Geramita and Wallis (1,2), Robinson (3), Robinson and Seberry (4) use (+1,0,-1) sequences satisfying a nonperiodic autocorrelation condition to construct new orthogonal designs. These constructions are based in part on the work of Turyn (5).

In this paper we consider (+1,-1) sequences of length n where n is a positive integer. Let A be such a sequence.

$$A = a_0, a_1, \ldots, a_{n-1}$$

For use in generating functions, it is convenient to number the elements of a sequence from 0 to n-1, rather than from 1 to n. Associated with the sequence A, we have its generating function a(x).

$$a(x) = a_0 + a_1 x + \ldots + a_{n-1} x^{n-1}$$

We define two operations on the sequence A as follows:

$$N(A) = -a_0, -a_1, \ldots, -a_{n-1}$$

$$R(A) = a_{n-1}, \ldots, a_1, a_0$$

The sequence R(A) has the generating function $x^{n-1}a(x^{-1})$. Observe that these operations commute; NR(A) = RN(A).

We define the *nonperiodic autocorrelation function*, NPAF, of A to be the following sequence:

$$\mathrm{NPAF}(A) = \mathrm{NPAF}_1(A),\mathrm{NPAF}_2(A),\ldots,\mathrm{NPAF}_{n-1}(A)$$

where $$\mathrm{NPAF}_k(A) = \sum_{i=0}^{n-k-1} a_i\, a_{i+k} .$$

For proving theorems, it is useful to express the NPAF(A) in terms of the generating function a(x).

$$a(x)a(x^{-1}) = n + \sum_{k=1}^{n-1} \mathrm{NPAF}_k(A)\ [x^k + x^{-k}]$$

In the statement of Theorem 8, we use the *nonperiodic correlation function*, NPCF, of two length n sequences A and B.

$$\mathrm{NPCF}(A,B) = \mathrm{NPCF}_{-(n-1)}(A,B),\mathrm{NPCF}_{-(n-2)}(A,B),\ldots,\mathrm{NPCF}_{n-1}(A,B)$$

where $$\mathrm{NPCF}_k(A,B) = \begin{cases} \displaystyle\sum_{i=0}^{n-1+k} a_i\, b_{i-k} & \text{for } -(n-1) \le k \le -1 \\ \displaystyle\sum_{i=0}^{n-1-k} a_{i+k}\, b_i & \text{for } 0 \le k \le n-1 \end{cases}$$

For example, if A = +1,+1,+1,+1,-1,-1 and B = -1,+1,-1,+1,+1,-1 , then

$$\mathrm{NPCF}(A,B) = -1,0,1,0,3,0,-3,0,-1,0,1 .$$

It is useful to express NPCF(A,B) in terms of the generating functions for A and B.

$$a(x)b(x^{-1}) = \sum_{k=-(n-1)}^{n-1} \mathrm{NPCF}_k(A,B)\, x^k$$

There are two binary operations which are used on sequences. The Hadamard (tensor) product of sequences A and B is denoted $A * B$.

$$A * B = a_1b_1,a_2b_1,\ldots,a_nb_1,a_1b_2,a_2b_2,\ldots,a_nb_2,\ldots,a_1b_m,a_2b_m,\ldots,a_nb_m$$

where A has length n and B has length m. The concatenation of sequences A and B is denoted $A|B$.

$$A|B = a_1, a_2, \ldots, a_n, b_1, b_2, \ldots, b_m$$

## 2. ELEMENTARY RESULTS

LEMMA 1: NPAF(A) = NPAF(N(A)) = NPAF(R(A)) = NPAF(NR(A)).

Proof: $a(x)a(x^{-1}) = -a(x)(-a(x^{-1})) \Rightarrow$ NPAF(A) = NPAF(N(A))

$a(x)a(x^{-1}) = x^{n-1}a(x)x^{-(n-1)}a(x^{-1}) \Rightarrow$ NPAF(A) = NPAF(R(A))

where A has length n.

LEMMA 2: For n odd, there are $2^{(n+1)/2}$ (+1,-1) sequences A of length n which satisfy R(A) = A. These sequences yield at most $2^{(n-1)/2}$ distinct NPAFs.

Proof: Let $n = 2m+1$. R(A) = A implies that $a_i = a_{2m-i}$ for $0 \le i \le m-1$. For $0 \le i \le m$, there are two choices for $a_i$: +1 and -1. Thus there are $2^{m+1} = 2^{(n+1)/2}$ sequences of length n which satisfy R(A) = A. These sequences yield at most $2^{(n-1)/2}$ distinct NPAFs because N(A) $\neq$ A and NPAF(A) = NPAF(N(A)) by Lemma 1.

LEMMA 3: For n even, there are $2^{n/2}$ (+1,-1) sequences A of length n which satisfy R(A) = A and $2^{n/2}$ (+1,-1) sequences A of length n which satisfy R(A) = N(A). These $2^{n/2+1}$ sequences yield at most $2^{n/2}$ distinct NPAFs.

Proof: Let $n = 2m$. R(A) = A implies that $a_i = a_{2m-1-i}$ for $0 \le i \le m-1$. For $0 \le i \le m-1$, there are two choices for $a_i$: +1 and -1. Thus there are $2^m = 2^{n/2}$ (+1,-1) sequences of length n which satisfy R(A) = A. Now R(A) = N(A) implies that $a_i = -a_{2m-1-i}$ for $0 \le i \le m-1$. Again for $0 \le i \le m-1$, there are two choices for $a_i$. Thus there are $2^m = 2^{n/2}$ (+1,-1) sequences of length n which satisfy R(A) = N(A). These $2^{n/2+1}$ sequences yield at most $2^{n/2}$ distinct NPAFs because N(A) $\neq$ A and NPAF(A) = NPAF(N(A)) by Lemma 1.

THEOREM 4: From the $2^n$ (+1,-1) sequences of length n, there are at most f(n) distinct NPAFs where

$$f(n) = \begin{cases} \frac{1}{4}(2^n - 2^{(n+1)/2}) + 2^{(n-1)/2} & \text{for } n \text{ odd} \\ \frac{1}{4}(2^n - 2^{n/2+1}) + 2^{n/2} & \text{for } n \text{ even}. \end{cases}$$

Proof: The remaining $2^n - 2^{(n+1)/2}$ (+1,-1) length n sequences not satisfying the conditions of Lemma 2, yield at most $\frac{1}{4}(2^n - 2^{(n+1)/2})$ distinct NPAFs. The remaining $2^n - 2^{n/2+1}$ (+1,-1) length n sequences not satisfying the conditions of Lemma 3, yield at most $\frac{1}{4}(2^n - 2^{n/2+1})$ distinct NPAFs. Thus Theorem 4 follows from Lemmas 1,2, and 3.

It is interesting to compare the upper bound, f(n), with the actual number, t(n), of distinct NPAFs obtained from the $2^n$ (+1,-1) sequences of length n. See Table 1.

Table 1

| n | 1 | 2 | 3 | 4 | 5 | 6 | 7 | 8 | 9 | 10 | 11 | 12 | 13 | 14 | 15 |
|---|---|---|---|---|---|---|---|---|---|---|---|---|---|---|---|
| f(n) - t(n) | 0 | 0 | 0 | 0 | 0 | 0 | 0 | 0 | 1 | 0 | 0 | 8 | 0 | 0 | 14 |

The values of f(n) - t(n) for $1 \le n \le 15$ were obtained by computer computations. In addition to obtaining these values, incomplete computations were done for n = 16 and n = 18. These computations helped the author to discover the theorems given in the next section.

## 3. MAIN RESULTS

A careful look at Table 1 suggests that the factorization of n influences t(n). This observation leads to the following theorem:

THEOREM 5: If $n = n_1 n_2 \cdots n_k$ where $n_1, n_2, \ldots, n_k \ge 3$ and $k \ge 1$, then there exists a set of $2^{k+1}$ (+1,-1) sequences of length n, all members of which have the same NPAF.

Proof: We prove this theorem by induction on k. For k = 1, let $A_1 = +1,\ldots,+1,-1$ be a sequence of length $n_1$, composed of $n_1 - 1$ positive ones and 1 negative one. Let $S_1 = \{A_1, N(A_1), R(A_1), NR(A_1)\}$. Since $n_1 \ge 3$ implies $n_1 - 1 \ge 2$, $A_1$ and $R(A_1)$ both have at least 2 positive ones. $A_1 \ne R(A_1)$ because $a_1 \ne a_{n_1}$. $N(A_1)$ and $NR(A_1)$ both have exactly 1 positive one. $N(A_1) \ne NR(A_1)$ because $a_1 \ne a_{n_1}$. Therefore $|S_1| = 4 = 2^{k+1}$ where k = 1. By Lemma 1, all members of $S_1$ have the same NPAF.

We make the following induction assumption for $k = m-1$: There exists a set $S_{m-1}$ such that $|S_{m-1}| = 2^{(m-1)+1}$ and all members of $S_{m-1}$ have length $n_1 n_2 \ldots n_{m-1}$ *and* the same NPAF. For each (+1,-1) sequence $X \in S_{m-1}$, we construct two sequences which are members of $S_m$, namely $\phi X$ and $\psi X$.

$$\phi X = X|\cdots|X|N(X) = X * A_m$$

$$\psi X = X|N(X)|\cdots|N(X) = X * NR(A_m)$$

Here $A_m = +1,\ldots,+1,-1$ is a sequence of length $n_m$, composed of $n_m - 1$ positive ones and 1 negative one. Let

$$S_m = \{\phi X \mid X \in S_{m-1}\} \cup \{\psi X \mid X \in S_{m-1}\} .$$

Based on our induction assumption, we shall prove that $|S_m| = 2^{m+1}$.

$$|\{\phi X \mid X \in S_{m-1}\}| = |\{\psi X \mid X \in S_{m-1}\}| = |S_{m-1}| = 2^m$$

Suppose that $\phi X = \psi Y$ for some $X, Y \in S_{m-1}$. Then $(\phi X)_i = (\phi Y)_i$ for $1 \le i \le n_1 n_2 \cdots n_{m-1}$ which implies that $X = Y$. Thus $\phi X = \psi X$ which leads to a contradiction since $X \neq N(X)$. Therefore $|S_m| = 2^{m+1}$.

To complete the proof of Theorem 5, we must show that for any $U, V \in S_m$, $NPAF(U) = NPAF(V)$. This result is an application of the following lemma:

LEMMA 6: Let A and B be (+1,-1) sequences of length j where $NPAF(A) = NPAF(B)$. Let C and D be (+1,-1) sequences of length k where $NPAF(C) = NPAF(D)$. Then $NPAF(A * C) = NPAF(B * D)$.

Proof: $NPAF(A) = NPAF(B)$ implies that $a(x)a(x^{-1}) = b(x)b(x^{-1})$.

$NPAF(C) = NPAF(D)$ implies that $c(x)c(x^{-1}) = d(x)d(x^{-1})$.

$NPAF(A * C)$ is generated by $a(x)c(x^j)a(x^{-1})c(x^{-j}) = b(x)d(x^j)b(x^{-1})d(x^{-j})$ which generates $NPAF(B * D)$.

As a result of Theorem 5, we have the fact that the upper bound f(n) is not achieved whenever $n = n_1 n_2 \ldots n_k$ where $n_1, n_2, \ldots, n_k \ge 3$ and $k \ge 2$. It seems extremely difficult to obtain sharp bounds on the number of order $2^{k+1}$ $(k \geq 2)$ sets of (+1,-1) sequences, all members of which have the same NPAF. The following two theorems give an indication of why this difficulty arises.

THEOREM 7: Let $A_1, A_2, \ldots, A_k$ be (+1,-1) sequences of length n and let T and U be (+1,-1) sequences of length m where NPAF(T) = NPAF(U). Then

$$\mathrm{NPAF}(A_1 * T \mid A_2 * T \mid \cdots \mid A_k * T) = \mathrm{NPAF}(A_1 * U \mid A_2 * U \mid \cdots \mid A_k * U).$$

Proof: Let $a_i(x)$ be the generating function for $A_i$ for $1 \leq i \leq k$; t(x) for T; u(x) for U. NPAF(T) = NPAF(U) implies that $t(x)t(x^{-1}) = u(x)u(x^{-1})$. The generating function for $\mathrm{NPAF}(A_1 * T \mid A_2 * T \mid \cdots \mid A_k * T)$ is

$$\left[\sum_{i=1}^{k} a_i(x)\,x^{(i-1)mn}\right] t(x^n)t(x^{-n}) \left[\sum_{i=1}^{k} a_i(x^{-1})\,x^{-(i-1)mn}\right]$$

$$= \left[\sum_{i=1}^{k} a_i(x)\,x^{(i-1)mn}\right] u(x^n)u(x^{-n}) \left[\sum_{i=1}^{k} a_i(x^{-1})\,x^{-(i-1)mn}\right]$$

which is the generating function for $\mathrm{NPAF}(A_1 * U \mid A_2 * U \mid \cdots \mid A_k * U)$.

Example: Let $A_1$ = +1,-1 and $A_2$ = +1,+1. Let T = +1,+1,-1 and U = +1,-1,-1. Since U = NR(T), NPAF(T) = NPAF(U). Therefore $\mathrm{NPAF}(A_1 * T \mid A_2 * T) = \mathrm{NPAF}(A_1 * U \mid A_2 * U)$ where

$$A_1 * T \mid A_2 * T \;=\; + - + - - + + + + + - -$$

and

$$A_1 * U \mid A_2 * U \;=\; + - - + - + + + - - - - \;.$$

Here "+" represents positive one and "-" represents negative one.

THEOREM 8: If NPCF(A,B) = -NPCF(B,A) where A and B are (+1,-1) sequences of length n, then any sequence composed of subsequences A,N(A),B, and N(B) has the same NPAF as the sequence where the subsequences appear in reverse order and where B is replaced by N(B) and N(B) is replaced by B.

Proof: Let S denote an arbitrary sequence of length mn, composed of subsequences

A,N(A),B, and N(B). Thus S has the following decomposition:

$$S = A * T \oplus B * U$$

where "$\oplus$" means componentwise sequence addition and where T and U are (+1,0,-1) sequences of length m. These sequences, T and U, have the following properties:

$$t_i u_i = 0 \qquad \text{for all } i \in \{0,1,2,\ldots,m-1\}$$

$$t_i + u_i \neq 0 \qquad \text{for all } i \in \{0,1,2,\ldots,m-1\} .$$

Let a(x) be the generating function for A; b(x) for B; t(x) for T; u(x) for U. Now NPCF(A,B) = -NPCF(B,A) implies that $a(x)b(x^{-1}) = -b(x)a(x^{-1})$. The generating function for NPAF(S) is

$$[a(x)t(x^n) + b(x)u(x^n)][a(x^{-1})t(x^{-n}) + b(x^{-1})u(x^{-n})]$$

$$= a(x)a(x^{-1})t(x^n)t(x^{-n}) + \underline{a(x)b(x^{-1})}t(x^n)u(x^{-n}) + \underline{b(x)a(x^{-1})}u(x^n)t(x^{-n}) + b(x)b(x^{-1})u(x^n)u(x^{-n})$$

$$= a(x)a(x^{-1})t(x^n)t(x^{-n}) - b(x)a(x^{-1})t(x^n)u(x^{-n}) - a(x)b(x^{-1})u(x^n)t(x^{-n}) + b(x)b(x^{-1})u(x^n)u(x^{-n})$$

$$= [a(x)x^{m(n-1)}t(x^{-n}) - b(x)x^{m(n-1)}u(x^{-n})][a(x^{-1})x^{-m(n-1)}t(x^n) - b(x^{-1})x^{-m(n-1)}u(x^n)]$$

which is the generating function for the NPAF(A * R(T) $\oplus$ B * R(U)).

Example: Let A = +1,+1,+1,+1,-1,-1 and B = -1,+1,-1,+1,+1,-1.

Thus $a(x) = 1 + x + x^2 + x^3 - x^4 - x^5$ and $b(x) = -1 + x - x^2 + x^3 + x^4 - x^5$.

Thus

$$a(x)b(x^{-1}) = -x^{-5} + x^{-3} + 3x^{-1} - 3x - x^3 + x^5$$

$$b(x)a(x^{-1}) = +x^{-5} - x^{-3} - 3x^{-1} + 3x + x^3 - x^5$$

Therefore $a(x)b(x^{-1}) = -b(x)a(x^{-1})$ which implies that NPCF(A,B) = -NPCF(B,A).

Let S = A|N(B)|B = A * (+1,0,0) $\oplus$ B * (0,-1,1).

Thus S = + + + + - - + - + - - + - + - + + - and

A * (0,0,+1) $\oplus$ B * (1,-1,0) = + - + - - + - + - + + - + + + + - - .

ACKNOWLEDGEMENTS

Jennifer Seberry and R.J. Turyn made helpful suggestions. Computer computations for this research were done on the University of Pittsburgh's DEC-10 system.

BIBLIOGRAPHY

(1) Anthony V. Geramita and Jennifer Seberry Wallis, "Orthogonal designs II", *Aequationes Math.* 13 (1975), 299-313.

(2) Anthony V. Geramita and Jennifer Seberry Wallis, "Orthogonal designs III", *Utilitas Math.* 6 (1974), 209-236.

(3) Peter J. Robinson, *Concerning the existence and construction of orthogonal designs*, Ph.D. Thesis, Australian National University, Canberra, 1977.

(4) Peter J. Robinson and Jennifer Seberry, "Orthogonal designs in powers of two", *Ars Combinatoria* (to appear).

(5) R.J. Turyn, "Hadamard matrices, Baumert-Hall units, four-symbol sequences, pulse compression, and surface wave encodings", *J. Combinatorial Th.* (Ser.A) 16 (1974), 313-333.

05C30

# TRIANGLES IN LABELLED CUBIC GRAPHS

N. C. Wormald
Department of Mathematics,
University of Newcastle,
New South Wales, 2308,
Australia

ABSTRACT

No cubic graph has an odd number of points. A method is found of calculating the number $t_p$ of labelled connected cubic graphs with 2p points rooted at a triangle. The method presupposes knowledge of the numbers $q_k$ of labelled connected cubic graphs with 2k points and $k < p$. Labelled connected cubic graphs have already been counted by Read, so this allows determination of the mean number $t_p/q_p$ of triangles in a labelled connected cubic graph with 2p points, for all $p > 1$. It is shown that $t_p/q_p \to \frac{4}{3}$ as $p \to \infty$.

Labelled cubic graphs were first counted by Read, who in [4] and [5] provides two different solutions to the problem. In the latter paper, the solution takes the form of a recurrence relation satisfied by the number of labelled cubic graphs with 2p points. A cubic graph has an even number of points, so this solution is complete. More recently [7], we derived the same recurrence relation using a different method which involves consideration of graphs which are cubic except for one or two points of degree 2. It will be shown that this technique can be used to count labelled connected cubic graphs rooted at a triangle. The numbers of such graphs with 2p points are tabulated for p up to 10.

This type of question has already been solved for graphs in general. The enumeration of labelled graphs rooted at a given subgraph is quite elementary, but the corresponding problem for unlabelled graphs requires the more advanced enumeration techniques of Harary [1] and Robinson [6]. A full discussion of the problem is to be found in [3, Sections 4.4 and 4.5].

In [5], Read also shows how to calculate the number $q_p$ of labelled connected cubic graphs on 2p points. Once we have calculated the number $t_p$ of labelled connected cubic graphs on 2p points rooted at a triangle, division by $q_p$ yields the mean number of triangles in a labelled connected cubic graph on 2p points. Numerical values of the quantity $t_p/q_p$ are tabulated for selected values of p up to p = 29. The rate of growth of the numbers $q_p$ is then used to show that $t_p/q_p \to \frac{4}{3}$

as $p \to \infty$. Also discussed is the problem of finding the average number of triangles in a labelled cubic graph which is not necessarily connected. All basic graph theoretic notation can be found in the book [2].

We begin by quoting a result derived in [7] that will be useful for our present purposes. A *closely cubic* graph is a graph in which one point has degree 2 and the rest have degree 3. Each such graph has an odd number of points, so for $p > 0$ let $c_p$ denote the number of labelled connected closely cubic graphs with $2p + 1$ points. Next, let $q = q(x)$ be the exponential generating function for labelled connected cubic graphs, that is, $q(x) = \sum_{p=0}^{\infty} \frac{q_p}{(2p)!} x^{2p}$. Similarly define $t = t(x) = \sum_{p=0}^{\infty} \frac{t_p}{(2p)!} x^{2p}$ and $c = c(x) = \sum_{p=0}^{\infty} \frac{c_p}{(2p+1)!} x^{2p+1}$. The result required from [7, Section 3] is the equation

$$(1) \qquad 3x^2q' = (2 - 2x^2 - x^4)c,$$

where $q'$ denotes the derivative of $q$ with respect to $x$.

The first step in enumerating labelled connected cubic graphs rooted at a triangle is to express the exponential generating function $t(x)$ in terms of $c(x)$ and $q(x)$. Equation (1) is then used to eliminate $c(x)$. Equating coefficients of $x^{2p}$ in the resultant equation expresses $t_p$ in terms of the numbers $q_k$ for which $k < p$.

THEOREM 1. The generating functions $t$, $q$ and $c$ are related by

$$(2) \qquad t = x^3q' + (x^3 + \tfrac{1}{2}x^5)c + \frac{1}{6}x^4,$$

where $q'$ denotes the derivative of $q$ with respect to $x$.

PROOF. Let $G$ be a labelled connected cubic graph on $2p$ points rooted at a triangle, called the root triangle. The various possibilities for $G$ are enumerated by the exponential generating function $t(x)$. The proof is effected by counting the number of choices for $G$ in a different way. To do this we consider four mutually exclusive possibilities for the structure of that part of $G$ which is near the root triangle.

A point $u$ of a graph $F$ is *adjacent* to a subgraph $E$ of $F$ if $u$ is adjacent to one of the points of $E$ but is not a point of $E$.

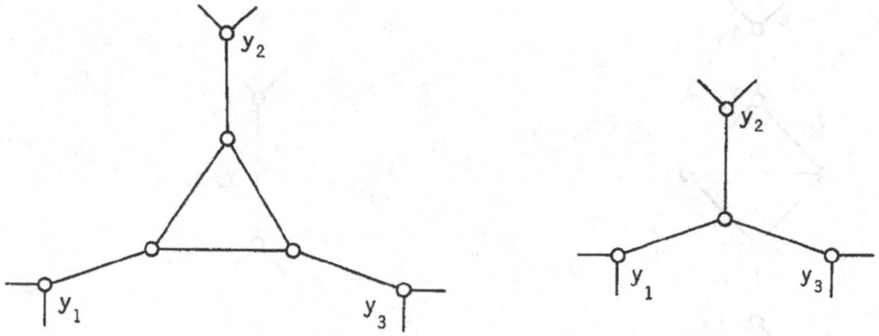


Figure 1. The process of shrinking a triangle in a cubic graph.

*CASE* 1. *There are three distinct points adjacent to the root triangle of G.* The root triangle of G is then of the type shown in the left side of Figure 1. Distinguish one of the points, called the root point, of the root triangle of G to obtain the graph G'. Denote the other points of the root triangle of G' by u and v. Next, obtain the rooted graph H' from G' by shrinking the root triangle to a single point, which is made the root point of H' and inherits the label of the root point of G'. The process of shrinking the triangle is represented in Figure 1. The labels of the points of H' lie in the range 1,2,...,2p with two integers missing, so H' can be converted to a labelled graph H on 2p-2 points by compressing the labels into the range 1,2,...,2p-2.

H is clearly a labelled rooted cubic graph on 2p-2 points. Furthermore, H is connected because G is connected. Thus, there are $(2p-2)q_{p-1}$ possibilities for H, one for each point in a labelled connected cubic graph on 2p-2 points. Each of these possibilities corresponds to $\binom{2p}{2}$ possibilities for H', one for each choice of the two missing labels. Then, for each graph H' there are 6 possibilities for G', corresponding to the 6 permutations of the labels on the points in the root triangle of G'. Thus, there are $6\binom{2p}{2}(2p-2)q_{p-1}$ possibilities for the graph G'. Finally, in the formation of G' from G there are just 3 choices for the root point. Hence, there are $2p(2p-1)(2p-2)q_{p-1}$ possibilities for G which satisfy the condition that three distinct points be adjacent to the root triangle. Note that this number is (2p)! times the coefficient of $x^{2p}$ in $x^3q'(x)$. Therefore, $x^3q'(x)$ is the exponential generating function for the possibilities for G in this case.

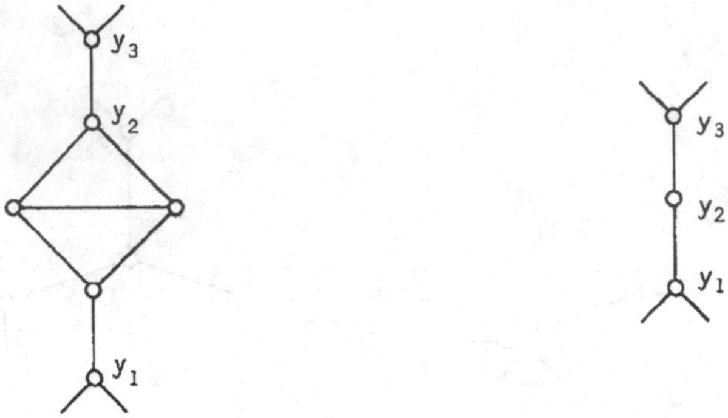


Figure 2. The removal of a triangle from a cubic graph.

*CASE 2. There are just two distinct points adjacent to the root triangle of G, these two points being non-adjacent to each other.*

The points around the root triangle of G are then as shown in the left side of Figure 2. Let $y_1$ and $y_2$ denote the points adjacent to the root triangle of G. Form the graph H from G by removing the three points in the root triangle of G and adding the line $\{y_1, y_2\}$, as shown in Figure 2. It is clear that H is a connected closely cubic graph, and by an argument similar to that used in case 1, it can be seen that there are $2p(2p-1)(2p-2)c_{p-2}$ possibilities for the graph G in the present case. Thus, the exponential generating function corresponding to G is $x^3c(x)$.

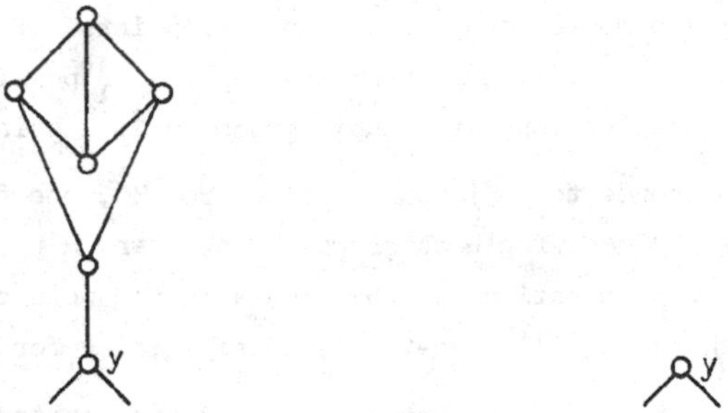


Figure 3. Another removal process sometimes applicable to cubic graphs.

*CASE 3. There are just two distinct points adjacent to the root triangle of G, these two points being adjacent to each other.*

In this case the portion of G near the root triangle is represented by the left side of Figure 3. If the root triangle and the two points adjacent to it are removed from G, as shown in the figure, the resultant graph is closely cubic and connected. When this process is analysed in the same manner as used in the previous cases, it is found that there are $p(2p-1)(2p-2)(2p-3)(2p-4)c_{p-3}$ possibilities for G. The corresponding exponential generating function is therefore $\frac{1}{2}x^5c(x)$.

*CASE 4. There is just one point adjacent to the root triangle of G.*
In this case, since G is connected it is necessarily a labelled version of $K_4$ rooted at a triangle. Hence, p must be 2, and there are four possibilities for G. As G has 4 points, the corresponding exponential generating function is $\frac{1}{6}x^4$.

Since just one of the four cases must hold, it follows that $t(x)$ is the sum of the exponential generating functions corresponding to G in each of the four cases, and we have the theorem.

On multiplying (2) by $(2-2x^2-x^4)$ and using (1) to eliminate $c(x)$ we obtain

$$(3)\quad (2 - 2x^2 - x^4)(t - \frac{1}{6}x^4) = (2x^3 + x^5 + \tfrac{1}{2}x^7)q'.$$

It is convenient at this point to ignore the cubic graph with 4 points. Accordingly, define $t^* = t - \frac{1}{6}x^4$ and for all p let $t^*_p$ be $(2p)!$ times the coefficient of $x^{2p}$ in $t^*$. Equating coefficients of $(2p)!x^{2p}$ in (3) now produces

$$(4)\quad t^*_p = 2\binom{2p}{2}t^*_{p-1} + 12\binom{2p}{4}t^*_{p-2} + 6\binom{2p}{3}q_{p-1} + 60\binom{2p}{5}q_{p-2} + 1260\binom{2p}{7}q_{p-3}.$$

The numbers $q_k$ are easily calculated using the recursive method of Read as appears in [5], and a table of these numbers for small values of k appears in [7]. Once $q_k$ is known for all k up to p-1, equation (4) together with the initial conditions $t^*_0 = t^*_1 = 0$ can be used to find $t^*_k$ for $2 \le k \le p$. The numbers $t_k$ are identical to the $t^*_k$ except for $t_2 = 4$. Table 1 lists these numbers for k up to 10.

The fraction $t_p/q_p$ is the mean number of triangles in a labelled connected cubic graph on 2p points. Approximate values of this quantity are supplied in Table 2 for selected values of p up to 29. *

---

* The author is grateful to the Australian Research Grants Committee for providing the support necessary for the programming of all the numerical work, by way of a grant to Professor R.W. Robinson. Programming was performed by Albert Nymeyer.

Table 1. The number $t_p$ of labelled connected cubic graphs on 2p points rooted at a triangle.

| $t_p$ | p |
|---:|---:|
| 0 | 1 |
| 4 | 2 |
| 120 | 3 |
| 33 600 | 4 |
| 18 446 400 | 5 |
| 18 361 728 000 | 6 |
| 30 199 104 936 000 | 7 |
| 76 326 119 565 696 000 | 8 |
| 280 889 824 362 219 072 000 | 9 |
| 1 443 428 429 045 578 335 360 000 | 10 |

Table 2. Approximate values of the mean number $t_p/q_p$ of triangles in a labelled connected cubic graph with 2p points.

| $t_p/q_p$ | p |
|---|---:|
| 4 | 2 |
| 1.714 | 3 |
| 1.739 | 4 |
| 1.652 | 5 |
| 1.591 | 6 |
| 1.549 | 7 |
| 1.519 | 8 |
| 1.497 | 9 |
| 1.479 | 10 |
| 1.428 | 15 |
| 1.403 | 20 |
| 1.389 | 25 |
| 1.381 | 29 |

From Table 2 it appears that $t_p/q_p$ approaches a constant as p increases. Our next aim is to prove this.

THEOREM 2. The mean number $t_p/q_p$ of triangles in a labelled connected cubic graph on 2p points approaches $\frac{4}{3}$ as $p \to \infty$.

PROOF. The notation $O(\ )$ refers to the passage of p to infinity. Let $d_p$ denote the number of labelled cubic graphs on 2p points, with the convention that $d_0 = 1$. Then $1 - q_p/d_p$ is the fraction of labelled cubic graphs on 2p points which are disconnected. In [7] it was shown that

$$p(2p+1)(3p+3) < d_{p+1}/d_p < p(2p+1)(3p+5) \text{ for } p \geq 3, \text{ and that}$$

$$(5) \qquad 1 - q_p/d_p = O(p^{-2}).$$

As direct consequences we have that $d_{p+1}/d_p = 6p^3 + O(p^2)$ and $q_p/d_p = 1 + O(p^{-2})$, and hence

$$(6) \qquad q_{p+1}/q_p = 6p^3 + O(p^2).$$

Division of (4) by $q_p$ yields

$$t^*_p/q_p = O(p^2)t^*_{p-1}/q_p + O(p^4)t^*_{p-2}/q_p$$
$$+ 6\binom{2p}{3}q_{p-1}/q_p + O(p^5)q_{p-2}/q_p + O(p^7)q_{p-3}/q_p.$$

In view of (6) this implies

$$t^*_p/q_p = \frac{4}{3} + O(p^{-1})t^*_{p-1}/q_{p-1} + O(p^{-2})t^*_{p-2}/q_{p-2}.$$

It follows that $t^*_p/q_p = \frac{4}{3} + O(p^{-1})$. Since $t^*_p = t_p$ for $p > 2$, the theorem is established.

Now that we have counted labelled connected cubic graphs rooted at a triangle, it is a simple matter to count these graphs when the restriction of connectedness is removed. Let $d(x) = \sum_{p=0}^{\infty} \frac{d_p}{(2p)!} x^{2p}$, so that d(x) is the exponential generating function for labelled cubic graphs. If G is a cubic graph rooted at a triangle, then G is the disjoint union of a labelled connected cubic graph rooted at a triangle and a cubic graph, where the latter is taken as the null-graph if G is connected. Thus, by the Labeled Counting Lemma [3,p.8] the exponential generating function for labelled cubic graphs rooted at a triangle is given by t(x)d(x). Alternatively, the number $\overline{t}_p$ of labelled cubic graphs on 2p points rooted at a triangle can be calculated by a method which is analogous to that used in the connected case but which allows the graphs under consideration to be disconnected.

From Table 2 it is apparent that, on the average, the more points a connected cubic graph has the fewer triangles it will contain, although this trend is only slight if the number of points is large. Consequently, the mean number $t_p/q_p$ of triangles in a labelled connected cubic graph with 2p points tends to be less than the corresponding number $\overline{t}_p/d_p$ for all labelled cubic graphs with 2p points. However, the following result demonstrates that the difference between these numbers is negligible when the number of points is large.

THEOREM 3. The mean number $\overline{t}_p/d_p$ of triangles in a labelled cubic graph on 2p points approaches $\frac{4}{3}$ as $p \to \infty$.

PROOF. Since a cubic graph on 2p points contains at most 2p triangles, there are at most $2p(d_p - q_p)$ triangles amongst the labelled disconnected cubic graphs on 2p points. On the other hand, the total number of triangles amongst the labelled connected cubic graphs on 2p points is $t_p$. Hence, the mean number $\overline{t}_p/d_p$ of triangles in a cubic graph on 2p points is at most

$$\frac{t_p + 2p(d_p - q_p)}{d_p} = \frac{t_p}{d_p} + 2p\left(1 - \frac{q_p}{d_p}\right)$$

which is just $t_p/q_p + O(p^{-1})$ in view of (5). Theorem 2 completes the proof.

The calculation of the mean number of triangles in a labelled graph with 2p points and 3p lines provides an interesting comparison to Theorem 3. Let G be one such graph rooted at a triangle. There are $\binom{2p}{3}$ possibilities for the labels of the root triangle of G, and then $\binom{q-3}{3p-3}$ ways to arrange the remaining $3p - 3$ lines of G, where $q = \binom{2p}{2}$. On the other hand, there are just $\binom{q}{3p}$ labelled graphs with 2p points and 3p lines. Hence, the average number of triangles in such a graph is $\binom{2p}{3}\binom{q-3}{3p-3}/\binom{q}{3p}$, which is $\frac{9}{2} + O(p^{-1})$. It is therefore seen that the average number of triangles in a labelled graph with 2p points and 3p lines is severely reduced when the restriction is imposed that each point have degree 3.

The methods of the present paper can also be used to find the numbers of labelled cubic graphs rooted at subgraphs other than triangles. However, if this more general problem is to be considered, it is often simpler to find asymptotic rather than exact answers. Elsewhere we perform the asymptotic enumeration of labelled cubic graphs rooted at a subgraph isomorphic to any given graph. Unfortunately, our present methods do not seem to be applicable to the analogous problems concerning unlabelled cubic graphs.

REFERENCES

[1] F. Harary, The number of dissimilar supergraphs of a linear graph, *Pacific J. Math.* 7 (1957) 903-911.

[2] F. Harary, *Graph Theory*. Addison-Wesley, Reading, Mass., 1969.

[3] F. Harary and E.M. Palmer, *Graphical Enumeration*. Academic Press, New York, 1973.

[4] R.C. Read, The enumeration of locally restricted graphs II, *J. London Math. Soc.* 35 (1960) 344-351.

[5] R.C. Read, Some unusual enumeration problems, *Annals N.Y. Acad. Sci.* 175 (1970) 314-326.

[6] R.W. Robinson, Enumeration of colored graphs, *J. Combinatorial Theory* 4 (1968) 181-190.

[7] N. Wormald, Enumeration of labelled graphs II: Cubic graphs with a given connectedness, to appear.

## PROBLEMS

1. A problem on duality (Blanche Descartes)

Once a jolly graphman drew himself a dual graph,
Mate of a graph he referred to as G.
And he prayed as he rubbed an edge away from each of them
Send us a problem, Blanche Descartes and me.
Send us a problem, send us a problem,
Send us a problem, Blanche Descartes and me.
The edges he rubbed were corresponding dual ones,
Up jumped the graphman, shouting with glee,
And he sang as he noted the remnants isomorphism,
Send us a problem, Blanche Descartes and me.
Send us a problem, etc.,
Up came the problem; must such isomorphism,
Always extend to the mate and to G?
And he cried as he tore his hoary hair in puzzlement.
We've got a problem, Blanche Descartes and me!
We've got a problem, etc.,

2. Perfect matroid designs (M. Deza)

A matroid of rank r on a finite set V is a *perfect matroid design* (PMD) if for some integer i, $0 \le i \le r$, the cardinality $|F^i| = \alpha_i$ of any flat $F^i$ of rank i depends only on i. We suppose $\alpha_0 = 0$, $\alpha_1 = 1$ and denote $\alpha_{r-1} = k$, $\alpha_r = |V| = r$. Steiner systems S(t, k, r), affine and projective geometries and their truncations are PMD's.

(i) Conjecture: the hyperplane family of any PMD is *maximal*, i.e., for any subset A such that $A \subseteq V$, $|A| = k$, A is not an $F^{r-1}$, there exists a hyperplane $A' = F^{r-1}$ with $|A \cap A'| \neq \alpha_0, \alpha_1, \ldots, \alpha_{r-2}$.

(ii) The family $\{F^i\}$ of all i-flats is *complete* if for any j, $0 \le j \le i$, there exists two i-flats $F_1^i$, $F_2^i$ with $|F_1^i \cap F_2^i| = \alpha_j$; $\{F^i\}$ is *maximum* if for any family $\{A_j\}$ with $A_j \in V$, $|A_j| = \alpha_i$, $|A_{j_1} \cap A_{j_2}| \in \{\alpha_0\ \alpha_1, \ldots, \alpha_i\}$ we have $|\{A_j\}| \le |\{F^i\}|$. Find minimal i such that $0 < i \le r - 1$. Find minimal i such that $1 \le i \le r - 1$ such that $\{F^i\}$ is complete (or maximum) if the PMD is AG(r, q)(or PG(2, q)). It has to be close to ½r.

(iii) A PMD is a *PMD-scheme* if the hyperplane family $\{F^{r-1}\}$ is a subscheme of a Johnson association scheme (we define $F_1^{r-1}$ and $F_2^{r-1}$ to be i-associated if $F_1^{r-1} \cap F_2^{r-2}$ is an (r - i)-flat). Examples of PMD-schemes are any PMD of rank 3 (BIBD corresponding to a strongly regular graph), any tight $S(t, k, r)$, any Möbius plane $S(3, k, k^2 - 2k + 2)$. Describe PMD-schemes. (The number of these PMD's has to be finite with respect to k.

(iv) Does there exist a PMD of rank 4 with $\alpha_i = 0, 1, 3, 13, 313$ for $i = 0, 1, 2, 3, 4$? (This is equivalent to a problem posed ten years ago by H. Hanani.)

3. Permutation graphs (R.B. Eggleton and A. Hartman)

The *permutation graph* $G(n, \mu)$ of degree n and distance $\mu$ is the graph with vertex set $S_n$, the permutations of $\{1, 2, \ldots, n\}$, two vertices being adjacent just if they differ on precisely $\mu$ elements (that is, their Hamming distance is $\mu$). This is the Cayley graph on $S_n$ generated by the permutations which derange $\mu$ elements.

It is known that $G(n, 2)$ and $G(n, 3)$ are edge transitive, whereas $G(n, 4)$ and $G(n, 5)$ are not. For any fixed $\mu \geq 6$, we know that $G(n, \mu)$ is not edge transitive when n is sufficiently large. Are there any permutation graphs with $\mu \geq 6$ which are edge transitive?

4. Tiling (P. Erdős)

(i) Given a, b, c, d, m, n natural numbers with $(a, b, c, d) = 1$, is it possible to tile an $m \times n$ rectangle using $a \times b$, $b \times c$ and $c \times d$ tiles, for all m, n sufficiently large?

(ii) Is it possible to tile an $m \times n$ rectangle using $a \times a$ and $b \times b$ tiles, for all m, n sufficiently large?

(During the conference E.G. Straus showed that $a \times a$, $b \times b$, $c \times c$ tiles will tile every sufficiently large $m \times n$ rectangle. This can be used to solve (i) above.

Immediately after the conference Straus also showed that (ii) is not possible.

The results of Straus generalise in the obious way to higher dimensions.)

5. Groups (P. Erdős and E.G. Straus)

Given a group G with finite commutator subgroup G', then there exist n (not necessarily distinct) elements $x_1, x_2, \ldots, x_n$ of G such that

$$G' = \{x_{i_1} x_{i_2} \ldots x_{i_n} : (i_1, \ldots, i_n) \in S_n\}.$$

That is, G' consists of the product of all the $x_i$ in all possible order. We know that we can choose $n < 3|G'|$. What better bounds for n in terms of $|G'|$ can be obtained? Is it possible to choose distinct $x_1, x_2, \ldots, x_n$?

6. Graphs (D.A. Holton)

If G is a graph on n vertices, determine $t_n$, the maximum number of triangles in G such that no two triangles have a common edge.

Remarks: (i) Godsil has shown that $t_n \le \frac{1}{6}n(1 + [\frac{n}{3}])$ and that this upper bound is actually obtained only for $n = 3$ $(G = K_3)$, $9(G = K_3 \times K_3)$, $15(G = \overline{L(K_6)})$ and a strongly regular graph of degree 10 on 27 vertices.

(ii) Erdős has pointed out that the problem is equivalent to the following. Consider a triple system of n elements where no two triples are allowed to have an edge in common and no 6 vertices can contain three triples. In other words the systems below are excluded. If $t_n$ is the number of such triples,

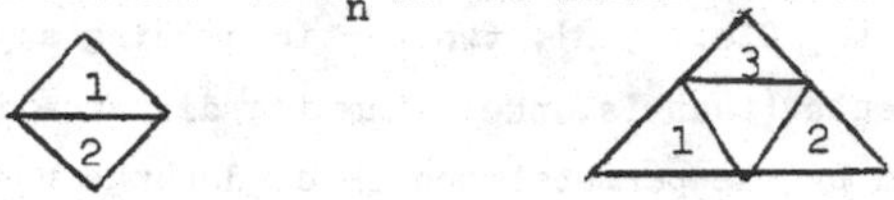


Ruzsa and Szemerédi have proved that $\frac{t_n}{n^2} \to 0$ and $t_n > \frac{Kn^2}{e^{c\sqrt{\log n}}}$. (This result will appear soon.)

Generalisation 1: (P. Erdős) Let G be a graph on n vertices and let K(m) be the complete graph on m vertices. Determine f(n, m, r), the largest number of copies of K(m) that can be contained in G, so that no two copies of K(m) have r vertices in common.

Generalisation 2: (P. Erdős) Let G be a graph on n vertices and let H be a graph on fewer than n vertices. Determine f(n; H), the largest number of copies of H that can be contained in G so that no two copies of H have a common edge.

7. Matrices (D. Nason, Capricornia Inst of Adv. Ed.)

Consider an $m \times n$ (0, 1) matrix which has precisely k or its entries equal to 1. Such a matrix is said to be *permitted* if none of its rows or columns contains any sequence of the form 010 or 101. Let $M_p$ be the number of permitted (0, 1) matrices and M the total number of (0, 1) matrices for given k, m and n. For m, $n \ge 10^7$, find (to withing 30% accuracy), $\log(M_p/M)$ as a function of k/mn.

Comment (E.G. Straus) If an m n matrix of 0's and 1's contains k ones, then the probability of an entry 1 is $\frac{k}{mn} = p$ and of an entry 0 is $q = 1 - p$. The probability that a cross 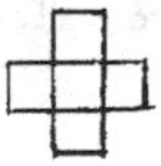 is not forbidden is $1 - p^2q - pq^2 - 2p^3q^2 - 2p^2q^3 = 1 - pq - 2p^2q^2 = r$.

Thus, under the assumption that these probabilities are independent (which is not quite correct) we would get $\sim r^{mn+o(mn)}M$ permitted matrices where the term o(mn) comes from conditions on the boundary rows and columns which are different. So $mn \log (1 - pq - 2p^2q^2)$ is probably a reasonably good estimate of $\log \frac{M_p}{M}$. Since I am not a statistician I will not estimate the correlations, but this should not be a difficult task.

Comment (A.P. Street) Note also that if we consider a partition of an $m \times n$ matrix A, with two rows and two columns removed as shown and the submatrices $A_i$ of approximately equal size, then provided that the submatrices are permitted, we can complete them to a permitted, $m \times n$ matrix. This shows an intuitive argument for the exponential nature of the problem.

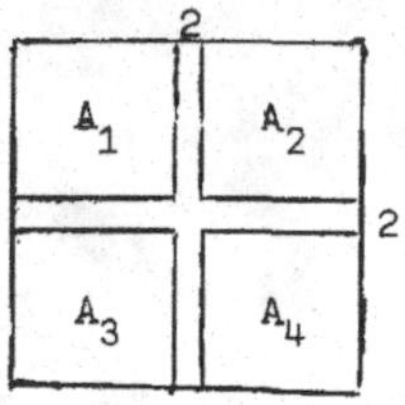


8. Weighted Graphs (E.G. Straus)

Motzkin and I proved: If we assign weights $x_1, \ldots, x_v$ to the vertices of a graph with $x_i \geq 0$, $\sum x_i = 1$ and form $f(x) = \sum x_i x_j$ over all edges $(i, j)$ of the graph, then $\max f(x) = \frac{1}{2}(1 - \frac{1}{k})$ is attained if we assign equal weights $\frac{1}{k}$ to the vertices of a maximal complete subgraph $K_k$ and 0 to all other vertices.

Generalize this as follows: Assign weights $x_1, \ldots, x_m$ with $x_i \geq 0$, $\sum x_i^{\ell+1} = 1$ to all subgraphs $K_\ell$ of a graph (if this is possible) and form $f(x) = \sum x_{i_1} x_{i_2} \ldots x_{i_{\ell+1}}$, summed over all subgraphs $K_{\ell+1}$ where the $x_{i_j}$ are the weights of the $K_\ell$ subgraphs. Then $f_\ell(x)$ attains its maximum if we assign equal weights to all the $K_\ell$ subgraphs of a maximal $K_k$ and zero weights to all other $K_\ell$

9. Latin cubes (E.G. Straus)

Three $n \times n \times n$ Latin cubes are orthogonal if the superposition yields all ordered triples. What is the maximal number of Latin cubes so that every three are orthogonal?

It is known that there exist 4 for every order $n \neq 2, 6$, $n + 2$ for $n = 2^m$ and $n + 1$ for $n = p^m$, p an odd prime.

## PARTICIPANTS

Mr. M. Adena, Australian National University, Canberra.
Professor B. Alspach, Simon Fraser University, Canada.
Professor R. Baxter, Australian National University, Canberra.
Dr. I. Beaman, University of Newcastle.
Mr. D. Billington, University of Melbourne.
Professor J.A. Bondy, University of Waterloo, Canada.
Dr. A. Brace, Canberra College of Advanced Education.
Dr. D. Breach, University of Canterbury, New Zealand.
Assoc. Professor W. Brisley, University of Newcastle.
Mr. J. Campbell, Macquarie, ACT.
Dr. R. Casse, University of Adelaide.
Dr. Chuan-Chong Chen, Nanyang University, Singapore.
Mr. K. Chidzey, University of Melbourne.
Dr. J.A. Cooper, University of Newcastle.
Professor M. Deza, C.N.R.S., Paris, France.
Mr. C. Dibley, University of Newcastle.
Dr. R. Duke, Open University, U.K.,
Mr. P. Eades, Australian National University, Canberra.
Dr. R.G. Eggleton, University of Newcastle.
Dr. I. Enting, Australian National University, Canberra.
Professor P. Erdös, Hungarian Academy of Sciences.
Mr. D. Glynn, University of Adelaide.
Mr. C. Godsil, University of Melbourne.
Mr. I.P. Goulden, University of Manitoba, Canada.
Professor M. Hall Jr., Californian Institute of Technology, U.S.A.
Mr. W.B. Hall, C.S.I.R.O., South Australia.
Professor H. Hanani, Technion, Israel.
Professor G. Harary, University of Michigan, U.S.A.
Mr. A. Hartman, University of Newcastle.
Ms. K. Heinrich, University of Newcastle.
Professor M. Herzog, Tel Aviv University, Israel.
Dr. D. Holton, University of Melbourne.
Ms. R. Hubbard, Queensland Institute of Technology.
Professor D. Hughes, Westfield College, London, U.K.
Dr. D. Hunt, University of New South Wales.
Mrs. N. Karot, University of Tasmania.
Mr. M.T. Kelly, Queensland Institute of Technology.

Mr. W.L. Kocay, University of Manitoba, Canada.
Dr. J-L. Lassez, University of Melbourne.
Mr. R. Levingston, University of Sydney.
Professor R. Lidl, University of Tasmania.
Dr. C. Little, R.M.I.T., Melbourne.
Professor C.F. Miller, University of Melbourne.
Ms. E.J. Morgan, University of Queensland.
Professor R.C. Mullin, University of Waterloo, Canada.
Dr. S.O. MacDonald, University of Queensland.
Mr. B. McKay, University of Melbourne.
Professor T.V. Narayana, University of Alberta, Canada.
Dr. B.H. Neumann, Australian National University, Canberra.
Dr. D.J. Newman, Australian National University, Canberra.
Mr. P. O'Halloran, Canberra College of Advanced Education.
Dr. D. Parrott, University of Adelaide.
Dr. C.S. Praeger, University of Western Australia.
Professor N.J. Pullman, Queens University, Canada.
Mr. S. Quinn, Marks Point, New South Wales.
Dr. A. Rahilly, Gippsland I.A.E., Victoria.
Dr. R. Razen, University of New South Wales.
Dr. B. Richmond, University of Newcastle.
Dr. P.J. Robinson, Australian National University, Canberra.
Professor R.W. Robinson, University of Newcastle.
Mr. D. Rogers, Mathematical Institute, Oxford. U.K.
Dr. D. Row, University of Tasmania.
Dr. J. Seberry, University of Sydney.
Dr. G.J. Simmons, Sandia Laboratories, U.S.A.
Mr. D. Skillicorn, Eastwood, New South Wales.
Dr. N. Smythe, Australian National University, Canberra.
Dr. E. Sonenberg, Monash University.
Professor T.P. Speed, University of Western Australia.
Professor J.N. Srivastava, Colorado State University, U.S.A.
Professor R.G. Stanton, University of Manitoba, Canada.
Professor E.G. Straus, University of California at Los Angeles, U.S.A.
Dr. A.P. Street, University of Queensland.
Ms. D.J. Street, St. Lucia, Queensland.
Ms. M. Sved, University of Adelaide.

Ms. Esther Szekeres, Macquarie University.

Professor G. Szekeres, University of New South Wales.

Dr. D. Taylor, University of Sydney.

Professor H.N.V. Temperley, University College, Swansea, U.K.

Miss Shiu-Kuen Tsang, Australian National University, Canberra.

Professor W. Tutte, University of Waterloo, Canada.

Professor J.M. Van Buggenhaut, University of Brussels, Belgium.

Professor S.A. Vanstone, University of Waterloo, Canada.

Dr. W. Wallis, University of Newcastle.

Professor M.E. Watkins, Syracuse University, U.S.A.

Mr. J. Weadon, University of Adelaide.

Dr. K. Wehrhahn, University of Sydney.

Dr. E.G. Whitehead Jr., University of Queensland.

Dr. E. Williams, C.S.I.R.O., Canberra.

Mr. N. Wormald, University of Newcastle.

Vol. 521: G. Cherlin, Model Theoretic Algebra – Selected Topics. IV, 234 pages. 1976.

Vol. 522: C. O. Bloom and N. D. Kazarinoff, Short Wave Radiation Problems in Inhomogeneous Media: Asymptotic Solutions. V. 104 pages. 1976.

Vol. 523: S. A. Albeverio and R. J. Høegh-Krohn, Mathematical Theory of Feynman Path Integrals. IV, 139 pages. 1976.

Vol. 524: Séminaire Pierre Lelong (Analyse) Année 1974/75. Edité par P. Lelong. V, 222 pages. 1976.

Vol. 525: Structural Stability, the Theory of Catastrophes, and Applications in the Sciences. Proceedings 1975. Edited by P. Hilton. VI, 408 pages. 1976.

Vol. 526: Probability in Banach Spaces. Proceedings 1975. Edited by A. Beck. VI, 290 pages. 1976.

Vol. 527: M. Denker, Ch. Grillenberger, and K. Sigmund, Ergodic Theory on Compact Spaces. IV, 360 pages. 1976.

Vol. 528: J. E. Humphreys, Ordinary and Modular Representations of Chevalley Groups. III, 127 pages. 1976.

Vol. 529: J. Grandell, Doubly Stochastic Poisson Processes. X, 234 pages. 1976.

Vol. 530: S. S. Gelbart, Weil's Representation and the Spectrum of the Metaplectic Group. VII, 140 pages. 1976.

Vol. 531: Y.-C. Wong, The Topology of Uniform Convergence on Order-Bounded Sets. VI, 163 pages. 1976.

Vol. 532: Théorie Ergodique. Proceedings 1973/1974. Edité par J.-P. Conze and M. S. Keane. VIII, 227 pages. 1976.

Vol. 533: F. R. Cohen, T. J. Lada, and J. P. May, The Homology of Iterated Loop Spaces. IX, 490 pages. 1976.

Vol. 534: C. Preston, Random Fields. V, 200 pages. 1976.

Vol. 535: Singularités d'Applications Differentiables. Plans-sur-Bex. 1975. Edité par O. Burlet et F. Ronga. V, 253 pages. 1976.

Vol. 536: W. M. Schmidt, Equations over Finite Fields. An Elementary Approach. IX, 267 pages. 1976.

Vol. 537: Set Theory and Hierarchy Theory. Bierutowice, Poland 1975. A Memorial Tribute to Andrzej Mostowski. Edited by W. Marek, M. Srebrny and A. Zarach. XIII, 345 pages. 1976.

Vol. 538: G. Fischer, Complex Analytic Geometry. VII, 201 pages. 1976.

Vol. 539: A. Badrikian, J. F. C. Kingman et J. Kuelbs, Ecole d'Eté de Probabilités de Saint Flour V-1975. Edité par P.-L. Hennequin. IX, 314 pages. 1976.

Vol. 540: Categorical Topology, Proceedings 1975. Edited by E. Binz and H. Herrlich. XV, 719 pages. 1976.

Vol. 541: Measure Theory, Oberwolfach 1975. Proceedings. Edited by A. Bellow and D. Kölzow. XIV, 430 pages. 1976.

Vol. 542: D. A. Edwards and H. M. Hastings, Čech and Steenrod Homotopy Theories with Applications to Geometric Topology. VII, 296 pages. 1976.

Vol. 543: Nonlinear Operators and the Calculus of Variations, Bruxelles 1975. Edited by J. P. Gossez, E. J. Lami Dozo, J. Mawhin, and L. Waelbroeck, VII, 237 pages. 1976.

Vol. 544: Robert P. Langlands, On the Functional Equations Satisfied by Eisenstein Series. VII, 337 pages. 1976.

Vol. 545: Noncommutative Ring Theory. Kent State 1975. Edited by J. H. Cozzens and F. L. Sandomierski. V, 212 pages. 1976.

Vol. 546: K. Mahler, Lectures on Transcendental Numbers. Edited and Completed by B. Diviš and W. J. Le Veque. XXI, 254 pages. 1976.

Vol. 547: A. Mukherjea and N. A. Tserpes, Measures on Topological Semigroups: Convolution Products and Random Walks. V, 197 pages. 1976.

Vol. 548: D. A. Hejhal, The Selberg Trace Formula for PSL (2,ℝ). Volume I. VI, 516 pages. 1976.

Vol. 549: Brauer Groups, Evanston 1975. Proceedings. Edited by D. Zelinsky. V, 187 pages. 1976.

Vol. 550: Proceedings of the Third Japan – USSR Symposium on Probability Theory. Edited by G. Maruyama and J. V. Prokhorov. VI, 722 pages. 1976.

Vol. 551: Algebraic K-Theory, Evanston 1976. Proceedings. Edited by M. R. Stein. XI, 409 pages. 1976.

Vol. 552: C. G. Gibson, K. Wirthmüller, A. A. du Plessis and E. J. N. Looijenga. Topological Stability of Smooth Mappings. V, 155 pages. 1976.

Vol. 553: M. Petrich, Categories of Algebraic Systems. Vector and Projective Spaces, Semigroups, Rings and Lattices. VIII, 217 pages. 1976.

Vol. 554: J. D. H. Smith, Mal'cev Varieties. VIII, 158 pages. 1976.

Vol. 555: M. Ishida, The Genus Fields of Algebraic Number Fields. VII, 116 pages. 1976.

Vol. 556: Approximation Theory. Bonn 1976. Proceedings. Edited by R. Schaback and K. Scherer. VII, 466 pages. 1976.

Vol. 557: W. Iberkleid and T. Petrie, Smooth $S^1$ Manifolds. III, 163 pages. 1976.

Vol. 558: B. Weisfeiler, On Construction and Identification of Graphs. XIV, 237 pages. 1976.

Vol. 559: J.-P. Caubet, Le Mouvement Brownien Relativiste. IX, 212 pages. 1976.

Vol. 560: Combinatorial Mathematics, IV, Proceedings 1975. Edited by L. R. A. Casse and W. D. Wallis. VII, 249 pages. 1976.

Vol. 561: Function Theoretic Methods for Partial Differential Equations. Darmstadt 1976. Proceedings. Edited by V. E. Meister, N. Weck and W. L. Wendland. XVIII, 520 pages. 1976.

Vol. 562: R. W. Goodman, Nilpotent Lie Groups: Structure and Applications to Analysis. X, 210 pages. 1976.

Vol. 563: Séminaire de Théorie du Potentiel. Paris, No. 2. Proceedings 1975–1976. Edited by F. Hirsch and G. Mokobodzki. VI, 292 pages. 1976.

Vol. 564: Ordinary and Partial Differential Equations, Dundee 1976. Proceedings. Edited by W. N. Everitt and B. D. Sleeman. XVIII, 551 pages. 1976.

Vol. 565: Turbulence and Navier Stokes Equations. Proceedings 1975. Edited by R. Temam. IX, 194 pages. 1976.

Vol. 566: Empirical Distributions and Processes. Oberwolfach 1976. Proceedings. Edited by P. Gaenssler and P. Révész. VII, 146 pages. 1976.

Vol. 567: Séminaire Bourbaki vol. 1975/76. Exposés 471–488. IV, 303 pages. 1977.

Vol. 568: R. E. Gaines and J. L. Mawhin, Coincidence Degree, and Nonlinear Differential Equations. V, 262 pages. 1977.

Vol. 569: Cohomologie Etale SGA 4½. Séminaire de Géométrie Algébrique du Bois-Marie. Edité par P. Deligne. V, 312 pages. 1977.

Vol. 570: Differential Geometrical Methods in Mathematical Physics, Bonn 1975. Proceedings. Edited by K. Bleuler and A. Reetz. VIII, 576 pages. 1977.

Vol. 571: Constructive Theory of Functions of Several Variables, Oberwolfach 1976. Proceedings. Edited by W. Schempp and K. Zeller. VI. 290 pages. 1977

Vol. 572: Sparse Matrix Techniques, Copenhagen 1976. Edited by V. A. Barker. V, 184 pages. 1977.

Vol. 573: Group Theory, Canberra 1975. Proceedings. Edited by R. A. Bryce, J. Cossey and M. F. Newman. VII, 146 pages. 1977.

Vol. 574: J. Moldestad, Computations in Higher Types. IV, 203 pages. 1977.

Vol. 575: K-Theory and Operator Algebras, Athens, Georgia 1975. Edited by B. B. Morrel and I. M. Singer. VI, 191 pages. 1977.

Vol. 576: V. S. Varadarajan, Harmonic Analysis on Real Reductive Groups. VI, 521 pages. 1977.

Vol. 577: J. P. May, $E_\infty$ Ring Spaces and $E_\infty$ Ring Spectra. IV, 268 pages. 1977.

Vol. 578: Séminaire Pierre Lelong (Analyse) Année 1975/76. Edité par P. Lelong. VI, 327 pages. 1977.

Vol. 579: Combinatoire et Représentation du Groupe Symétrique, Strasbourg 1976. Proceedings 1976. Edité par D. Foata. IV, 339 pages. 1977.

Vol. 580: C. Castaing and M. Valadier, Convex Analysis and Measurable Multifunctions. VIII, 278 pages. 1977.

Vol. 581: Séminaire de Probabilités XI, Université de Strasbourg. Proceedings 1975/1976. Edité par C. Dellacherie, P. A. Meyer et M. Weil. VI, 574 pages. 1977.

Vol. 582: J. M. G. Fell, Induced Representations and Banach *-Algebraic Bundles. IV, 349 pages. 1977.

Vol. 583: W. Hirsch, C. C. Pugh and M. Shub, Invariant Manifolds. IV, 149 pages. 1977.

Vol. 584: C. Brezinski, Accélération de la Convergence en Analyse Numérique. IV, 313 pages. 1977.

Vol. 585: T. A. Springer, Invariant Theory. VI, 112 pages. 1977.

Vol. 586: Séminaire d'Algèbre Paul Dubreil, Paris 1975–1976 (29ème Année). Edited by M. P. Malliavin. VI, 188 pages. 1977.

Vol. 587: Non-Commutative Harmonic Analysis. Proceedings 1976. Edited by J. Carmona and M. Vergne. IV, 240 pages. 1977.

Vol. 588: P. Molino, Théorie des G-Structures: Le Problème d'Equivalence. VI, 163 pages. 1977.

Vol. 589: Cohomologie l-adique et Fonctions L. Séminaire de Géométrie Algébrique du Bois-Marie 1965–66, SGA 5. Edité par L. Illusie. XII, 484 pages. 1977.

Vol. 590: H. Matsumoto, Analyse Harmonique dans les Systèmes de Tits Bornologiques de Type Affine. IV, 219 pages. 1977.

Vol. 591: G. A. Anderson, Surgery with Coefficients. VIII, 157 pages. 1977.

Vol. 592: D. Voigt, Induzierte Darstellungen in der Theorie der endlichen, algebraischen Gruppen. V, 413 Seiten. 1977.

Vol. 593: K. Barbey and H. König, Abstract Analytic Function Theory and Hardy Algebras. VIII, 260 pages. 1977.

Vol. 594: Singular Perturbations and Boundary Layer Theory, Lyon 1976. Edited by C. M. Brauner, B. Gay, and J. Mathieu. VIII, 539 pages. 1977.

Vol. 595: W. Hazod, Stetige Faltungshalbgruppen von Wahrscheinlichkeitsmaßen und erzeugende Distributionen. XIII, 157 Seiten. 1977.

Vol. 596: K. Deimling, Ordinary Differential Equations in Banach Spaces. VI, 137 pages. 1977.

Vol. 597: Geometry and Topology, Rio de Janeiro, July 1976. Proceedings. Edited by J. Palis and M. do Carmo. VI, 866 pages. 1977.

Vol. 598: J. Hoffmann-Jørgensen, T. M. Liggett et J. Neveu, Ecole d'Eté de Probabilités de Saint-Flour VI – 1976. Edité par P.-L. Hennequin. XII, 447 pages. 1977.

Vol. 599: Complex Analysis, Kentucky 1976. Proceedings. Edited by J. D. Buckholtz and T. J. Suffridge. X, 159 pages. 1977.

Vol. 600: W. Stoll, Value Distribution on Parabolic Spaces. VIII, 216 pages. 1977.

Vol. 601: Modular Functions of oneVariableV, Bonn1976. Proceedings. Edited by J.-P. Serre and D. B. Zagier. VI, 294 pages. 1977.

Vol. 602: J. P. Brezin, Harmonic Analysis on Compact Solvmanifolds. VIII, 179 pages. 1977.

Vol. 603: B. Moishezon, Complex Surfaces and Connected Sums of Complex Projective Planes. IV, 234 pages. 1977.

Vol. 604: Banach Spaces of Analytic Functions, Kent, Ohio 1976. Proceedings. Edited by J. Baker, C. Cleaver and Joseph Diestel. VI, 141 pages. 1977.

Vol. 605: Sario et al., Classification Theory of Riemannian Manifolds. XX, 498 pages. 1977.

Vol. 606: Mathematical Aspects of Finite Element Methods. Proceedings 1975. Edited by I. Galligani and E. Magenes. VI, 362 pages. 1977.

Vol. 607: M. Métivier, Reelle und Vektorwertige Quasimartingale und die Theorie der Stochastischen Integration. X, 310 Seiten. 1977.

Vol. 608: Bigard et al., Groupes et Anneaux Réticulés. XIV, 334 pages. 1977.

Vol. 609: General Topology and Its Relations to Modern Analysis and Algebra IV. Proceedings 1976. Edited by J. Novák. XVIII, 225 pages. 1977.

Vol. 610: G. Jensen, Higher Order Contact of Submanifolds of Homogeneous Spaces. XII, 154 pages. 1977.

Vol. 611: M. Makkai and G. E. Reyes, First Order Categorical Logic. VIII, 301 pages. 1977.

Vol. 612: E. M. Kleinberg, Infinitary Combinatorics and the Axiom of Determinateness. VIII, 150 pages. 1977.

Vol. 613: E. Behrends et al., $L^p$-Structure in Real Banach Spaces. X, 108 pages. 1977.

Vol. 614: H. Yanagihara, Theory of Hopf Algebras Attached to Group Schemes. VIII, 308 pages. 1977.

Vol. 615: Turbulence Seminar, Proceedings 1976/77. Edited by P. Bernard and T. Ratiu. VI, 155 pages. 1977.

Vol. 616: Abelian Group Theory, 2nd New Mexico State University Conference, 1976. Proceedings. Edited by D. Arnold, R. Hunter and E. Walker. X, 423 pages. 1977.

Vol. 617: K. J. Devlin, The Axiom of Constructibility: A Guide for the Mathematician. VIII, 96 pages. 1977.

Vol. 618: I. I. Hirschman, Jr. and D. E. Hughes, Extreme Eigen Values of Toeplitz Operators. VI, 145 pages. 1977.

Vol. 619: Set Theory and Hierarchy Theory V, Bierutowice 1976. Edited by A. Lachlan, M. Srebrny, and A. Zarach. VIII, 358 pages. 1977.

Vol. 620: H. Popp, Moduli Theory and Classification Theory of Algebraic Varieties. VIII, 189 pages. 1977.

Vol. 621: Kauffman et al., The Deficiency Index Problem. VI, 112 pages. 1977.

Vol. 622: Combinatorial Mathematics V, Melbourne 1976. Proceedings. Edited by C. Little. VIII, 213 pages. 1977.

Vol. 623: I. Erdelyi and R. Lange, Spectral Decompositions on Banach Spaces. VIII, 122 pages. 1977.

Vol. 624: Y. Guivarc'h et al., Marches Aléatoires sur les Groupes de Lie. VIII, 292 pages. 1977.

Vol. 625: J. P. Alexander et al., Odd Order Group Actions and Witt Classification of Innerproducts. IV, 202 pages. 1977.

Vol. 626: Number Theory Day, New York 1976. Proceedings. Edited by M. B. Nathanson. VI, 241 pages. 1977.

Vol. 627: Modular Functions of One Variable VI, Bonn 1976. Proceedings. Edited by J.-P. Serre and D. B. Zagier. VI, 339 pages. 1977.

Vol. 628: H. J. Baues, Obstruction Theory on the Homotopy Classification of Maps. XII, 387 pages. 1977.

Vol. 629: W. A. Coppel, Dichotomies in Stability Theory. VI, 98 pages. 1978.

Vol. 630: Numerical Analysis, Proceedings, Biennial Conference, Dundee 1977. Edited by G. A. Watson. XII, 199 pages. 1978.

Vol. 631: Numerical Treatment of Differential Equations. Proceedings 1976. Edited by R. Bulirsch, R. D. Grigorieff, and J. Schröder. X, 219 pages. 1978.

Vol. 632: J.-F. Boutot, Schéma de Picard Local. X, 165 pages. 1978.

Vol. 633: N. R. Coleff and M. E. Herrera, Les Courants Résiduels Associés à une Forme Méromorphe. X, 211 pages. 1978.

Vol. 634: H. Kurke et al., Die Approximationseigenschaft lokaler Ringe. IV, 204 Seiten. 1978.

Vol. 635: T. Y. Lam, Serre's Conjecture. XVI, 227 pages. 1978.

Vol. 636: Journées de Statistique des Processus Stochastiques, Grenoble 1977, Proceedings. Edité par Didier Dacunha-Castelle et Bernard Van Cutsem. VII, 202 pages. 1978.

Vol. 637: W. B. Jurkat, Meromorphe Differentialgleichungen. VII, 194 Seiten. 1978.

Vol. 638: P. Shanahan, The Atiyah-Singer Index Theorem, An Introduction. V, 224 pages. 1978.

Vol. 639: N. Adasch et al., Topological Vector Spaces. V, 125 pages. 1978.